Linear Algebra

Textbooks in Mathematics

Series editors:

Al Boggess, Kenneth H. Rosen

Linear Algebra

James R. Kirkwood, Bessie H. Kirkwood

Real Analysis

With Proof Strategies

Daniel W. Cunningham

Train Your Brain

Challenging Yet Elementary Mathematics

Bogumil Kaminski, Pawel Pralat

Contemporary Abstract Algebra, Tenth Edition

Joseph A. Gallian

Geometry and Its Applications

Walter J. Meyer

Linear Algebra

What You Need to Know

Hugo J. Woerdeman

Introduction to Real Analysis, 3rd Edition

Manfred Stoll

Discovering Dynamical Systems Through Experiment and Inquiry

Thomas LoFaro, Jeff Ford

Functional Linear Algebra

Hannah Robbins

Introduction to Financial Mathematics

With Computer Applications

Donald R. Chambers, Qin Lu

Linear Algebra

An Inquiry-based Approach

Jeff Suzuki

https://www.routledge.com/Textbooks-in-Mathematics/book-series/CANDHTEX-BOOMTH

Linear Algebra

An Inquiry-Based Approach

Jeff Suzuki
Brooklyn College

CRC Press is an imprint of the
Taylor & Francis Group, an **informa** business
A CHAPMAN & HALL BOOK

First edition published 2021
by CRC Press
6000 Broken Sound Parkway NW, Suite 300, Boca Raton, FL 33487-2742

and by CRC Press
2 Park Square, Milton Park, Abingdon, Oxon, OX14 4RN

CRC Press is an imprint of Taylor & Francis Group, LLC

Library of Congress Cataloging-in-Publication Data

Names: Suzuki, Jeff, author.
Title: Linear algebra : an inquiry-based approach / Jeff Suzuki.
Description: First edition. | Boca Raton, FL : CRC Press, 2021. | Series: Textbooks in mathematics | Includes bibliographical references and index.
Identifiers: LCCN 2020050821 (print) | LCCN 2020050822 (ebook) | ISBN 9780367248963 (hardback) | ISBN 9780429284984 (ebook)
Subjects: LCSH: Algebras, Linear.
Classification: LCC QA184.2 .S89 2021 (print) | LCC QA184.2 (ebook) | DDC 512/.5--dc23
LC record available at https://lccn.loc.gov/2020050821
LC ebook record available at https://lccn.loc.gov/2020050822

ISBN: 9780367248963 (hbk)
ISBN: 9780367754877 (pbk)
ISBN: 9780429284984 (ebk)

To Lauren Rose: For introducing me to whole new worlds of teaching and mathematics.

Contents

Introduction and Features

What you hold in your hand or—this *is* the 21st century—view on screen shouldn't exist.

Let me explain. Mathematics is often taught by presenting students with a broad concept, like linear independence, and then a set of tools for identifying, using, or creating it. More rigorous courses might include proofs, like the theorem that the eigenvectors for a given eigenvalue form a vector space. Understandably, this approach leaves most students with the impression that mathematics is a collection of results and algorithms. But these are merely the products of mathematics, and are about as interesting to the mathematician as last week's bread to the baker or last year's canvas to the artist. The theorem has been proved, the bread baked, the portrait painted, and the important question is: What's *next*?

Mathematics itself is the *creation* of these concepts and algorithms. In a perfect world, students would begin with a simple concept, and then develop the tools necessary to handle this concept. This is the idea behind **inquiry based learning** (IBL).

The value of IBL should be clear. In the modern world, if an existing method can solve a problem, then a clever programmer can code that method, and a computer can implement the method more rapidly, more accurately, and more cheaply than any human being. So what's really important is not *how* to solve a problem, but how to *create* the solution to a problem that *no one* has solved. Thus, a 21st century mathematics textbook should focus on teaching students to be creative.

Herein lies the paradox. It is, by definition, impossible to teach someone to be creative. Consequently, this book shouldn't exist.

Yet it obviously does. The reason is that while no one can teach you how to be creative, they can foster an environment that supports the creative process. This can be done by:

- Providing opportunities to be creative,
- Teaching "ways of thinking" that will make it easier for them to be creative.

These are the twin goals of this textbook. It's important to understand these goals, because they result in a textbook that is probably unlike any you've ever worked with before.

First, creativity is a *habit.* The more you create, the easier it is for you to create even more. In the context of problem solving, this means creating a solution to a problem you've never solved before. But you only ever get one chance to solve a problem for the first time, and this chance is forever lost by being *told* how to solve the problem. Thus, as much as feasible, solutions to problems will *not* be presented as ready-made algorithms. Instead, the reader will have a chance to create the solution through a sequence of activities that build up the necessary components for a solution.

Second, creativity is a *process.* It is rare that someone comes up with *the* solution to a problem at a single sitting. More often than not, a solution is found in stages: first to a drastically simplified problem, perhaps even a concrete example; then to gradually more sophisticated problems. Oftentimes this means solving the same problem several times, to get a feel for what solving the general problem entails. If it seems like many of the problems in this book are asking "the same thing"—they are, though with the expectation that successive solutions will display greater insight and understanding.

Third, creativity is *introspective.* Careful and constant re-examination of the creative process helps build the scaffolding for the *next* act of a creation. The original meaning of "proof" was to test, and it's often useful to test our solutions to the breaking point.

And fourth, creativity is *messy.* I tell my students that one way you can gauge the importance of a concept is the number of *different* terms and symbols we have for it. That's because an important concept is usually invented independently by many different people or groups of researchers, and they choose a name that's based on their own idiosyncrasies. What this means, particularly in emerging fields, is that the notation and terminology are not standardized. Most linear algebra texts take great pains to be consistent so that a vector is always indicated using boldface type ($\mathbf{v}$) or using an arrow ($\overrightarrow{v}$). While we won't go so far as to randomly select a different convention every time we introduce a concept, we will not take pains to be consistent in our notation, to prepare students for their *next* encounter with linear algebra—which might not be in a mathematics course.

For the Student . . . and Teacher

In an IBL course, the instructor lays out some broad principles, and allows the student leeway to explore them. To do this effectively, the student must understand the underlying concepts and be ready and willing to work with them. At the same time, the instructor has to resist the all-too-natural urge to give the student the algorithm required to solve "the problem." This dynamic is important to understand: The goal of the teacher is not to impart knowledge to the student, but to guide them so they can create their own understanding.

This can be a challenge. For example, simply presenting the standard definition of vector addition and scalar multiplication and going through a few examples can be done in a few minutes. But it will take considerably more time for students to work through Activities 1.3, 1.5, and 1.6, especially if this is the first time they've had to develop a mathematical concept from first principles.

There are some ways to approach this. The first is a more traditional approach: a standard lecture, with the activities assigned as homework. While some of the activities could be assigned as homework, particularly those that focus on the application of an algorithm or a concept, assigning the bulk of them to be done outside of class will negate the whole purpose of teaching an IBL course. The main hurdle faced by students in an IBL course is that they will encounter obstacles on the path to enlightenment. If the activities are done as homework, students might not be able to overcome the obstacle or—worse—use the Internet to find an answer, short-circuiting the entire process of creating their own solution.

A more effective way to implement an IBL course is to limit the time the instructor speaks, then let the students work problems during the rest of the class period. Then, if a student encounters an insurmountable obstacle, the instructor can provide guidance or useful suggestions for overcoming the obstacle. One way to do this is to use a "flipped" or "inverted" classroom approach, where all lectures are done outside of class, usually by having students watch a video.

I use the following structure for my own class:

- Before class, students watch some lecture videos on a topic in linear algebra. The videos I use are at `https://www.youtube.com/jeffsuzuki1`, but any of thousands of others could be used.

- Class begins with a short "overview" lecture, summarizing the main concepts students of the assigned lectures.

- The rest of the class is spent having students work, alone or in groups, to solve problems that build understanding or develop key ideas. These problems form the core of the following text.

Fair warning: As an instructor, you will be present for many an "Aha!" moment. Since, for most of us, those "Aha!" moments are the reason we got into teaching, the IBL approach is addictive, and you'll want to do it in every class you teach. Meanwhile, as a student, you'll find yourself working much more closely with your instructor and your fellow students. Since this is the very definition of "collegiality," you'll come to expect it from every class you take. If this approach leads to a revolution in the college experience from both faculty and students ... so be it.

Prerequisites

Most introductory linear algebra courses have calculus as a prerequisite. This is a baffling requirement, as there is *nothing* in an introductory linear algebra course that requires calculus. Indeed, one could make an argument that there is *nothing* in linear algebra, introductory *or* advanced, that requires calculus, outside of applications to calculus and postcalculus problems.

So why do most introductory linear algebra courses require calculus as a prerequisite? The only justification for the calculus prerequisite is the expectation that students possess some mystical quality called "mathematical maturity." Introductory linear algebra typically includes mathematical proofs, and the assumption is that students who are not "mathematically mature" won't be able to meet the expectations of the course. While this is true, it begs the question: Where does one *acquire* mathematical maturity?

My contention is that mathematical maturity is acquired through courses where proof is a regular and required part of the course. But this means that *some* course must be the first time students are exposed to mathematical proof. Why not make linear algebra that course? With this idea in mind, this text is written with an eye toward introducing proofs to students who have *not* taken calculus. More broadly, I've tried to ease the transition to abstraction that is the hallmark of higher mathematics. Thus, many concepts might be introduced using a very concrete example or application, but as soon as feasible, the concrete is replaced with the abstract: "Math ever generalizes."

Almost everything in this text should be accessible to students who've taken "college algebra," and with the exception of problems that originate from calculus questions, *everything* should be within the scope of a student who has taken a standard precalculus course.

Suggested Sequences

This material grew from a linear algebra course I teach at Brooklyn College, a large public institution in the northeast. In fact, it grew so much that there's enough material here for a two (or even three)-semester course on linear algebra, so selecting the material for your own course can be challenging. A few suggestions follow.

Our own course begins with vector arithmetic and ends with eigenvalues and eigenvectors; I typically use the following:

- All of Chapter 1;
- All of Chapter 2 except Section 2.5 (on computational cost);
- All of Chapter 3 except Section 3.8 (applications to cryptography);
- All of Chapter 4, except Section 4.7 (which deals with complex numbers);
- All of Chapter 5, except Section 5.10 (least squares) and Section 5.7 (on normed vector spaces).
- All of Chapter 6, except Section 6.8 (computational cost) and Section 6.11 (the permutation definition of the determinant).
- Sections 7.2 and 7.5 from Chapter 7.

While the Activities were designed to be done during class, some will necessarily have to be done as homework. I've indicated (with an H) those activities that are better suited to be done outside of class. There are also a number of activities that are more peripheral (marked with a P) to the understanding of the concepts and applications of elementary linear algebra: for example, Activity 1.13 helps students visualize vectors in more than three dimensions, while other activities develop applications or more advanced topics. These might also be relegated to homework status.

Vector Arithmetic

The basic properties of vector arithmetic, as well as a number of applications, are covered in Chapter 1.

- Activities 1.1 (H), 1.5 (H), 1.6 (H), 1.20, 1.25 introduce and motivate vector addition, scalar multiplication, and the dot product.

- Activities 1.2, 1.3, 1.4 provide nongeometric examples of vectors, while Activities 1.11, 1.12, 1.13 (P), 1.14, 1.15, 1.16, and 1.31 build up geometric concepts of vectors and vector arithmetic.
- Activities 1.17, 1.18, 1.19 (P) build up to the vector equations of lines, planes, and hyperplanes.
- Activities 1.7 (P), 1.23 (P), 1.24 (P) are more explanatory of why we *don't* do certain things: e.g., why we don't define componentwise multiplication.
- Activities 1.32 (H) and 1.35 (H) provide some examples of applied vector arithmetic.
- If you're teaching a proofs-based course, Activities 1.8, 1.9, 1.10, 1.22, 1.26, 1.27, 1.30, 1.33, and 1.34 are designed to help students write good mathematical proofs.
- The remaining activities in the chapter, namely Activities 1.21, 1.28, 1.29, require a more advanced background (complex arithmetic in this case).

Linear Equations

The use of matrices to represent and solve systems of linear equations is covered in Chapter 2. We include an approach, based on the Chinese *fang cheng shu* method, that avoids fractional coefficients; this allows us to introduce new problems, like finding integer solutions to indeterminate equations.

- Activities 2.1 (P), 2.2 (H), 2.3, 2.4, 2.5, 2.6, 2.7 (H), and 2.8 cover basic row operations.
- Activities 2.9, 2.10 (P), 2.11, 2.12 introduce and develop the concepts of free and basic variables, and the rank of a matrix.
- Activities 2.13 (P), 2.14 (H), 2.15 (H), 2.16 (H), and 2.17 (H) provide a number of examples of applied problems that can be solved using matrix algebra.

Linear Transformations

The concept of matrices as linear transformations is developed in Chapter 3. We begin by considering geometric transformations.

- Activities 3.1, 3.2 (P), 3.3, 3.4 (P), 3.5 (P), 3.6 (P), 3.17, 3.19, and 3.20 develop a student's understanding of linear transformations and their representation in matrix form.
- Activities 3.10, 3.11, 3.12, 3.13 (H), 3.14 (H) develop more abstract transformations, such as discrete-time models and state transition models.

- Activities 3.22 (P), 3.23 (P), and 3.24 (P) are focused on cryptographic uses of transformation matrices.
- The other Activities in this chapter (3.7, 3.8, 3.9, 3.15, 3.16, 3.18, and 3.21) develop standard topics of domain, codomain, range, and the identity transformation.

Matrix Algebra

Matrix algebra is developed in Chapter 4. Here we develop the definitions of matrix arithmetic based on our desire to identify a matrix with a linear transformation. The idea of equivalent definitions is introduced.

- The rationale behind scalar multiplication, matrix addition, and matrix multiplication is developed by Activities 4.1, 4.3, 4.5.
- An introduction to the idea of equivalent definitions is given in Activities 4.2 (P), 4.4 (P) and 4.6 (P).
- The properties of matrix products are developed in Activities 4.7 (P), 4.10, 4.11, 4.17 (P), 4.18, 4.19 (P), 4.20, and 4.24; and for complex matrices, Activity 4.25 (P) and 4.26 (P).
- The transpose and its properties are developed in 4.12 (P), 4.13 (H), 4.14 (H), 4.15, and 4.16.
- Activities 4.21, 4.22, 4.23 (H) develop an algorithm for finding matrix inverses.
- Practical aspects and applications of matrix arithmetic are explored in Activities 4.8 (H) and 4.9 (P).

Vector Spaces

The theory of vector spaces occupies Chapter 5.

- Basic properties of vector spaces are developed in Activities 5.1, 5.2, 5.3, 5.4, 5.5.
- Coordinates, span, independence, and basis are developed in Activities 5.6 (P), 5.7, 5.8, 5.9, 5.10, 5.11, 5.12, 5.13, 5.14, 5.15 (P), 5.16 (P), 5.17 (P).
- Changing basis, leading up to the Gram-Schmidt process, is motivated by Activities 5.18 (H), 5.20 (H), 5.21 (P), and developed in Activities 5.19 and 5.22.
- Activity 5.23 (P) and 5.24 (P) develop some applications vector spaces; although these activities are included in the section on normed vector space, the only thing they really require from that section is the 1-norm.

- More advanced topics (normed vector spaces, inner product spaces) are developed in Activities 5.25 (P), 5.26 (P), 5.27 (P), 5.28 (P), 5.29 (P), 5.30 (P), 5.31 (P), 5.32 (P).
- Other applications are developed in 5.33 (P), 5.34 (P), 5.35 (P), 5.36 (P), 5.37 (P), 5.38 (P).
- Activities 5.39 (P), 5.40 (P), 5.41 (P), 5.42 (P), 5.43 (P), 5.44 (P), and 5.45 (P) develop the least squares method, which is likely one of the most important and impactful uses of linear algebra.

Determinants

Chapter 6 develops the theory of determinants.

There are some, notably Sheldon Axler, who feel the determinant should be eliminated from elementary linear algebra. There is a justification for this viewpoint: every problem in elementary linear algebra that can be solved using the determinant can be solved more quickly using an alternative method.

At the same time, the determinant, with its completely unintuitive computational formula, is useful as a case study in the creation of mathematics. So even though I strongly counsel students against using the determinant for any purpose, the irony of mathematics is that the chapters on determinants is the longest in this book!

- Activities 6.1, 6.2, 6.3 (P), 6.4 (P), 6.5, develop the determinant of a 2×2 matrix is identified as a quantity of interest.
- The determinant as a multilinear function of the rows and columns of a matrix is established through Activities 6.6 (P), 6.7, 6.8 (H), 6.9 (P), 6.10, and 6.11.
- These properties are then used to derive the Laplace expansion through Activities 6.12, 6.13 (H), 6.14 (H), 6.15, 6.16, 6.17, 6.18, 6.19 (H), 6.20 (P), 6.21 (P).
- Further properties of the determinant are derived from the Laplace expansion in Activities 6.22 (H), 6.23 (H), 6.24 (H), 6.25 (H), 6.26 (H), 6.27 (H), 6.28 (H), 6.29 (H), and 6.30 (H).
- In point of fact, the Laplace expansion is terribly inefficient. This is established in Activity 6.31 (P). A more efficient approach is presented in 6.32 (P) and 6.33 (P) (which rely on Activities 4.11, 6.26, 6.27, and 6.29). Activities 6.34 (P), 6.35 (P), 6.36 (P), 6.37 (P), and 6.38 (P) explore some of the ramifications.
- Practical uses of the determinant, when the computational cost is taken into account, are presented in Activities 6.39 (P), 6.40 (P), 6.41 (P), and 6.42 (P, requires calculus).

- Finally, we present a third approach to the determinant, using permutations, in Activities 6.43 (P), 6.44 (P), 6.45 (P), 6.46 (P), and 6.47 (P).

Eigenvalues and Eigenvectors

The theory of eigenvalues and eigenvectors is developed in Chapter 7.

Finding eigenvalues and eigenvectors is the one common use of the determinant in elementary linear algebra. As it turns out, we can find them using a nondeterminant method; this chapter develops this method. It also presents the more traditional approach.

- A geometric approach to the topic is built through Activities 7.1 (H) and 7.2 (H).
- Properties of eigenvalues and eigenvectors are introduced and developed in Activities 7.3 (P), 7.4 (note that these are placed *before* any eigenvalues and eigenvectors are found); and Activities 7.7, 7.26, 7.28, and 7.29.
- Several different approaches to finding eigenvalues and eigenvectors are introduced. Activity 7.5 (H), 7.6 (H) build on the algebraic nature of the eigenvalues and eigenvectors as solutions to a system of equations; Activity 7.8 (P), 7.9 (P), 7.10 (P), and 7.30 (P) develop a numerical approach (and identify some potential problems with such an approach).
- A determinant-free approach, which reinforces basic concepts of vector spaces, is developed in Activities 7.18, 7.19, 7.20, and 7.21. Finally, the special case of 2×2 matrices is considered in Activity 7.27.
- A more traditional approach, which relies on computing determinants, is introduced by Activities 7.11, 7.12, 7.13, 7.14, 7.15 (P).
- Uses of eigenvalues and eigenvectors are explored in Activities 7.16 (P), 7.17 (P), 7.33 (P), and 7.34 (P).
- Generalized eigenvectors are developed by Activities 7.22 (P), 7.23 (P), 7.24 (P), 7.25 (P), 7.31 (P), and 7.32 (P).

Matrix Decompositions

Finally, Chapter 8 lays the groundwork for various matrix decompositions, including Singular Value Decomposition; the material would be suitable for projects or independent study by advanced students.

1

Tuples and Vectors

1.1 Tuples

We'll begin our journey into linear algebra by considering a ubiquitous feature of life in the 21st century: personal information forms. For example, a medical record might look something like this:

Name: Jeff Suzuki	Gender: M
Height: 5'6"	Weight: 135
BP: 130 over 70	Age: 37

This form *appears* to contain information. But does it? In the 1940s, information theory pioneer Claude Shannon (1916–2001) made an important realization: If you already know the answer, the answer isn't information. Thus the answer to "What is the person's name?" is information: If we picked out a random form, we wouldn't know the name of the person the form belongs to. But the answer to "What goes in the first box?" is *not* information: Regardless of which form we picked out, the *same* thing goes in the first box, so in some sense the label "Name" in the first box, as well as the other labels, are unnecessary.

To be sure, we couldn't just use a blank form; we have to agree in advance that the name will go in the first box, the gender in the second, and so on. But once we make that agreement, it's not necessary to keep the labels; we could simply record the answers to the questions:

Jeff Suzuki, M, 36, 5'6", 135, 130 over 70

This allows us to represent information spread over a sheet of paper into a compact list. In fact, if you think about it, we can take any arrangement of information and write it as a list, as long as we agree on the order of the elements. This leads to:

Definition 1.1 (Tuple). *A tuple (or n-tuple, if more specificity is required) is an ordered list of n values. We say the n-tuple has length n.*

The preceding are 6-tuples, because each contains 6 values: the patient's name, gender, age, height, weight, and blood pressure. We don't always specify n: thus we might introduce the 6-tuple of patient medical information, but later on we might simply speak of it as a tuple.

Example 1.1. *Identify which of the following lists are tuples. Of the tuples, what are the lengths?*

- *A grocery list, "Ham, bread, gallon of milk."*
- *The coordinates of a point* $(5, 8)$.
- *The map directions "Turn left, go five blocks, turn right, go three blocks."*

Solution. *The grocery list is not a tuple, since the list is not ordered.*

The set of coordinates is a 2-tuple: there are two elements, and the point $(8, 5)$ *is different from the point* $(5, 8)$.

The map directions form a 4-tuple: following the directions in a different order could take you to a different place.

If we have more than one tuple, we'll need to introduce some notation to keep them from running together. One way to do that is to begin and end a tuple with a **grouping symbol**. Thus, when we give coordinates, we enclose them in a set of parentheses: $(2, -7)$. A more common grouping symbol for tuples is a set of angle brackets: we begin with a $\langle$ and end with a $\rangle$. Thus one medical record might be:

⟨Jeff Suzuki, M, 36, 5'6", 135, 130 over 70⟩

We note that this isn't the only way to represent tuples; we can use parentheses:

(Jeff Suzuki, M, 36, 5'6", 135, 130 over 70)

or square brackets

[Jeff Suzuki, M, 36, 5'6", 135, 130 over 70]

and sometimes we omit the commas

(Jeff Suzuki M 36 5'6" 135 130 over 70)

or

[Jeff Suzuki M 36 5'6" 135 130 over 70]

The only grouping symbols we avoid are braces, which are generally used for *un*ordered list.

It's important to remember that once we've made an agreement on the order of the data, we have to stick to it. Thus, if the first entry of the tuple is the name of the person, then the tuple ⟨36, M, 135, 5'6", Jeff Suzuki, 130 over 70⟩ would be the medical record for a male named "36" who is 135 years of age.

One of the first questions a mathematician asks when they are introduced to a new object is: How can we compare these objects? The most basic comparison is that of equality: two things are either equal, or they are not. In the following activity, we'll build towards a definition of the equality of two tuples.

Activity 1.1: Equality

A1.1.1 Let s be an English sentence with exactly five words. Are 5-word sentences tuples? Why/why not?

A1.1.2 Consider the sentences, "The dog bit the child" and "The child bit the dog."

a) Since the two sentences have the same words, are they the same sentence?

b) What does this suggest about how to determine if two sentences are equal?

Activity 1.1 suggests the following definition:

Definition 1.2 (Equality of tuples). *Two tuples are equal when they have the same components in the same order.*

So the tuple

$$\langle \text{Red}, 15, \text{Green}, 87, \text{Blue}, 54 \rangle$$

and the tuple

$$\langle 87, 15, \text{Red}, \text{Green}, \text{Blue}, 54 \rangle$$

are not equal: they have the same components (red, green, blue, 15, 87, 54) but they are in different orders.

Before continuing, we'll want to say a few words about definitions. Any mathematics book is filled with definitions; we've introduced two in just a few pages. In real life, we often view definitions as secondary: If you want information, you go to an encyclopedia, not a dictionary. But in mathematics, definitions are supremely important. *Everything* in mathematics is derived from the definitions. If you forget a theorem, formula, or algorithm, you can reproduce it by going back to the definitions. But if you forget a definition, you can't reason your way toward it. In fact, we go so far as to say:

Strategy. *Definitions are the whole of mathematics; all else is commentary.*

1.2 Vectors

As a general rule, the more you can do with an object, the more useful it is. At this point, we can compare two tuples and decide if they're equal. But if they're not equal, we can't say whether one is greater or lesser, nor can we add or multiply two tuples. To do that, we need to define an arithmetic of tuples.

To begin with, we might consider whether we can do arithmetic with the *components* of the tuples. In our medical records example, we might not be able to say what it means to multiply two names or to add two genders. But we can do arithmetic with quantities like age and weight. That's because these values are represented by real numbers.

Let's look into these properties. If a, b, c are real numbers, we have the following properties:

- There are two **binary operations** defined: addition $a + b$ and multiplication ab. (There are more operations, of course, but we'll focus on these two for now)

- The sum or product of two real numbers is a real number. This is known as **closure under addition** and **closure under multiplication**.

- In a sum or product of two real numbers, the order they're listed doesn't matter: $a + b = b + a$, and $ab = ba$. This is known as **commutativity of addition** and **commutativity of multiplication**.

- The sum or product of three numbers can be done "two at a time," and it doesn't matter how we group them: $(a + b) + c = a + (b + c)$, and $(ab)c = a(bc)$. This is known as **associativity of addition** and **associativity of multiplication**.

- There are real numbers 0 and 1, with the property that $a + 0 = a$, and $1a = a$: in other words, if you operate with these numbers, you don't change what you started with. We say that we have an **additive identity** 0 and a **multiplicative identity** 1.

- For any real number a, there is a real number designated $-a$ where $a + (-a) = 0$, the additive identity. Likewise, provided $a \neq 0$, there is a real number a^{-1}, where $aa^{-1} = 1$, the multiplicative identity. We say that every real number has an **additive inverse**, and every real number (except 0) has a **multiplicative inverse**.

- For real numbers a, b, and c, we have $a(b + c) = ab + ac$. We say that the real numbers have **the distributive property of multiplication over addition**.

"Math ever generalizes," so a mathematician could conceive that there might be *other* things that satisfy all these properties. To that end, we'll introduce a mathematical object called a *field*, which has all the properties we associate with the real numbers. We define:

Definition 1.3 (Field). *A field $\mathbb{F}$ consists of some set and two operations, which we call addition $a + b$ and multiplication ab, where the following properties hold for any elements a, b, c in $\mathbb{F}$:*

- ***Closure under addition and multiplication:*** *$a + b$ and ab are in $\mathbb{F}$*

- ***Associativity of addition and multiplication:*** $(a+b)+c = a+(b+c)$ *and* $(ab)c = a(bc)$.
- ***Commutativity of addition and multiplication:*** $a + b = b + a$ *and* $ab = ba$.
- ***Additive and multiplicative identities:*** *There is an additive identity* 0 *and a multiplicative identity* 1 *where* $a + 0 = a$ *and* $1a = a$.
- ***Additive and multiplicative inverses:*** *For any* a*, there is additive inverse* $-a$ *where* $a + (-a) = 0$ *and, for* $a \neq 0$*, there is a multiplicative inverse* a^{-1} *where* $aa^{-1} = 1$.

The most familiar fields are the field of real numbers, designated $\mathbb{R}$, and the field of complex numbers, designated $\mathbb{C}$. For a refresher on complex numbers and their arithmetic, see Section 9.2.

Notice that our field only has two operations: addition and multiplication. What about the other two operations of basic arithmetic? These are defined in terms of the field operations:

Definition 1.4 (Subtraction and Division in a Field)**.** *For* a, b *in field* $\mathbb{F}$*, we define*

$$a - b = a + (-b)$$

and

$$a \div b = ab^{-1}$$

provided $b \neq 0$.

Thus, subtraction is addition of the additive inverse, while division is multiplication by the multiplicative inverse.

It's important to keep in mind that there are only so many symbols: You've seen expressions like $+$, $-a$, and a^{-1}, which have very specific meanings when a is a real number. But if you're not dealing with real numbers, $+$, $\times$, $-a$, and a^{-1} *could mean something entirely different.* To borrow a phrase from the study of foreign languages, these are *faux amis*: "false friends," that *appear* to mean something familiar, but in fact mean something very different. Thus in France, you couldn't buy something with a *coin*; in Spain, people would be horrified if you ate a *pie*; and in Boston, a *regular coffee* has cream and sugar in addition to the caffeine.

With the idea of a field idea in mind, we define:

Definition 1.5 (Vectors)**.** *A vector is a* n*-tuple whose components come from a field* $\mathbb{F}$.

We say the vector is in (or from) the set $\mathbb{F}^n$. In this book, we will almost always be considering vectors in $\mathbb{R}^n$: vectors with n components, all of which are real numbers. Occasionally, we will consider vectors from the set $\mathbb{C}^n$, whose components are complex numbers.

There are several common notations for vectors, but the two most important are to use boldface type ($\mathbf{v}$) or an arrow ($\vec{v}$). Since different disciplines have different preferences, we'll use both and make no attempt to be consistent. Note that we can write a vector as a single object ($\mathbf{v}$ or $\vec{v}$), or write it in **component form** ($\langle v_1, v_2, v_3 \rangle$).

Example 1.2. *If* $\mathbf{v} = \langle 4, -3, 1, 5 \rangle$, *then the components are* 4, −3, 1, 5.

Activity 1.2: Feature Vectors

As we will see, linear algebra provides powerful tools for analyzing and using vectors. This means it can be applied to anything that can be interpreted as a vector.

A1.2.1 A group of students take an exam, and the results of the exam are tabulated as follows:

A	B	C	D	F
8	15	23	3	1

a) Explain why the exam results can be represented by a 5-tuple. Identify the significance of each component of the tuple.

b) Is the 5-tuple a vector? If it is, what set is it from? If not, why not?

A1.2.2 In a digital image, each pixel has a color that can be specified by the intensity of the red, green, and blue light emitted (RGB color).

a) Explain why the color of a pixel can be represented by a 3-tuple. What is the significance of each component of the tuple?

b) Is the 3-tuple a vector? If it is, what set is it in? If not, why not?

Feature vectors are somewhat anomalous, in the sense that their components are usually whole numbers: thus, no class would have −3 As, or 5.7 Cs. RGB intensities have further restrictions, in that they must be between 0 and 255. Since the whole numbers do not form a field (no number has an additive inverse, and only 1 has a multiplicative inverse), then strictly speaking, feature vectors are not vectors. In practice, this distinction is often ignored, and "feature vectors" are generally considered to be vectors.

Activity 1.3: Vectors

A1.3.1 Suppose we represent a polynomial $a_3x^3 + a_2x^2 + a_1x + a_0$, where the a_is are real numbers, as a tuple $\langle a_3, a_2, a_1, a_0\rangle$.

a) Explain why these tuples are vectors.
b) What set are these vectors from?
c) Write as a vector: $3x^3 - 7x^2 + 8x + 5$.
d) Write as a vector: $x^3 + 7x + 5$.
e) Write as a vector: $3x - 7x^2 + 8x^3 + 5$
f) Write as a vector: $8x^2 + 12x + 7$.
g) Verify your answers in Activity A1.3.1c through Activity A1.3.1f by interpreting the vector you wrote as a polynomial and confirming it's the same polynomial you started with.
h) Interpret as a polynomial: $\langle 5, 1, -4, 6\rangle$.
i) Interpret as a polynomial: $\langle 0, 0, 0, 5\rangle$.
j) Interpret as a polynomial: $\langle 1, 0, 0, 1\rangle$.
k) Verify your answers in Activity A1.3.1h through Activity A1.3.1j by writing the polynomial you wrote as a vector and confirming it's the same vector you started with.

A1.3.2 We can represent a linear equation in three variables $ax+by+cz = d$, where a, b, c, d are real numbers, as a tuple $\langle a, b, c, d\rangle$.

a) Explain why these tuples are vectors.
b) What set are these vectors from?
c) Write as a vector: $3x - 5y + 2z = 11$.
d) Write as a vector: $2x + 5y + 8z = 3$.
e) Write as a vector: $2x + 5y + 8 = 3z$.
f) Write as a vector: $2z + 5y + 8x = 3$.
g) Interpret as a linear equation: $\langle 2, 5, 8, 3\rangle$.
h) Interpret as a linear equation: $\langle 2, 5, -4, 0\rangle$.
i) Verify your answers to Activities A1.3.2c through A1.3.2h: If you wrote a vector, interpret it as a linear equation and confirm that it's the equation you started with; if you wrote a linear equation, write it as a vector and confirm it's the vector you started with.

One particularly important type of vector is known as a **document vector**.

Activity 1.4: Document Vectors

Given a library of documents, which can be anything from a short search engine query ("What is linear algebra?") to a multivolume encyclopedia, a **document vector** is created as follows: First, indexers create a list of keywords. Then a document is converted into a vector whose ith component is the number of times the ith keyword appears in the document.

A1.4.1 Suppose we choose the following six keywords: *the*, *arctic*, *bear*, *is*, *platypus*, *Australia*.

a) Find the document vector for the sentence: "The polar bear is a type of animal that lives in the arctic, where there is snow and ice."
b) Find the document vector for the sentence: "The continent of Australia is the home to the platypus, which is one of the very few poisonous mammals."
c) Find the document vector for the sentence: "The koala bear and the platypus are not native to the arctic."
d) Write a proper English sentence with document vector $\langle 2, 1, 1, 0, 0 \rangle$.
e) Suppose $\mathbf{u}, \mathbf{v}$ are two document vectors, with $\mathbf{u} = \mathbf{v}$. Are the documents the same? Why/why not?

Because the components of a vector are field elements, and we can perform arithmetic with field elements, we might try to perform arithmetic on vectors. Thus, we might ask what $\mathbf{u} + \mathbf{v}$ is equal to.

The standard practice is to provide a definition for vector addition. But before we do that, remember: Definitions are the whole of mathematics; all else is commentary. This means our definitions affect *everything*. A badly conceived definition could make it difficult or even impossible to proceed past a certain point, or—worse yet—may lead to contradictory mathematical results.

So instead of asking how we should *define* vector addition, let's start by asking what we want vector addition to *do*. When we find $\mathbf{u} + \mathbf{v}$, what do we want the result to represent?

Activity 1.5: Vector Addition

A1.5.1 Suppose

$$\langle a_3, a_2, a_1, a_0 \rangle \text{ represents } a_3x^3 + a_2x^2 + a_1x + a_0$$
$$\langle b_3, b_2, b_1, b_0 \rangle \text{ represents } b_3x^3 + b_2x^2 + b_1x + b_0$$

where the a_is and b_is are real numbers.

a) Add the polynomials.
b) To what vector will the sum of the polynomials be equal?
c) What does this suggest about how to define

$$\langle a_3, a_2, a_1, a_0 \rangle + \langle b_3, b_2, b_1, b_0 \rangle$$

d) Subtract the polynomials.
e) To what vector will the difference of the polynomials be equal?
f) What does this suggest about how to define vector subtraction

$$\langle a_3, a_2, a_1, a_0 \rangle - \langle b_3, b_2, b_1, b_0 \rangle$$

A1.5.2 Suppose the vector $\langle a_{11}, a_{12}, b_1 \rangle$ corresponds to the equation $a_{11}x + a_{12}y = b_1$.

a) Consider the system of equations

$$\begin{aligned} a_{11}x + a_{12}y &= b_1 \\ a_{21}x + a_{22}y &= b_2 \end{aligned}$$

What equation do you get when you add the two equations together?
b) What is the corresponding vector?
c) What does this suggest about how to define $\langle a_{11}, a_{12}, b_1 \rangle + \langle a_{21}, a_{22}, b_2 \rangle$?
d) Returning to the system of equations, what equation do you get when you subtract one equation from the other?
e) What is the corresponding vector?
f) What does this suggest about how to define $\langle a_{11}, a_{12}, b_1 \rangle - \langle a_{21}, a_{22}, b_2 \rangle$?

A1.5.3 Suppose a document is formed by joining together ("concatenating") several documents; for example, the articles in a magazine or an encyclopedia. How would you find the document vector for the concatenated document?

A1.5.4 Suppose $\langle a_1, a_0 \rangle$ represents $a_1x + a_0$, while $\langle b_2, b_1, b_0 \rangle$ represents $b_2x^2 + b_1x + b_0$.

a) What should $\langle 0, a_1, a_0 \rangle + \langle b_2, b_1, b_0 \rangle$ equal?
b) Are the *vectors* $\langle 0, a_1, a_0 \rangle$ and $\langle a_1, a_0 \rangle$ equal? Why/why not? Base your answer on the definition of the equality of two vectors.
c) Can we write the *vector* equation $\langle 0, a_1, a_0 \rangle + \langle b_2, b_1, b_0 \rangle = \langle a_1, a_0 \rangle + \langle b_2, b_1, b_0 \rangle$? Why/why not?

Activity 1.5 shows that if we want the sum of two vectors to represent the sum of the two quantities they represent, then we should define:

Definition 1.6 (Vector Addition). *Let* $\mathbf{u}, \mathbf{v} \in \mathbb{F}^n$, *with* $\mathbf{u} = \langle u_1, u_2, \dots u_n \rangle$, $\mathbf{v} = \langle v_1, v_2, \dots, v_n \rangle$. *Then*

$$\mathbf{u} + \mathbf{v} = \langle u_1 + u_2, v_1 + v_2, \dots, u_n + v_n \rangle$$

and

$$\mathbf{u} - \mathbf{v} = \langle u_1 - u_2, v_1 - v_2, \dots, u_n - v_n \rangle$$

where $u_i - v_i = u_i + (-v_i)$.

Since we are defining the sum and difference of two vectors in terms of the sum of the components, we say that addition and subtraction are performed **componentwise**.

Additionally, Activity 1.5 suggests defining the addition of two vectors with *different* numbers of components may be problematic. So we choose *not* to define the addition of two vectors if they have different numbers of components.

Example 1.3. *Find the following:*

- $\langle 5, 3, 1 \rangle + \langle 2, 1, -3 \rangle$.
- $\langle a, b, c \rangle + \langle 3, 5, -1 \rangle$.
- $\langle 1, 1, 0 \rangle - \langle 2, 5, x \rangle$.

Solution. *Operating componentwise:*

$$\begin{aligned} \langle 5, 3, 1 \rangle + \langle 2, 1, -3 \rangle &= \langle 5 + 2, 3 + 1, 1 + (-3) \rangle \\ &= \langle 7, 4, -2 \rangle \\ \langle a, b, c \rangle + \langle 3, 5, -1 \rangle &= \langle a + 3, b + 5, c + (-1) \rangle \\ &= \langle a + 3, b + 5, c - 1 \rangle \\ \langle 1, 1, 0 \rangle - \langle 2, 5, x \rangle &= \langle 1 - 2, 1 - 5, 0 - x \rangle \\ &= \langle -1, -4, -x \rangle \end{aligned}$$

Another operation we might consider is multiplication. However, multiplication of vectors is more complicated than addition, because there are two different *types* of multiplication. First, consider an expression like $\frac{1}{3} \cdot 11$ or $\sqrt{2}x$ or $5\mathbf{u}$. In this type of multiplication, the multiplier ($\frac{1}{3}$ or $\sqrt{2}$ or 5) is a real number, but the multiplicand (11 or x or $\mathbf{u}$) might not be. This is called **scalar multiplication**.

Activity 1.6: Scalar Multiplication

In the following, assume all variables represent real numbers.

A1.6.1 Suppose the vector $\langle a_2, a_1, a_0\rangle$ corresponds to the polynomial $a_2x^2 + a_1x + a_0$.

a) To what vector does $c(a_2x^2 + a_1x + a_0)$ correspond?
b) What does this suggest about how to define $c\langle a_2, a_1, a_0\rangle$?
c) To what vector does $(a_2x^2 + a_1x + a_0)c$ correspond?
d) What does this suggest about how to define $\langle a_2, a_1, a_0\rangle c$?
e) How should $-\langle a_2, a_1, a_0\rangle$ be defined?

A1.6.2 Suppose the vector $\langle a_{11}, a_{12}, b_1\rangle$ corresponds to the equation $a_{11}x + a_{12}y = b_1$.

a) If both sides of the equation are multiplied by c, what vector corresponds to the new equation?
b) What does this suggest about how to define scalar multiplication $c\langle a_{11}, a_{12}, b_1\rangle$?
c) What about $\langle a_{11}, a_{12}, b_1\rangle c$?
d) How should $-\langle a_{11}, a_{12}, b_1\rangle$ be defined?

Activity 1.6 suggests the following definition:

Definition 1.7 (Scalar Multiplication). *Let* $\mathbf{v} = \langle v_1, v_2, \ldots, v_n\rangle$ *be a vector in* $\mathbb{F}^n$*, and let* a *be any element of* $\mathbb{F}$*. We say that* a *is a* ***scalar****, and define the* ***multiplication of*** $\mathbf{v}$ ***by scalar*** a

$$a\mathbf{v} = \langle av_1, av_2, \ldots, av_n\rangle$$

Additionally $-\mathbf{v} = -1\mathbf{v}$.

As with addition, scalar multiplication is done componentwise.

We note that while we could write an expression like $\mathbf{v}a$ and have it understood (Activity A1.6.1d and A1.6.2c), we typically write the scalar multiplier first: $a\mathbf{v}$, not $\mathbf{v}a$.

Example 1.4. *Find the products:* $5\langle 2, -1, 4\rangle$, $s\langle 1, 4, 3\rangle$.

Solution. *Multiplying componentwise:*

$$\begin{aligned} 5\langle 2, -1, 4\rangle &= \langle 5\cdot 2, 5(-1), 5\cdot 4\rangle \\ &= \langle 10, -5, 20\rangle \\ s\langle 1, 4, 3\rangle &= \langle s\cdot 1, s\cdot 4, s\cdot 3\rangle \\ &= \langle s, 4s, 3s\rangle \end{aligned}$$

We introduce one more useful strategy for thinking mathematically:

Strategy. *How you speak influences how you think.*

This is an extension of the importance of definitions: using the right words helps to think about a topic in the right way. In this case, you should always refer to $a\mathbf{v}$ as a *scalar multiple.* What's important to observe here is that we have defined the product of a *scalar* and a vector, and *not* defined the product of two vectors.

What if we try to multiply two vectors?

Activity 1.7: Componentwise Multiplication?

Suppose we define the product of two vectors componentwise, as

$$\langle u_1, u_2, u_3\rangle \times \langle v_1, v_2, v_3\rangle = \langle u_1v_1, u_2v_2, u_3v_3\rangle$$

A1.7.1 Let $\langle a_2, a_1, a_0\rangle$ correspond to the polynomial $a_2x^2 + a_1 + a_0$,and $\langle b_2, b_1, b_0\rangle$ correspond to the polynomial $b_2x^2 + b_1x + b_0$

a) Use the definition above to find $\langle a_2, a_1, a_0\rangle \times \langle b_2, b_1, b_0\rangle$.

b) Does the resultant vector correspond to the product of the polynomials $(a_2x^2 + a_1 + a_0)(b_2x^2 + b_1x + b_0)$?

A1.7.2 Suppose $\langle a_{11}, a_{12}, b_1\rangle$ corresponds to the equation $a_{11}x + a_{12}y = b_1$, and $\langle a_{21}, a_{22}, b_2\rangle$ corresponds to the equation $a_{21}x + a_{22}y = b_2$.

a) Use our the definition above find $\langle a_{11}, a_{12}, b_1\rangle \times \langle a_{21}, a_{22}, b_2\rangle$.

b) Does the resultant vector correspond to the product of the two equations?

Activity 1.7 suggests that the componentwise product of two vectors doesn't seem to mean anything useful. So for now, we'll leave expressions like $\mathbf{u} \times \mathbf{v}$ undefined.

In contrast to the seeming uselessness of componentwise multiplication, one of the most important definitions in linear algebra (and possibly all of mathematics) is:

Definition 1.8. *A linear combination is the sum of scalar multiples of vectors.*

So if $\mathbf{u}$, $\mathbf{v}$, and $\mathbf{w}$ are vectors in $\mathbb{R}^n$, then $5\mathbf{u} - 2\mathbf{v} + 11\mathbf{w}$ is a linear combination of vectors in $\mathbb{R}^n$.

Example 1.5. *Let* $\mathbf{u}, \mathbf{v}$ *be vectors in* $\mathbb{R}^4$, *where* $\mathbf{u} = \langle 3, -4, 2, 7\rangle$, $\mathbf{v} = \langle 4, 1, -8, -6\rangle$. *Find* $3\mathbf{u}$, $2\mathbf{v}$, $3\mathbf{u} + 2\mathbf{v}$; *and* $\mathbf{u} - 3\mathbf{v}$.

Solution. $3\mathbf{u}$ *is a scalar multiple of* $\mathbf{u}$, *so we find it by multiplying each component of* $\mathbf{u}$ *by the scalar* 3*:*

$$\begin{aligned} 3\mathbf{u} &= 3\langle 3, -4, 2, 7\rangle \\ &= \langle 3\cdot 3, 3\cdot(-4), 3\cdot 2, 3\cdot 7\rangle \\ &= \langle 9, -12, 6, 21\rangle \end{aligned}$$

Similarly:

$$\begin{aligned} 2\mathbf{v} &= 2\langle 4, 1, -8, -6\rangle \\ &= \langle 2\cdot 4, 2\cdot 1, 2\cdot(-8), 2\cdot(-6)\rangle \\ &= \langle 8, 2, -16, -12\rangle \end{aligned}$$

We add the vectors componentwise:

$$\begin{aligned} 3\mathbf{u} + 2\mathbf{v} &= \langle 9, -12, 6, 21\rangle + \langle 8, 2, -16, -12\rangle \\ &= \langle 9+8, -12+2, 6+(-16), 21+(-12)\rangle \\ &= \langle 17, -10, -10, 9\rangle \end{aligned}$$

We subtract the vectors componentwise:

$$\begin{aligned} \mathbf{u} - 3\mathbf{v} &= \mathbf{u} - (3\mathbf{v}) \\ &= \langle 3, -4, 2, 7\rangle - \langle 3\cdot 4, 3\cdot 1, 3\cdot(-8), 3\cdot(-6)\rangle \\ &= \langle 3, -4, 2, 7\rangle - \langle 12, 3, -24, -18\rangle \\ &= \langle 3-12, -4-3, 2-(-24), 7-(-18)\rangle \\ &= \langle -9, -7, 26, 25\rangle \end{aligned}$$

Mathematicians also like to break things all the way down. This requires identifying the simplest possible instance of a concept. In the case of vectors, we define:

Definition 1.9 (Elementary Vectors). *Let* $\mathcal{V}$ *be vectors in* $\mathbb{F}^n$, *where* $\mathbb{F}$ *is a field. The* ***elementary vector*** $\mathbf{e}_i$ *is the vector where the ith component is* 1 *(the multiplicative identity for the field) and the remaining components are* 0 *(the additive identity for the field).*

Example 1.6. *Find the elementary vectors for* $\mathbb{R}^3$.

Solution. *Since* $\mathbb{R}$ *is the set of real numbers, the additive identity is* 0 *and the multiplicative identity is* 1, *so our elementary vectors will be*

$$\mathbf{e}_1 = \langle 1, 0, 0\rangle \qquad \mathbf{e}_2 = \langle 0, 1, 0\rangle \qquad \mathbf{e}_3 = \langle 0, 0, 1\rangle$$

For historical reasons, these elementary vectors are sometimes called **i**, **j**, **k**. However, this risks confusing the *vector* **i** with the imaginary number $i = \sqrt{-1}$, and it limits our scope to three dimensions, so we'll avoid this practice.

What could be simpler than an elementary vector?

Definition 1.10 (Zero Vector). *The zero vector in $\mathbb{F}^n$, designated* $\mathbf{0}$*, is the vector with n components, all of which are* 0 *(the additive identity in* $\mathbb{F}$*).*

For example, the zero vector in $\mathbb{R}^5$ is the vector with 5 components, all of which are 0: $\langle 0, 0, 0, 0, 0 \rangle$.

1.3 Proofs

For many students, an introductory course in linear algebra is their first encounter with the idea of a rigorous mathematical proof. In this section, we'll introduce a few proof ideas.

Activity 1.8: Evidence Collection

In the following, let $\vec{v}_1 = \langle 3, 5 \rangle$, $\vec{v}_2 = \langle 1, 8 \rangle$, $\vec{v}_3 = \langle 6, -5 \rangle$.

A1.8.1 Consider the addition of two vectors.

a) "There are only so many symbols." Explain why the $+$ in a statement like $5 + 3$ is **NOT** the same as the $+$ in a statement like $\vec{v}_1 + \vec{v}_2$.
b) Find $\vec{v}_1 + \vec{v}_2$.
c) Find $\vec{v}_2 + \vec{v}_1$.
d) Does vector addition appear to be commutative? Why/why not?
e) Find three more examples of vector addition that support the conclusion you reached in Activity A1.8.1d.

A1.8.2 Consider the addition of three vectors. Remember parentheses indicate an expression that must be evaluated first.

a) Find $(\vec{v}_1 + \vec{v}_2) + \vec{v}_3$.
b) Find $\vec{v}_1 + (\vec{v}_2 + \vec{v}_3)$.
c) Does vector addition appear to be associative?
d) Find three more examples of vector addition that support the conclusion you reached in Activity A1.8.2c.

In Activity A1.8.1 and A1.8.2, you gathered evidence that vector addition is commutative and associative. Gathering evidence is an important part of the scientific method: after you've gathered enough evidence, you're entitled to claim a result as a scientific *theory.*

Mathematicians work a little differently. After we've collected enough evidence to convince ourselves that something is true, we try to construct a proof. The most important thing to remember about proofs:

An example is NOT a proof!

That doesn't mean examples are useless in higher mathematics, and it's often possible to create a proof *from* an example. A useful strategy:

Strategy. *Replace the specific with the generic.*

For example, our evidence for the commutativity of vector addition came from our observation that when we added specific vectors, we got the same result no matter what order we added the vectors. Perhaps you added $\langle 5, 7\rangle$ and $\langle 2, 3\rangle$. You might have written down the addition $\langle 5, 7\rangle + \langle 2, 3\rangle = \langle 7, 10\rangle$. While there's nothing wrong with this as a mathematical statement, it will be more difficult to generalize our example. That's because the numbers in our result, $\langle 7, 10\rangle$ are *specific* numbers. In particular, the 7 came from the specific numbers 5 and 2, while the 10 came from the specific numbers 7 and 3.

Instead, it's better to write $\langle 5, 7\rangle + \langle 2, 3\rangle = \langle 5 + 2, 7 + 3\rangle$. The reason this is more useful is that it's easier to switch out the specific values (5, 7, 2, 3) with generic values (x_1, x_2, y_1, y_2) to get

$$\langle x_1, x_2\rangle + \langle y_1, y_2\rangle = \langle x_1 + y_1, x_2 + y_2\rangle$$

This suggests another useful strategy:

Strategy. *Don't do arithmetic.*

You might notice that this addition of two arbitrary vectors exactly mimics the definition of vector addition. Indeed, we could have begun at this point. Remember: Definitions are the whole of mathematics; all else is commentary. Consequently, another important proof strategy:

Strategy. *If you're not invoking one or more definitions, you're probably not doing a proof.*

We wanted to see if vector addition was commutative, so we need to add the two vectors "the other way." By our definition, we'll have:

$$\langle y_1, y_2\rangle + \langle x_1, x_2\rangle = \langle y_1 + x_1, y_2 + x_2\rangle$$

Now the definition of a vector (Definition 1.5) tells us that the components of a the vector come from a field. And our definition of a field (Definition 1.3) tells us that addition of field elements is commutative. This means that

$$x_1 + y_1 = y_1 + x_1 \text{ and } x_2 + y_2 = y_2 + x_2$$

which means the *components* of the two vector sums are the same. But our definition of the equality of two tuples (Definition 1.2) tells us that if two tuples have the same components in the same order, they are equal. Consequently we have

$$\langle x_1, x_2 \rangle + \langle y_1, y_2 \rangle = \langle y_1, y_2 \rangle + \langle x_1, x_2 \rangle$$

Let's summarize our results, with the reasons behind them:

- By the definition of vector addition, $\langle x_1, x_2 \rangle + \langle y_1, y_2 \rangle = \langle x_1 + y_1, x_2 + y_2 \rangle$
- By the definition of vector addition, $\langle y_1, y_2 \rangle + \langle x_1, x_2 \rangle = \langle y_1 + x_1, y_2 + x_2 \rangle$
- By the definition of a field (which requires commutativity of addition), $x_1 + y_1 = y_1 + x_1$ and $x_2 + y_2 = y_2 + x_2$ so the components of the two vectors are the same.
- By the definition of the equality of two vectors,

$$\langle x_1 + y_1, x_2 + y_2 \rangle = \langle y_1 + x_1, y_2 + x_2 \rangle$$

- Since the left hand side is $\langle x_1, x_2 \rangle + \langle y_1, y_2 \rangle$, and the right hand side is $\langle y_1, y_2 \rangle + \langle x_1, x_2 \rangle$, we have

$$\langle x_1, x_2 \rangle + \langle y_1, y_2 \rangle = \langle y_1, y_2 \rangle + \langle x_1, x_2 \rangle$$

It's worth pointing out that in this five-step proof, we've used definitions no less than four times. We emphasize again: If you're not invoking one or more definitions, you're probably not doing a proof.

At this point, we introduce an important idea:

The last line of the proof is what you've actually proven.

So it appears we've proven that the addition of two vectors is commutative.

However, never promise more than you've delivered. In this case, what we *actually* proved is that the addition of two vectors in $\mathbb{F}^2$ is commutative; we've said nothing about the addition of two vectors in $\mathbb{F}^3$, or $\mathbb{F}^{5000}$.

To prove commutativity of vector addition in general, we need to consider an arbitrary vector in $\mathbb{F}^n$. Such a vector might be represented as $\langle x_1, x_2, \ldots, x_n \rangle$, where the "..." (known as an *ellipsis*) is shorthand for "more like what came before and what followed."

It's risky to use an ellipsis in a proof, because it's not always clear what they represent. For example, if we write "3, 5, 7, ...," it's not clear what follows the 7: on the one hand, we might be writing down odd numbers; on the other hand, we might be writing down odd *primes*. Other interpretations are possible for the sufficiently creative.

While it's risky, it's sometimes necessary, or at the very least convenient. Thus if we want to prove that vector addition in $\mathbb{F}^n$ is commutative, we want to end with the equation

$$\langle x_1, x_2, \ldots, x_n \rangle + \langle y_1, y_2, \ldots, y_n \rangle = \langle y_1, y_2, \ldots, y_n \rangle + \langle x_1, x_2, \ldots, x_n \rangle$$

To get there, we'll offer a useful strategy:

Strategy. *You can always write down* **one** *side of an equality.*

So we can start our proof by writing down

$$\langle x_1, x_2, \ldots, x_n\rangle + \langle y_1, y_2, \ldots, y_n\rangle =$$

To determine what goes on the other side of the equals, we can invoke a few definitions and see if we can get to the desired result.

Activity 1.9: Properties of Vector Arithmetic

A1.9.1 In the following, you'll prove: The addition of vectors in $\mathbb{F}^n$ is commutative.

a) Write down

$$\langle x_1, x_2, \ldots, x_n\rangle + \langle y_1, y_2, \ldots, y_n\rangle =$$

Complete the sentence: "According to the definition of vector addition, this is equal to ..."

b) Write down

$$\langle y_1, y_2, \ldots, y_n\rangle + \langle x_1, x_2, \ldots, x_n\rangle =$$

Complete the sentence: "According to the definition of vector addition, this is equal to ..."

c) Prove/disprove the two sums are the same. Justify every step by referring to a specific definition, e.g., "According to the definition of vector addition ..."

A1.9.2 Prove: The addition of vectors in $\mathbb{F}^n$ is associative. In other words, given three vectors $\vec{u}$, $\vec{v}$, $\vec{w}$, we have

$$(\vec{u} + \vec{v}) + \vec{w} = \vec{u} + (\vec{v} + \vec{w})$$

A1.9.3 Prove: If a, b are scalars in $\mathbb{F}$, and $\vec{v}$ is a vector in $\mathbb{F}^n$, then $(ab)\vec{v} = a(b\vec{v})$. (This is sometimes called *associativity of scalar multiplication*)

A1.9.4 Prove: If a is a scalar in $\mathbb{F}$ and $\vec{u}$, $\vec{v}$ are vectors in $\mathbb{F}^n$, then $a(\vec{u} + \vec{v}) = a\vec{u} + a\vec{v}$. (This is sometimes called *left distributivity*).

A1.9.5 Prove: If a, b are scalars in $\mathbb{F}$ and $\vec{u}$ is a vector in $\mathbb{F}^n$, then $(a+b)\vec{u} = a\vec{u} + b\vec{u}$. (This is sometimes called *right distributivity*)

In Activity 1.9, you proved several important vector properties:

Theorem 1.1 (Vector Properties). *Let* $\mathbf{u}$*,* $\mathbf{v}$*,* $\mathbf{w}$ *be vectors in* $\mathbb{F}^n$*, and let* a, b *be scalars from* $\mathbb{F}$*. Then:*

- *Vector addition is commutative, so* $\mathbf{u} + \mathbf{v} = \mathbf{v} + \mathbf{u}$*,*
- *Vector addition is associative, so* $(\mathbf{u} + \mathbf{v}) + \mathbf{w} = \mathbf{v} + (\mathbf{u} + \mathbf{w})$*,*
- *Scalar multiplication is associative, so* $a(b\mathbf{u}) = (ab)\mathbf{u}$*,*
- *Scalar multiplication is left distributive, so* $a(\mathbf{u} + \mathbf{v}) = a\mathbf{u} + a\mathbf{v}$*,*
- *Scalar multiplication is right distributive, so* $(a + b)\mathbf{u} = a\mathbf{u} + b\mathbf{u}$*.*

One type of proof you might not have seen before is a **proof by mathematical induction**. As an example of mathematical induction works, consider the following. A field requires a binary operation: $a + b$ has to have a value. But fields do not require trinary operations: it's conceivable that $a+b+c$ might be undefined!

Of course, if $a + b + c$ was undefined, we wouldn't be able to do very much with the field, so we use the following convention: when there are multiple summands, we add from left to right. So $a + b + c$ is the same as $(a + b) + c$, where we have an implied parentheses around the first two terms. Similarly, $a + b + c + d$ would be found by adding $a + b + c$ first (which we do by finding $(a + b) + c$), and then adding d:

$$a + b + c + d = ((a + b) + c) + d$$

and so on.

Now consider associativity requirement of a field: $(a + b) + c = a + (b + c)$. This tells us we can add *three* terms by adding the first pair, or by adding the second pair, and we'll get the same result either way.

But what if we have four terms: Does $(a + b + c) + d = a + (b + c + d)$? We *can't* invoke the field definition, because the definition only tells us what happens with *three* terms.[1] We need to prove that $(a+b+c)+d = a+(b+c+d)$.

Now suppose we do that. "Never promise more than you've delivered," so a proof that $(a + b + c) + d = a + (b + c + d)$ tells us *nothing* about what happens when there are *five* terms—let alone five hundred. So what can we do?

That's where mathematical induction comes in. In general, you might want to consider a proof by induction when you can break the statement to be proven into an orderly sequence of statements, where no possibility would be missed. For example, we can break associativity down into the cases dealing with:

- The sum of *three* terms: $(a_1 + a_2) + a_3 = a_1 + (a_2 + a_3)$.

[1]There's a reason law schools like mathematics majors: mathematicians tend to be very literal, and if it's not written down, we don't allow it.

- The sum of *four* terms: $(a_1 + a_2 + a_3) + a_4 = a_1 + (a_2 + a_3 + a_4)$.
- The sum of *five* terms: $(a_1 + a_2 + a_3 + a_4) + a_5 = a_1 + (a_2 + a_3 + a_4 + a_5)$

and so on. This sequence is ordered: we know what the next statement will be about (the sum of six terms). Moreover, every statement of the form "The sum of n terms" will be somewhere on the list.

The general structure of a proof by mathematical induction is the following:

- First, we prove a *base case*: the first case that has to be true.
- Next, we prove that *if* the result holds for some generic case, *then* it holds for the next case. This is known as the *induction step*.

In this case, the first case is the sum of three terms:

$$(a_1 + a_2) + a_3 = a_1 + (a_2 + a_3)$$

To prove the base case, remember: Definitions are the whole of mathematics; all else is commentary. In this case:

Proof. According to the definition of a field, $(a_1+a_2)+a_3 = a_1+(a_2+a_3)$. □

Now for the induction step: We *assume* the statement holds for k terms, so

$$(a_1 + a_2 + \ldots + a_{k-1}) + a_k = a_1 + (a_2 + \ldots + a_{k-1} + a_k)$$

We want to show that the statement holds for $k + 1$ terms:

$$(a_1 + a_2 + \ldots + a_{k-1} + a_k) + a_{k+1} = a_1 + (a_2 + \ldots + a_{k-1} + a_k + a_{k+1})$$

Again, since this is what we want to prove, this has to be the last line we write.

Remember: You can always write down one side of an equality. So let's write down the left hand side of the equality we want to prove:

$(a_1 + a_2 + \ldots + a_{k-1} + a_k) + a_{k+1} =$

It's important to recognize that, at this point, we *don't* know what we can write on the right hand side. However, notice that inside the parentheses we have a sum of many terms. Remember that this has to be evaluated by adding the first terms together: in other words, there's an implied set of parentheses around the first set of terms. This means we can write:

$$= ((a_1 + a_2 + \ldots + a_{k-1}) + a_k) + a_{k+1}$$

If you look at what we've written, you'll see a somewhat familiar expression: $(a_1+a_2+\ldots+a_{k-1})+a_k$. And we assumed something about this expression: namely it was the same as $a_1+(a_2+\ldots+a_{k-1}+a_k)$. So we can replace it, giving us:

$$= (a_1 + (a_2 + \ldots + a_{k-1} + a_k)) + a_{k+1}$$

The thing to recognize here is that because it's in parentheses, $(a_2 + \ldots + a_{k-1} + a_k)$ is a *single* quantity. To emphasize this, we'll cover it up:

$$= (a_1 + \blacksquare\blacksquare\blacksquare\blacksquare\blacksquare) + a_{k+1}$$

Notice we have the sum of *three* things, so we can apply associativity:

$$= a_1 + (\blacksquare\blacksquare\blacksquare\blacksquare\blacksquare + a_{k+1})$$

Now we'll apply the "kindergarten rule": put things back where you found them. Remember the $\blacksquare\blacksquare\blacksquare\blacksquare\blacksquare$ covered up the sum $(a_2+\ldots+a_{k-1}+a_k)$, so uncovering it:

$$= a_1 + ((a_2 + \ldots + a_{k-1} + a_k) + a_{k+1})$$

Now remember that when we add, we add from left to right, which means the inner set of parentheses are redundant: we *have* to do the sum first, whether or not the parentheses are there. So we drop the inner set of parentheses:

$$= a_1 + (a_2 + \ldots + a_{k-1} + a_k + a_{k+1})$$

Since we've maintained equality throughout, we can join the beginning to the end to obtain:

$$(a_1 + a_2 + \ldots + a_k) + a_{k+1} = a_1 + (a_2 + \ldots + a_{k-1} + a_k + a_{k+1})$$

which completes our induction step.

How does this prove our result? First (base step), we know that associativity holds for the sum of three terms. Next (induction step), we know that if associativity holds for the sum of three terms, it holds for the "next" sum: namely, the sum of four terms.

Now:

Strategy. *Lather, rinse, repeat.*

Since associativity holds for four terms, we know (induction step) it holds for "next" sum: namely, the sum five terms. And since it holds for five terms, we know (induction step) it holds for "next" sum: namely, the sum of six terms. And since it holds for six terms, it holds for seven, and so on.

Activity 1.10: More Vector Properties

A1.10.1 In the following, you'll prove that the sum of any number of vectors in $\mathbb{R}^n$ is associative.

a) Prove the base step: $(\mathbf{u} + \mathbf{v}) + \mathbf{w} = \mathbf{u} + (\mathbf{v} + \mathbf{w})$.
b) Prove the induction step: if the sum of k vectors in $\mathbb{R}^n$ is associative, then the sum of $k + 1$ vectors is associative.
c) Explain why Activities A1.10.1a and A1.10.1b prove the sum of any number of vectors in $\mathbb{R}^n$ is associative.

A1.10.2 In the following, you'll prove that left distributivity holds for any number of vectors in $\mathbb{R}^n$ (see Activity A1.9.4 for the meaning of left distributivity).

a) Prove the base step: $a(\mathbf{u} + \mathbf{v}) = a\mathbf{u} + a\mathbf{v}$.
b) Prove the induction step: If $a(\mathbf{u}_1 + \mathbf{u}_2 + \ldots + \mathbf{u}_k) = a\mathbf{u}_1 + a\mathbf{u}_2 + \ldots + a\mathbf{u}_k$, then $a(\mathbf{u}_1 + \mathbf{u}_2 + \ldots + \mathbf{u}_k + \mathbf{u}_{k+1}) = a\mathbf{u}_1 + a\mathbf{u}_2 + \ldots + a\mathbf{u}_k + a\mathbf{u}_{k+1}$.
c) Explain why the base step and the induction step prove that left distributivity holds for any number of vectors.

A1.10.3 Prove: Right distributivity of vectors in $\mathbb{R}^n$ holds for any number of scalars. (See Activity A1.9.5 for the meaning of right distributivity).

1.4 Directed Distances

A useful thing to do in mathematics (and in life) is to view a single concept from several different perspectives. In mathematics, there are two important viewpoints: the algebraic and the geometric. Algebra concerns itself with numbers, formulas, and equations. Geometry focuses on figures, drawings, and graphs. For millennia, mathematics was based on geometry. That began to change during the 16th century, and today mathematics is almost entirely based on algebra. Thus our definition of a vector is a purely algebraic definition: it's a n-tuple of field elements.

However, it's often useful to have a geometric picture of a mathematical concept: an idea of what it "looks like." To develop that geometric picture, suppose P and Q are two points in space. To navigate *from* P *to* Q, we to follow a sequence of steps that tell us which *direction* to travel, and what *distance* to travel. We define:

Definition 1.11 (Directed Distance). *Let P, Q be two points in $\mathbb{R}^n$. $\overrightarrow{PQ}$ is the* ***directed distance*** *from P to Q.*

Activity 1.11: Directed Distances

In the following, let the coordinates of P, Q, R be $(2, 5)$, $(1, 3)$, and $(3, 8)$, respectively; let O be the origin of our coordinate system.

A1.11.1 Consider the directed distances from O to the points P, Q, or R.

a) Express $\overrightarrow{OP}$ by stating how far horizontally and how far vertically you must go in order to travel from O to P. (Be sure to identify whether you are going to the right or left, upward or downward)

b) Express $\overrightarrow{OQ}$ and $\overrightarrow{OR}$ in an analogous fashion.

A1.11.2 Consider a directed distance, like "Go x units horizontally, then y units vertically."

a) Explain why we can give this directed distance as an ordered 2-tuple. What does the first component correspond to? What does the second component correspond to?

b) Explain why $x, y \in \mathbb{R}$.

c) Why does this mean that directed distances are vectors?

A1.11.3 Consider a directed distance given using the cardinal directions (north, south, east, and west); for example, "Go n units north, then w units west." Explain why we can give this directed distance as an ordered 4-tuple. What does each component correspond to?

Activity 1.11 suggests the following:

Definition 1.12. *Let the coordinates of a point X in $\mathbb{R}^n$ be $(x_1, x_2, x_3, \ldots, x_n)$. Then the directed distance from the origin O to the point X is the vector*

$$\overrightarrow{OX} = \langle x_1, x_2, x_3, \ldots, x_n \rangle$$

where $\overrightarrow{OX}$ is a vector in $\mathbb{R}^n$.

Conversely, if $\vec{v} = \langle v_1, v_2, \ldots, v_n \rangle$ is a vector in $\mathbb{R}^n$, then we can interpret $\vec{v}$ as a directed distance from the origin to the point $(v_1, v_2, \ldots, v_n)$.

What about the directed distance between two points, neither of which is the origin? We might draw this using an arrow running from P to Q:

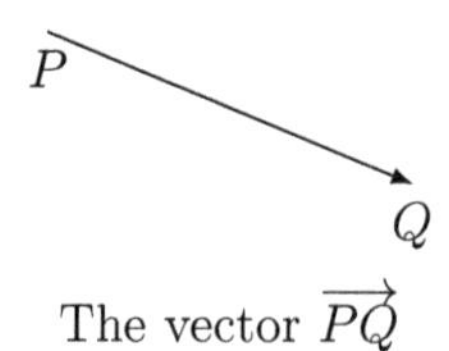

The vector $\overrightarrow{PQ}$

Activity 1.12: More Directed Distances

A1.12.1 For the following, let the points $P = (7, 1)$, $Q = (5, 4)$, $R = (3, 2)$, $S = (1, 3)$; also let O be the origin of the coordinate system.

a) Find the vectors: $\overrightarrow{OP}$, $\overrightarrow{OQ}$, $\overrightarrow{OR}$, $\overrightarrow{OS}$.

b) Express $\overrightarrow{PQ}$ as a directed distance, by stating how far horizontally and how far vertically you must go to travel *from P to Q*. (Use the convention that positive distances correspond to movement right or upward, and negative distances correspond to movement left or downward)

c) Use these horizontal and vertical distances to express $\overrightarrow{PQ}$ as a vector.

d) Express $\overrightarrow{PR}$ as a vector.

A1.12.2 Suppose $\overrightarrow{OX} = \langle 3, 5 \rangle$ and $\overrightarrow{XY} = \langle 1, 4 \rangle$.

a) Explain why $\overrightarrow{XO} = \langle -3, -5 \rangle$.

b) Find $\overrightarrow{YX}$.

c) Assuming O is the origin, what are the coordinates of Y?

d) Find $\overrightarrow{OY}$.

A1.12.3 Suppose M, N are points in $\mathbb{R}^3$ with coordinates (x_1, x_2, x_3) and (y_1, y_2, y_3), respectively. Find $\overrightarrow{MN}$ and $\overrightarrow{NM}$.

Activity 1.12 suggests the following:

Definition 1.13. *Suppose the coordinates of P and Q are $(x_1, x_2, \ldots, x_n)$ and $(y_1, y_2, \ldots, y_n)$. Then directed distance from P to Q is the vector*

$$\overrightarrow{PQ} = \langle y_1 - x_1, y_2 - x_2, \ldots, y_n - x_n \rangle$$

Note that this definition can be applied to vectors with any number of components.

Example 1.7. *Suppose $X = (1, 5, 3, 8)$, $Y = (2, 7, -3, -5)$ are points in $\mathbb{R}^4$. Find $\overrightarrow{XY}$.*

Solution.

$$\overrightarrow{XY} = \langle 2 - 1, 7 - 5, -3 - 3, -5 - 8 \rangle = \langle 1, 2, -6, -13 \rangle$$

Since a vector can have any number of components, but we can't effectively visualize objects with more than three dimensions, it might seem that the geometric viewpoint is too limited to be useful. But is it?

Activity 1.13: Vectors in $\mathbb{R}^{5000}$

Let P, Q, R, S be points in $\mathbb{R}^{5000}$, and let O be the origin. Note that this means each point will have 5000 coordinates. All of your answers to the following should be based on geometry.

A1.13.1 Explain why the points P, Q, R can be represented in a single plane.

A1.13.2 Is it possible to draw the vectors $\overrightarrow{PQ}$, $\overrightarrow{QR}$ in the same plane? Why/why not?

A1.13.3 Is it possible to draw the vectors $\overrightarrow{PQ}$, $\overrightarrow{PS}$ in the same plane (possibly different from the plane in Activity A1.13.2? Why/why not?

A1.13.4 Is it possible to draw the vectors $\overrightarrow{PQ}$, $\overrightarrow{QR}$, *and* $\overrightarrow{PS}$ in the same plane? Why/why not?

Activity 1.13 illustrates a useful point: Given *any* three points, regardless of whether they're in $\mathbb{R}^3$ or $\mathbb{R}^{5000}$, we can draw them as if all three points were in the same plane.

Activity 1.14: The Geometry of Vectors

For the following, let the points $P = (7, 1)$, $Q = (5, 4)$, $R = (3, 2)$, $S = (1, 5)$; also let O be the origin of the coordinate system.

A1.14.1 Graph P, Q, R, S, then answer the following questions.

a) Find $\overrightarrow{PQ}$, $\overrightarrow{PR}$, $\overrightarrow{PS}$, $\overrightarrow{QR}$, $\overrightarrow{QS}$, and $\overrightarrow{RS}$. Also, draw these as directed distances.
b) Explain why $\overrightarrow{PQ} = \overrightarrow{RS}$, even though their beginning and ending points are completely different.

A1.14.2 If possible, find the *coordinates* of the specified point.

a) X so $\overrightarrow{PX} = \overrightarrow{QR}$.
b) Y so $\overrightarrow{YQ} = \overrightarrow{RS}$.
c) Graph the points P, Q, R, S, X, Y.
d) Graph the vectors $\overrightarrow{PX}$, $\overrightarrow{QR}$, $\overrightarrow{YQ}$, $\overrightarrow{RS}$.
e) What *geometric* feature(s) of vectors can be used to determine whether two vectors are equal?

Activity 1.14 points to an unexpected feature of vectors: they have no *spatial* location. We can move a vector anywhere, and as long as we change neither its direction nor its length, it will be the same vector.

Activity 1.15: Direction and Magnitude

A1.15.1 The figure shows several vectors. Which vectors are equal?

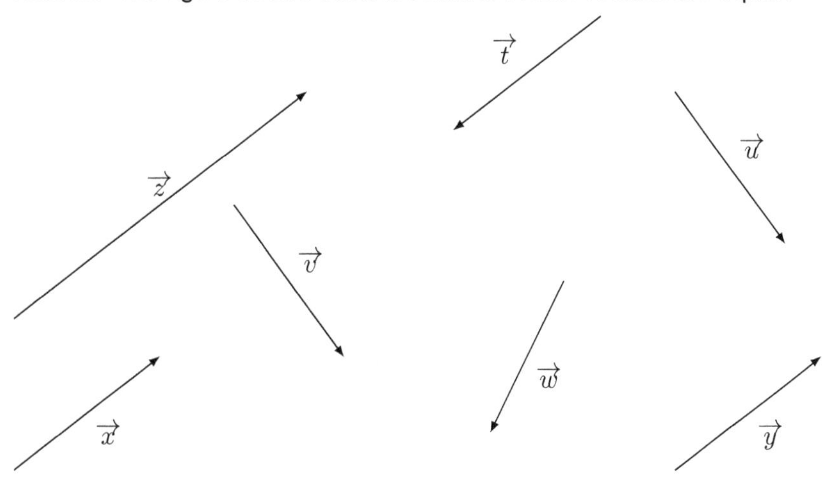

A1.15.2 Suppose P, Q, R, S are four points in $\mathbb{R}^{5000}$.

a) Explain why we can't, in general, draw all four points in the same plane. Base your answer on geometry.

b) Explain why this means we can't draw $\overrightarrow{PQ}$, $\overrightarrow{RS}$ in the same plane.

c) Explain why we can find a point X with $\overrightarrow{PX} = \overrightarrow{RS}$, where $\overrightarrow{PX}$, $\overrightarrow{PQ}$ can be drawn in the same plane.

d) Explain why we can find point a Y with $\overrightarrow{QY} = \overrightarrow{RS}$, where $\overrightarrow{PQ}$, $\overrightarrow{QY}$ can be drawn in the same plane.

Our ability to "move" vectors, as long as we retain their direction and length, has an important implication for visualizing vectors. Take two vectors $\vec{x}$, $\vec{y}$ in $\mathbb{R}^n$. Even if the vectors live in a higher dimensional space, we can move both vectors into the same two-dimensional plane, say the plane of the text or the chalkboard. There are two useful ways to do this: we can either move the vectors so they *start* at the same point; or move the vectors so that one begins where the other ends:

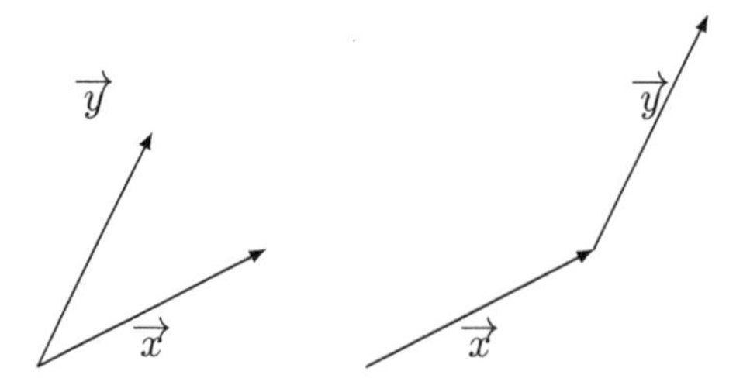

Vectors begin at same point   Second vector begins at end of first vector

Let's use this ability to represent *any* vectors in two-dimensional space to find a geometric interpretation of vector addition and scalar multiplication.

Activity 1.16: Vector Arithmetic

A1.16.1 Suppose $A = (2, 3)$, $B = (4, 1)$.

a) Graph A, B, $\overrightarrow{AB}$.

b) Find $\overrightarrow{AB}$, $3\overrightarrow{AB}$, and $-\overrightarrow{AB}$.

c) Find X so $\overrightarrow{AX} = 3\overrightarrow{AB}$.

d) Describe the relation between $\overrightarrow{AX}$ and $\overrightarrow{AB}$ *geometrically*. In particular, how do the length and direction of $\overrightarrow{AX}$ and $\overrightarrow{AB}$ compare?

e) Find Y so $\overrightarrow{AY} = -\overrightarrow{AB}$.

f) Describe the relations between $\overrightarrow{AY}$ and $\overrightarrow{AB}$ *geometrically*. In particular, how do the length and direction of $\overrightarrow{AY}$ and $\overrightarrow{A}B$ compare?

A1.16.2 Suppose $P = (1, 5)$, $Q = (3, 7)$, and $R = (2, 10)$.

a) Graph P, Q, R.

b) Find $\overrightarrow{PQ}$, $\overrightarrow{QR}$, and $\overrightarrow{PR}$ (expressing them as vectors). Also draw these vectors as directed distances.

c) Find $\overrightarrow{PQ} + \overrightarrow{QR}$.

d) What appears to be the relationship between $\overrightarrow{PQ} + \overrightarrow{QR}$ and $\overrightarrow{PR}$? (Write this relation as an *algebraic* equation involving the vectors)

e) How would you describe the *geometric* relationship between the vectors $\overrightarrow{PQ}$, $\overrightarrow{QR}$, and $\overrightarrow{PR}$?

A1.16.3 Let $K = (1, 4)$, $L = (2, 7)$, $M = (5, 1)$ be three vertices of parallelogram $KLMN$.

a) Graph K, L, M.

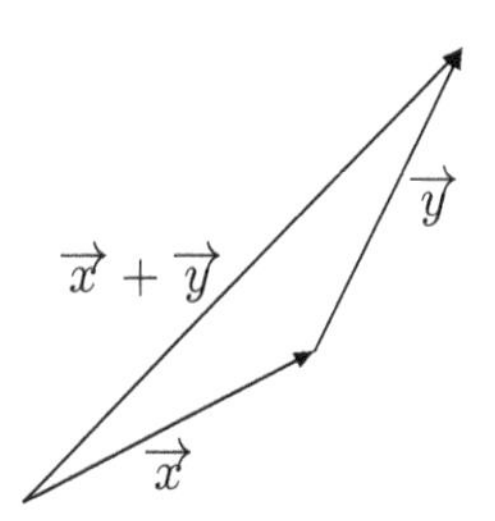

Tip-to-Tail Addition

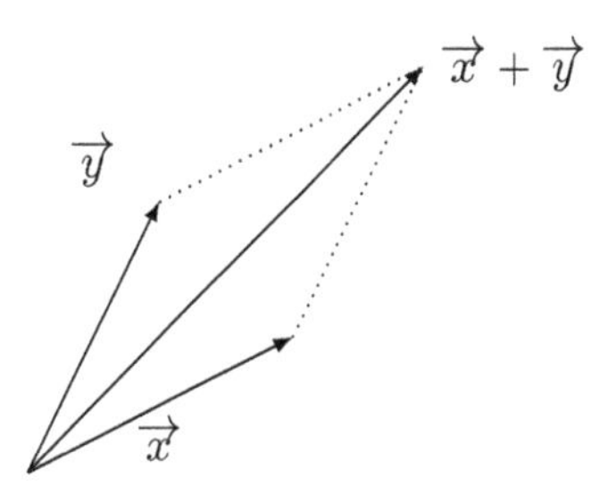

Parallelogram Addition

FIGURE 1.1
Addition of Vectors

b) Where must N be to complete the parallelogram?

c) Find $\overrightarrow{LK}$, $\overrightarrow{LM}$.

d) Find $\overrightarrow{LK} + \overrightarrow{LM}$.

e) Find X so $\overrightarrow{LX} = \overrightarrow{LK} + \overrightarrow{LM}$.

f) Describe the *geometric* relationship between $\overrightarrow{LK}$, $\overrightarrow{LM}$, and $\overrightarrow{LK} + \overrightarrow{LM}$.

Activity A1.16.1 suggests the following interpretation for scalar multiples of a vector: Given a vector $\overrightarrow{u}$ and scalar c, the vector $c\overrightarrow{u}$ is a vector whose length is increased by a factor of $|c|$. If $c > 0$, $c\overrightarrow{u}$ has the same direction; if $c < 0$, $c\overrightarrow{u}$ has the opposite direction.

Activity A1.16.2 suggests the following geometric interpretation for the sum of two vectors: Given two vectors $\overrightarrow{u}$, $\overrightarrow{v}$, if we place the starting point of $\overrightarrow{v}$ at the ending point of $\overrightarrow{u}$, then $\overrightarrow{u} + \overrightarrow{v}$ is the vector joining the starting point of $\overrightarrow{u}$ to the ending point of $\overrightarrow{v}$. This is called the **tip-to-tail addition** of vector addition.

Meanwhile, Activity A1.16.3 suggests a different interpretation: If we place the two vectors $\overrightarrow{u}$, $\overrightarrow{v}$ so they originate at the same point, then $\overrightarrow{u} + \overrightarrow{v}$ is the vector joining the common starting point of the vectors to the opposite vertex of the parallelogram whose sides correspond to $\overrightarrow{u}$ and $\overrightarrow{v}$. This is called the **parallelogram addition** of two vectors. We illustrate both in Figure 1.1.

This geometric interpretation suggests a way to describe lines and planes.

Activity 1.17: Vector Equation of a Line

A1.17.1 Let $P = (7, 1)$ and $Q = (3, 5)$.

a) Find the equation of the line $\overleftrightarrow{PQ}$. Also draw the line $\overleftrightarrow{PQ}$.

b) Find $\overrightarrow{PQ}$. Also draw the vector.

c) Suppose $\overrightarrow{PR} = 2\overrightarrow{PQ}$. Find R.

d) Is R on the line $\overleftrightarrow{PQ}$? Why/why not?

e) Suppose $\overrightarrow{PS} = -4\overrightarrow{PQ}$. Find S.

f) Is S on the line $\overleftrightarrow{PQ}$? Why/why not?

g) Suppose $\overrightarrow{PT} = \frac{1}{5}\overrightarrow{PQ}$. Find T.

h) Is T on the line $\overleftrightarrow{PQ}$? Why/why not?

i) Suppose c is a real number. What does the preceding suggest about X, where $\overrightarrow{PX} = c\overrightarrow{PQ}$?

A1.17.2 Suppose $A = (x_1, y_1)$, $B = (x_2, y_2)$.

a) Write the equation of the line $\overleftrightarrow{AB}$.

b) Find $\overrightarrow{AB}$.

c) Show that if t is a real number, then X, where $\overrightarrow{AX} = t\overrightarrow{AB}$, is on the line $\overleftrightarrow{AB}$. (You must show the *coordinates* of X will be the coordinates of a point on the line)

A1.17.3 Let O be the origin, and let $M = (2, 6)$ and $N = (-3, 8)$.

a) Find $\overrightarrow{OM}$, $\overrightarrow{ON}$, and $\overrightarrow{MN}$. Also draw the vectors.

b) Based on your drawing, what is the relationship between $\overrightarrow{OM}$, $\overrightarrow{ON}$, and $\overrightarrow{MN}$? Explain in terms of the geometric interpretation of vectors and vector arithmetic.

c) Let c be any real number. Sketch a figure showing X, where $\overrightarrow{OX} = \overrightarrow{OM} + c\overrightarrow{MN}$. What does this suggest about the equation of the line in terms of the vectors $\overrightarrow{OM}$, $\overrightarrow{MN}$?

Activity 1.17 suggests:

Theorem 1.2 (Vector Equation of a Line). *Let X be any point on $\overleftrightarrow{PQ}$, and O the origin of our coordinate system. Then*

$$\overrightarrow{OX} = \overrightarrow{OP} + t\overrightarrow{PQ}$$

for some scalar t.

Intuitively, we may view this result as follows: To get from the origin O to some point X on the line,

- First, use $\overrightarrow{OP}$ to go from the origin O to some point P on the line,

- Then follow $t\overrightarrow{PQ}$, which takes you from the point P on the line to another point X on the line.

Example 1.8. *Find the vector equation of the line through* $P = (1, 3, 5, 8)$ *and* $Q = (-4, 3, 7, 0)$.

Solution. *We find*

$$\overrightarrow{OP} = \langle 1, 3, 5, 8 \rangle$$
$$\overrightarrow{PQ} = \langle -5, 0, 2, -8 \rangle$$

so

$$\overrightarrow{OP} + t\overrightarrow{PQ} = \langle 1, 3, 5, 8 \rangle + t\langle -5, 0, 2, -8 \rangle$$

We can interpret our equation as follows:

- $\langle 1, 3, 5, 8 \rangle$ will take you from the origin to a point P on the line,
- From there, traveling any distance $t\langle -5, 0, 2, -8 \rangle$ will take you to another point on the line.

Activity 1.18: Vector Equation of a Plane

A1.18.1 Suppose P, Q, R are points on the plane PQR.

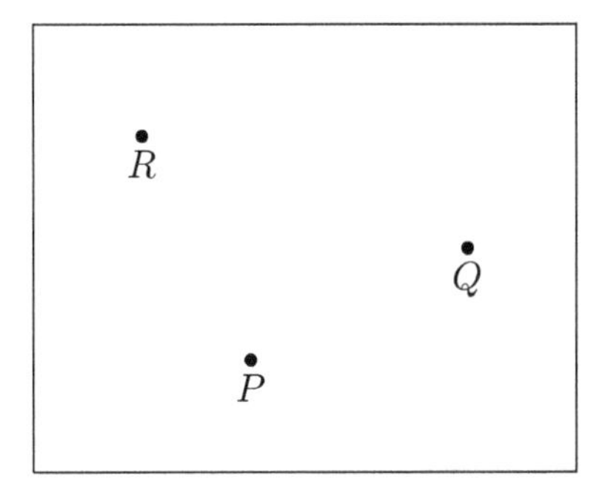

a) Draw vectors representing $\overrightarrow{PQ}$ and $\overrightarrow{PR}$.

b) Find X_1, where $\overrightarrow{PX_1} = \overrightarrow{PQ} + \overrightarrow{PR}$. Is X_1 on the plane? Why/why not?

c) Find X_2, where $\overrightarrow{PX_2} = 3\overrightarrow{PQ} + 2\overrightarrow{PR}$. Is X_2 on the plane? Why/why not?

d) Find X_3, where $\overrightarrow{PX_3} = -2\overrightarrow{PQ} + \overrightarrow{PR}$. Is X_3 on the plane? Why/why not?

e) Suppose X_0 is any point on the plane PQR. What do your answers suggest about a way to find another point on the plane?

f) Suppose $T = (-3, 5)$, $U = (1, 4)$, $V = (2, 7)$. Use the approach you described in Activity A1.18.1e to find another point X on the plane TUV.

g) Suppose $L = (4, 1, 7)$, $M = (1, 5, 3)$, and $N = (2, 1, -8)$. Use the approach you described in Activity A1.18.1e to find another point X on the plane LMN.

A1.18.2 Let $P = (x_1, y_1, z_1)$, $Q = (x_2, y_2, z_2)$, and $R = (x_3, y_3, z_3)$. Find a vector equation for X, where X is another point on the plane PQR.

Activity 1.18 suggests:

Theorem 1.3 (Vector Equation of a Plane). *Let X be any point on plane PQR, and O the origin of our coordinate system. Then*

$$\overrightarrow{OX} = \overrightarrow{OP} + s\overrightarrow{PQ} + t\overrightarrow{QR}$$

for any scalars s, t.

Again, we might interpret our equation as follows:

- $\overrightarrow{OP}$ is the directed distance that takes you from the origin to some point P on the plane,
- $s\overrightarrow{PQ}$ and $t\overrightarrow{QR}$ will move you from point P to another point X on the plane.

Example 1.9. *Let $P = (1, 5, 8)$, $Q = (-3, 1, 7)$, and $R = (0, 2, 4)$. Find the vector equation of the plane PQR.*

Solution. *We find:*

$$\overrightarrow{OP} = \langle 1, 5, 8 \rangle$$
$$\overrightarrow{PQ} = s\langle -4, -4, -1 \rangle$$
$$\overrightarrow{PR} = t\langle -1, -3, -4 \rangle$$

so

$$\overrightarrow{OP} + s\overrightarrow{PQ} + t\overrightarrow{PR} = \langle 1, 5, 8 \rangle + s\langle -4, -4, -1 \rangle + t\langle -1, -3, -4 \rangle$$

Note that in the vector equation of the line and the plane, the parameters can take on any value. We say they are "free variables," in the sense that there are no constraints on them. On the other hand, the coordinate values $x, y, z, \ldots$ are constrained: we compute their values from the free variables.

It's also worth pointing out that the vector equations don't depend on the number of components. This means we can generalize the concept of lines and planes into higher dimensional spaces.

Activity 1.19: Hyperspace

A1.19.1 The vector equations allow us to gain some insight into the concept of dimension.

a) We regard lines as one-dimensional objects. What feature of the vector equation of a line corresponds to this dimension of 1?

b) We regard planes a two-dimensional objects. What feature of the vector equation of a plane corresponds to this dimension of 2?

A1.19.2 Let $T = (1, 5, -3, 8)$ and $V = (5, 6, 4, 7)$. Write the equation of the line $\overleftrightarrow{TV}$.

A1.19.3 Let $A = (1, 5, 3, -2)$, $B = (6, 1, -8, 3)$, $C = (1, 1, 0, -5)$. Write the equation of the plane ABC.

A1.19.4 A **hyperplane** is a geometric object whose dimension is one less than the space it lives in. Consider a hyperplane in $\mathbb{R}^4$. What would you expect its vector equation to look like? How many points in $\mathbb{R}^4$ would you need to find this equation?

1.5 Magnitude

If a vector is a directed distance, the natural questions to ask are "What direction?" and "How far?" We'll deal with the second question first.

Activity 1.20: Length of a Vector

A1.20.1 Consider the points P, Q, and assume O is the origin.

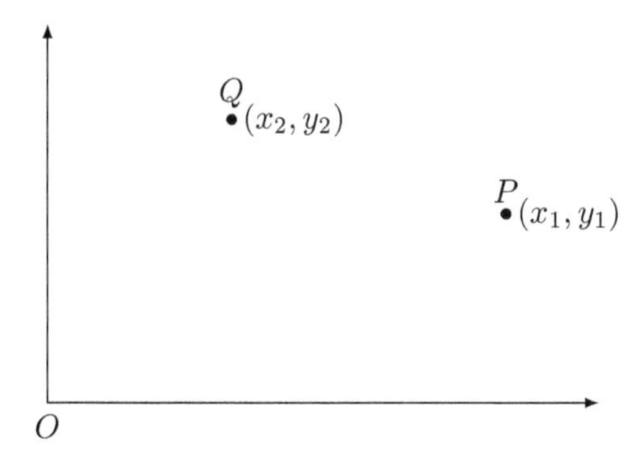

A1.20.2 Find the length of the line segment $\overline{OP}$ and $\overline{OQ}$.

A1.20.3 Remember one interpretation of a vector is as a "directed distance."

a) What distance should be associated with the vector $\overrightarrow{OQ}$?

b) What distance should be associated with the vector $\overrightarrow{OP}$?

c) What distance should be associated with the vector $\overrightarrow{QP}$?

Activity A1.20.3 suggests we can identify the length of a vector $\overrightarrow{XY}$ with the distance between the points X and Y, at least for vectors in $\mathbb{R}^2$. The Pythagorean Theorem generalizes to higher dimensional spaces; this is in fact a consequence of our ability to shift *any* three points into a plane (Activity 1.13). Consequently, we introduce the following definition for the length or **magnitude** of the vector:

Definition 1.14 (Magnitude). *Let $\vec{x} = \langle x_1, x_2, \ldots, x_n \rangle$, where the x_is are real numbers. The* **magnitude of** $\vec{x}$*, written $|\vec{x}|$ or $||\vec{x}||$, is defined to be*

$$\sqrt{x_1^2 + x_2^2 + x_3^2 + \ldots + x_n^2}$$

Because this corresponds to the geometric length of the directed line segment, $||\vec{x}||$ is sometimes called the **Euclidean norm** of the vector.

Example 1.10. *Find the magnitude of $\vec{u} = \langle 3, -1, 5, 8 \rangle$.*

Solution. *The magnitude is the square root of the sum of the squares of the vector components, so*

$$\begin{aligned} ||\vec{u}|| &= \sqrt{3^2 + (-1)^2 + 5^2 + 8^2} \\ &= \sqrt{9 + 1 + 25 + 64} \\ &= \sqrt{99} \end{aligned}$$

Our definition of magnitude presupposes the components of our vector are real numbers. But all mathematicians like to generalize, so the natural question to ask at this point is: Could we extend this definition to vectors in arbitrary fields?

Activity 1.21: Complex Magnitudes

In the following, suppose $\vec{v} = \langle v_1, v_2, \ldots, v_n \rangle$ is a vector in $\mathbb{C}^n$, where $\mathbb{C}$ is the set of complex numbers. Define

$$||\vec{v}|| = \sqrt{v_1^2 + v_2^2 + \ldots + v_n^2}$$

A1.21.1 Prove: For $\vec{0} \in \mathbb{C}^n$, $||\vec{0}|| = 0$.

A1.21.2 Consider the following vectors in $\mathbb{C}^2$, where the components are complex numbers.

a) Find $||\vec{v}||$, where $\vec{v} = \langle 3, i \rangle$.
b) Find $||\vec{v}||$, where $\vec{v} = \langle 4, 2i \rangle$.
c) Find $||\vec{v}||$, where $\vec{v} = \langle 3i, 1 \rangle$.
d) Find $||\vec{v}||$, where $\vec{v} = \langle 2i, 2 \rangle$.
e) What difficulties arise if we consider $||\vec{v}||$ to be a distance?

Because interpreting $||\vec{v}||$ gives us peculiar results when our vectors have complex components, we'll leave Definition 1.14 as it is (for now!), restricting its use to vectors in $\mathbb{R}^n$. However, we will return to this problem in Activity 5.28.

Another important question mathematicians consider is *interactions* between different concepts. For example, we have the concept of scalar multiplication and magnitude. What relationships can be found between these concepts?

Activity 1.22: Scaling Vectors

For the following, let $\vec{u} = \langle 1, 4, -4 \rangle$ and $\vec{v} = \langle 3, 4, 12 \rangle$.

A1.22.1 Find the following:

a) $||\vec{u}||$
b) $2\vec{u}$
c) $||2\vec{u}||$
d) $-3\vec{v}$
e) $||-3\vec{v}||$
f) $k\vec{v}$, where k is in $\mathbb{R}$.
g) $||k\vec{v}||$. Note: k might be negative!

A1.22.2 Suppose $\vec{x}$ is a vector in $\mathbb{R}^3$.

a) What appears to be the relationship between $||\vec{x}||$ and $||k\vec{x}||$, where k is in $\mathbb{R}$?
b) Prove your result.

Activity A1.22.2 leads to the following:

Theorem 1.4 (Scalar Multiple). *Let $\vec{v}$ be a vector in $\mathbb{R}^n$, and let c be a real number. Then*

$$||c\vec{v}|| = |c|||\vec{v}||$$

1.6 Direction

If the magnitude gives us the length of a vector, what will give us its direction?

Activity 1.23: Direction Angles

A1.23.1 Consider the vector $\langle 5, 12 \rangle$.

a) Draw the vector (assume its initial point is at the origin).
b) Find the magnitude of the vector.
c) Find the angle the vector makes with the positive x-axis.

A1.23.2 Suppose $\vec{v} = \langle v_1, v_2 \rangle$ is a vector with initial point at the origin. Find the angle θ the vector makes with the positive x-axis.

While Activity 1.23 suggests we could identify the direction of a vector with the angle it makes with the positive x-axis, we quickly run into problems if our vector has more than two components.

Activity 1.24: More Direction Angles

A1.24.1 Consider the *point* $Z = (3, 4, 12)$, and let O be the origin of the coordinate system in $\mathbb{R}^3$.

a) Find $\overrightarrow{OZ}$.
b) What point X is "directly beneath" Z in the xy-plane?
c) Find $\overrightarrow{OX}$. (This is called the **projection of $\overrightarrow{OZ}$ onto the xy-plane.**)
d) Find the angle $\overrightarrow{OX}$ makes with the positive x-axis.
e) Find the angle $\overrightarrow{OZ}$ makes with the xy-plane.
f) Find the corresponding angles for $\vec{v} = \langle x_1, x_2, x_3 \rangle$.

A1.24.2 Suppose $P = (x_1, x_2, x_3, x_4)$, and let O be the origin of the coordinate system $\mathbb{R}^4$

a) Find $\overrightarrow{OP}$.
b) What point Z in $\mathbb{R}^3$ could be "directly beneath" the point P in $\mathbb{R}^4$?
c) Find $\overrightarrow{OZ}$. This is the projection of $\overrightarrow{OP}$ onto the xyz-hyperplane.
d) Find X where X is "directly beneath" Z in the xy plane.

e) Find the angle $\overrightarrow{OX}$ makes with the x-axis.

f) Find the angle $\overrightarrow{OZ}$ makes with the xy-plane.

g) Find the angle $\overrightarrow{OP}$ makes with the xyz-hyperplane.

The difficulty of extending the direction angles into $\mathbb{R}^3$, let alone $\mathbb{R}^n$, encourages us to think about the direction of a vector in a different way. In this case, mathematicians use the time-honored strategy of procrastination: unless there's a need to talk about the direction of a vector, we won't worry about it. More generally, we use the vector itself as the direction: the vector "points" in the direction of the vector.

While it's not that important to talk about the direction of a *single* vector, it can be useful to *compare* the direction of two vectors. To make this comparison, we'll rely on the fact that any two vectors can be represented as vectors in the same plane (Activity 1.13).

Activity 1.25: The Angle Between Vectors

The following activities refer to the triangle shown:

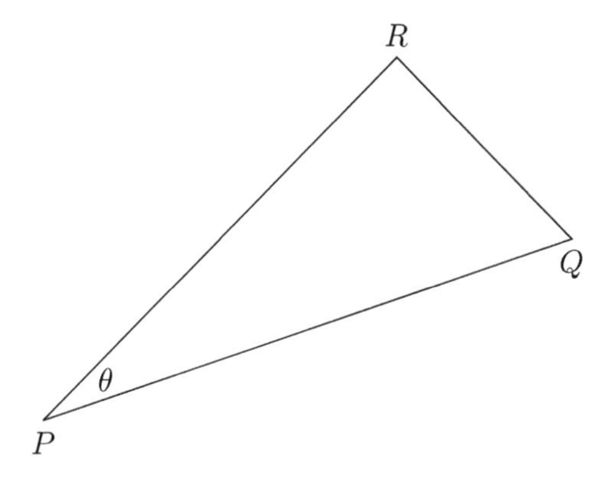

A1.25.1 Suppose $P = (1, 1, 3)$, $Q = (1, 4, 3)$, $R = (4, 6, 3)$.

a) Explain why, even though P, Q, R are points in $\mathbb{R}^3$, we can represent them as vertices of a triangle. (In other words, the figure above is a representation of the points P, Q, R in the plane of the page)

b) Find the lengths of $\overline{PQ}$, $\overline{PR}$, and $\overline{QR}$.

c) According to the law of cosines,

$$QR^2 = PQ^2 + PR^2 - 2(PQ)(PR)\cos\theta$$

where θ is the angle between $\overline{PQ}, \overline{PR}$. Use the law of cosines to find θ.

A1.25.2 Suppose X, Y, Z are points in $\mathbb{R}^n$

a) Express $\overrightarrow{YZ}$ in terms of $\overrightarrow{XY}$, $\overrightarrow{XZ}$. Suggestion: Draw a picture, then interpret the vectors $\overrightarrow{XY}$, $\overrightarrow{XZ}$, and $\overrightarrow{YZ}$ as a sum of vectors.

b) Express the lengths $\overline{XY}$, $\overline{XZ}$, $\overline{YZ}$ in terms of the vectors $\overrightarrow{XY}$, $\overrightarrow{XX}$.

c) Express the angle θ between the vectors $\overrightarrow{XY}$, $\overrightarrow{XZ}$ in terms of the vectors $\overrightarrow{XY}$, $\overrightarrow{XZ}$.

A1.25.3 Suppose $\overrightarrow{PQ} = \langle u_1, u_2\rangle$, $\overrightarrow{PR} = \langle v_1, v_2\rangle$.

a) Using the formula you found in Activity A1.25.2c, express the angle θ between the vectors $\overrightarrow{PQ}$, $\overrightarrow{PR}$ in terms of the *components* of the vectors.

b) Find the angle between the vectors $\langle 1, 0\rangle$ and $\langle 5, 12\rangle$.

c) In Activity A1.23.1, you found the angle between these vectors (because it was the angle the vector $\langle 5, 12\rangle$ makes with the positive x-axis). Verify that your results are the same.

A1.25.4 Suppose $\overrightarrow{PQ} = \langle u_1, u_2, u_3\rangle$, $\overrightarrow{PR} = \langle v_1, v_2, v_3\rangle$. Using the formula you found in Activity A1.25.2c, express the angle θ between the vectors $\overrightarrow{PQ}$, $\overrightarrow{PR}$ in terms of the *components* of the vectors.

A1.25.5 Find the angle between the given vectors.

a) $\langle 5, 12\rangle$ and $\langle 3, -4\rangle$.

b) $\langle 1, 8\rangle$ and $\langle 6, 2\rangle$.

c) $\langle 1, 5, 3\rangle$ and $\langle -2, 5, -2\rangle$.

d) $\langle 2, -4, 1\rangle$ and $\langle 5, 2, -2\rangle$.

A1.25.6 Suppose $\vec{u} = \langle 1, 3, -1, 7\rangle$ and $\vec{v} = \langle -2, 3, 5, 0\rangle$. Find the angle between $\vec{u}$ and $\vec{v}$.

In Activity 1.25 we saw that if $\vec{u} = \langle u_1, u_2\rangle$ and $\vec{v} = \langle v_1, v_2\rangle$ were vectors in $\mathbb{R}^2$, then the cosine of the angle between them could be expressed as

$$\cos\theta = \frac{u_1v_1 + u_2v_2}{\sqrt{u_1^2 + u_2^2}\sqrt{v_1^2 + v_2^2}}$$

Notice the numerator $u_1v_1 + u_2v_2$ is the sum of the componentwise products.

If they were vectors in $\mathbb{R}^3$, the cosine of the angle between could be expressed as

$$\cos\theta = \frac{u_1v_1 + u_2v_2 + u_3v_3}{\sqrt{u_1^2 + u_2^2 + u_3^2}\sqrt{v_1^2 + v_2^2 + v_3^2}}$$

Notice the numerator $u_1v_1 + u_2v_2 + u_3v_3$ is the sum of the componentwise products.

With a little effort, we can show that the formula generalizes to vectors in $\mathbb{R}^n$, so the cosine of the angle between two vectors in $\mathbb{R}^n$ can be found by

$$\cos\theta = \frac{u_1 v_1 + u_2 v_2 + \ldots + u_n v_n}{\sqrt{u_1^2 + u_2^2 + \ldots + u_n^2}\sqrt{v_1^2 + v_2^2 + \ldots + v_n^2}}$$

Notice that the numerator consists of the sum of the componentwise products.

There's an old saying: Once is an accident; twice is coincidence; but three times suggests something important is going on. The fact that the sum of the componentwise products keeps reappearing motivates the following definition:

Definition 1.15 (Dot Product). *Let* $\vec{u} = \langle u_1, u_2, \ldots, u_n \rangle$, $\vec{v} = \langle v_1, v_2, \ldots, v_n \rangle$ *be vectors in* $\mathbb{R}^n$. *We define the* ***dot product***

$$\vec{u} \cdot \vec{v} = u_1 v_1 + u_2 v_2 + \cdots + u_n v_n$$

Example 1.11. *Suppose* $\vec{u} = \langle 2, 1, 5, 4 \rangle$ *and* $\vec{v} = \langle 1, 1, 3, -8 \rangle$. *Find* $\vec{u} \cdot \vec{v}$.

Solution. *We find the sum of the componentwise products:*

$$\begin{aligned} \vec{u} \cdot \vec{v} &= \langle 2, 1, 5, 4 \rangle \cdot \langle 1, 1, 3, -8 \rangle \\ &= 2 \cdot 1 + 1 \cdot 1 + 5 \cdot 3 + 4 \cdot (-8) \\ &= 2 + 1 + 15 + (-32) \\ &= -14 \end{aligned}$$

We note that for any vector $\vec{u} = \langle u_1, u_2, \ldots, u_n \rangle$, we have

$$\begin{aligned} \vec{u} \cdot \vec{u} &= u_1^2 + u_2^2 + \cdots + u_n^2 \\ &= ||\vec{u}||^2 \end{aligned}$$

so we can even rewrite our magnitude formula in terms of the dot product:

Theorem 1.5 (Dot Product and Magnitude). *If* $\vec{u}$ *is a vector in* $\mathbb{R}^n$,

$$||\vec{u}|| = \sqrt{\vec{u} \cdot \vec{u}}$$

As long as we're dealing with vectors in $\mathbb{R}^n$, we can express the result from Activity 1.25 in a compact form:

Theorem 1.6. *For any vectors* $\vec{u}, \vec{v}$ *in* $\mathbb{R}^n$, *the angle* θ *between them satisfies*

$$\cos\theta = \frac{\vec{u} \cdot \vec{v}}{||\vec{u}||||\vec{v}||}$$

Since $\cos\theta = \cos(-\theta)$, it's possible for the angle between $\vec{u}$ and $\vec{v}$ to be positive or negative, depending on whether we have to rotate clockwise or counterclockwise. However, we typically disregard this distinction, and take the positive value as "the" angle between the two vectors

Example 1.12. *Find the angle between the vectors* $\vec{u} = \langle 2,5,1,4\rangle$ *and* $\vec{v} = \langle 1,1,3,8\rangle$.

Solution. *We find*

$$\begin{aligned}\vec{u}\cdot\vec{v} &= 42\\ ||\vec{u}|| &= \sqrt{46}\\ ||\vec{v}|| &= \sqrt{75}\end{aligned}$$

So

$$\begin{aligned}\cos\theta &= \frac{\vec{u}\cdot\vec{v}}{||\vec{u}||||\vec{v}||}\\ &= \frac{42}{\sqrt{46}\sqrt{75}}\\ &\approx 0.7151\end{aligned}$$

This gives $\theta \approx 44.35°$.

Activity 1.26: Properties of the Dot Product

A1.26.1 Consider the vectors $\vec{u} = \langle 0,5,1\rangle$, $\vec{v} = \langle 3,0,0\rangle$, $\vec{w} = \langle 0,3,-15\rangle$.

a) Find $\vec{u}\cdot\vec{v}$, $\vec{u}\cdot\vec{v}$, and $\vec{v}\cdot\vec{u}$.

b) Prove/disprove: The dot product of any two vectors in $\mathbb{R}^3$ is 0.

A1.26.2 Let $\vec{u} = \langle 1,2,5\rangle$, $\vec{v} = \langle 3,1,7\rangle$, $\vec{w} = \langle 1,5,0\rangle$. If possible, find the value indicated; if not possible, explain why not. (Assume $\cdot$ always refers to the dot product)

a) $\vec{u}\cdot\vec{v}$

b) $\vec{v}\cdot\vec{w}$

c) $(\vec{u}\cdot\vec{v})\cdot\vec{w}$

d) $\vec{u}+(\vec{v}\cdot\vec{w})$

e) $(\vec{u}+\vec{v})\cdot\vec{w}$

A1.26.3 Suppose $\vec{u}, \vec{v} \in \mathbb{R}^4$, with $a \in \mathbb{R}$.

a) Write down the component form of "generic" vectors $\vec{u}$, $\vec{v}$ in $\mathbb{R}^4$.

b) Find $a(\vec{u}\cdot\vec{v})$. **Note**: As usual, the operations inside the parentheses must be evaluated first.

c) Find $(a\vec{u})\cdot\vec{v}$.

d) Prove/disprove: $a(\vec{u}\cdot\vec{v}) = (a\vec{u})\cdot\vec{v}$.

e) Does your result hold if the vectors are in $\mathbb{R}^{500}$? Defend your conclusion.

A1.26.4 Let $\vec{u}$, $\vec{v}$, $\vec{w}$ be two vectors in $\mathbb{R}^n$.

a) Write down "generic" vectors $\vec{u}$, $\vec{v}$ in component form. Note: Using an ellipsis (...) is permissible in this case.
b) Find $\vec{u} \cdot \vec{v}$.
c) Find $\vec{u} \cdot \vec{w}$.
d) Find $\vec{u} \cdot \vec{v} + \vec{u} \cdot \vec{w}$.
e) Find $\vec{u} \cdot (\vec{v} + \vec{w})$. (Remember operations in parentheses must be done first)
f) Prove $\vec{u} \cdot (\vec{v} + \vec{w}) = \vec{u} \cdot \vec{v} + \vec{u} \cdot \vec{w}$.

A1.26.5 Let $\vec{u}$, $\vec{v}$, $\vec{w}$ be two vectors in $\mathbb{R}^n$.

a) Prove/disprove: $\vec{u} \cdot \vec{v} = \vec{v} \cdot \vec{u}$.
b) Prove/disprove: $(\vec{u} + \vec{v}) + \vec{w} = \vec{u} + (\vec{v} + \vec{w})$.
c) Prove $(\vec{u} + \vec{v}) \cdot (\vec{u} + \vec{v}) = ||\vec{u}||^2 + ||\vec{v}||^2 + 2\vec{u} \cdot \vec{v}$.

Activity 1.26 leads to several important properties of the dot product and vector arithmetic, which we summarize as:

Theorem 1.7. *Let $\vec{u}$, $\vec{v}$, $\vec{w}$ be vectors in $\mathbb{R}^n$, and let a, b be in $\mathbb{R}$. Then:*

- ***Commutativity:*** $\vec{u} \cdot \vec{v} = \vec{v} \cdot \vec{u}$
- ***Distributivity:*** $\vec{u} \cdot (\vec{v} + \vec{w}) = \vec{u} \cdot \vec{v} + \vec{u} \cdot \vec{w}$.
- ***Associativity of Scalar Multiplication:*** $(a\vec{u}) \cdot \vec{v} = a(\vec{u} \cdot \vec{v})$.

It should be clear that proof is an important component of higher mathematics. What's less often clear is that *re*proof is *also* an important component of higher mathematics. Let's see if we can construct a different proof of Theorem Theorem 1.6.

Activity 1.27: The Dot Product, Revisited

The following activities refer to the triangle shown:

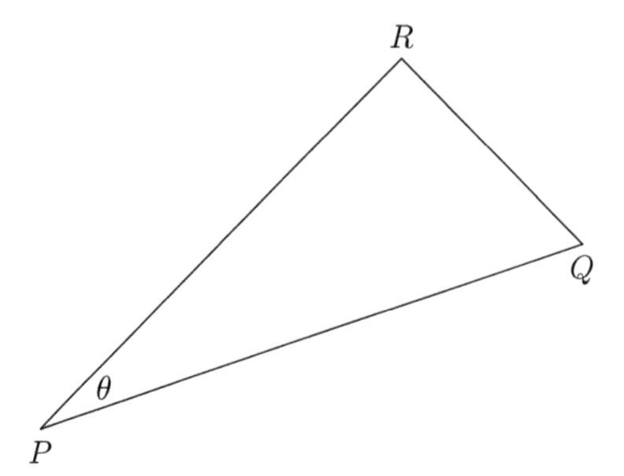

In the following, you may use Theorems 1.5 and 1.7, but **DO NOT** use Theorem 1.6 in any form.

Let $\mathbf{u} = \overrightarrow{PQ}$, $\mathbf{v} = \overrightarrow{PR}$.

A1.27.1 Express $\overrightarrow{QR}$ in terms of $\mathbf{u}$, $\mathbf{v}$.

A1.27.2 Use the law of cosines to write an equation to find $\cos\theta$.

A1.27.3 Use the properties of the dot product to prove Theorem 1.6.

Our geometric interpretations of $||\mathbf{v}||$ and $\mathbf{u} + \mathbf{v}$ suggest another useful property of vectors.

Activity 1.28: The Triangle Inequality, Part One

A1.28.1 Suppose $\vec{u}$, $\vec{v}$ are vectors in $\mathbb{R}^n$.

a) Draw a picture to represent the vectors $\vec{u}$, $\vec{v}$, and the sum $\vec{u} + \vec{v}$ (using the tip-to-tail representation of vector addition).
b) Interpret $||\vec{u}||$, $||\vec{v}||$, and $||\vec{u} + \vec{v}||$ *geometrically.*
c) Explain *geometrically* why it must be true that

$$||\vec{u}|| + ||\vec{v}|| \geq ||\vec{u} + \vec{v}||$$

A1.28.2 Prove: If a, b, c are nonnegative real numbers where $(b + c)^2 \geq a^2$, then $b + c \geq a$.

A1.28.3 In the following, assume u, v are real numbers.

a) Find a real number u where $\sqrt{u^2} \neq u$.
b) Find real numbers u, v where $\sqrt{(u+v)^2} \neq u + v$.
c) Will $\left(\sqrt{u^2}\right)^2 = u^2$? Why/why not?

d) Prove: $\sqrt{u^2} + \sqrt{v^2} \geq \sqrt{(u+v)^2}$.

Activity 1.28 suggests:

Theorem 1.8 (Triangle Inequality). *For* $\vec{a}, \vec{b} \in \mathbb{R}^n$,

$$||\vec{a} + \vec{b}|| \leq ||\vec{a}|| + ||\vec{b}||$$

However, it should be clear proving the triangle inequality, even where $\mathbf{u}, \mathbf{v} \in \mathbb{R}^1$, is extremely challenging. To prove it, we'll need to establish a lemma.

Activity 1.29: Cauchy-Bunyakovsky-Schwarz

A1.29.1 Let $\vec{u}$, $\vec{v}$ be vectors in $\mathbb{R}^n$, and define $f(x) = ||\vec{u} + x\vec{v}||^2$, where x is a real number.

a) Explain why $f(x) \geq 0$ for all x.
b) Show $f(x)$ is a quadratic function of x, and that the corresponding graph $y = f(x)$ is that of a parabola opening upwards.
c) Show that the minimum value of $f(x)$ will be $||\vec{u}||^2 - \dfrac{(\vec{u} \cdot \vec{v})^2}{||\vec{v}||^2}$.
d) Prove: $(\vec{u} \cdot \vec{v})^2 \leq ||\vec{u}||^2||\vec{u}||^2$.

Activity A1.29.1 proves an important relationship between the dot product and the magnitude:

Theorem 1.9 (Cauchy-Bunyakovsky-Schwarz). *For any vectors* $\vec{u}$, $\vec{v}$ *in* $\mathbb{R}^n$,

$$(\vec{u} \cdot \vec{v})^2 \leq ||\vec{u}||^2||\vec{v}||^2$$

Equivalently,

$$|\vec{u} \cdot \vec{v}| \leq ||\vec{u}||||\vec{v}||$$

Activity 1.30: The Triangle Inequality, Part Two

A1.30.1 Suppose $\vec{u}$, $\vec{v}$ are vectors in $\mathbb{R}^n$.

a) Show $||\vec{u} + \vec{v}||^2 = ||\vec{u}||^2 + ||\vec{v}||^2 + 2\vec{u} \cdot \vec{v}$.
b) Explain why $||\vec{u}||^2 + ||\vec{v}||^2 + 2\vec{u} \cdot \vec{v} \leq ||\vec{u}||^2 + ||\vec{v}||^2 + 2|\vec{u} \cdot \vec{v}|$.
c) Prove: $||\vec{u} + \vec{v}||^2 \leq ||\vec{u}||^2 + 2||\vec{u}||||\vec{v}|| + ||\vec{v}||^2$.

d) Prove; $||\vec{u} + \vec{v}|| \leq ||u|| + ||\vec{v}||$.

Activity A1.30.1 proves Theorem 1.8.

Another important question is to ask how similar two vectors are. If we focus on the direction of the two vectors, we might rule that the smaller the angle θ between two vectors, the more similar they are. This leads to:

Definition 1.16 (Cosine Similarity). *The **cosine similarity** of two vectors is the cosine of the angle θ between them.*

Activity 1.31: Cosine Similarity

A1.31.1 Write an expression for the cosine of the angle between two vectors $\vec{u}$ and $\vec{v}$.

A1.31.2 Let $\vec{u} = \langle 1, 5, 2 \rangle$, $\vec{v} = \langle 1, 5, 3 \rangle$, $\vec{w} = \langle 2, 9, 5 \rangle$.

a) Use cosine similarity to determine which vector is most similar to $\vec{u}$.
b) Use cosine similarity to determine which vector is most similar to $\vec{w}$.
c) Suppose $\vec{x}$, $\vec{y}$, and $\vec{z}$ are vectors. If $\vec{y}$ is the vector most similar to $\vec{y}$, is $\vec{y}$ the vector most similar to $\vec{x}$?

A1.31.3 Suppose $\vec{v}_1, \vec{v}_2, \vec{v}_3, \ldots, \vec{v}_n$ are all the same length. Prove: If $\vec{v}_i \cdot \vec{v}_j > \vec{v}_i \cdot \vec{v}_k$, then the angle between $\vec{v}_i$ and $\vec{v}_j$ is smaller than the angle between $\vec{v}_i$ and $\vec{v}_k$.

A1.31.4 The picture shows several vectors of the same length.

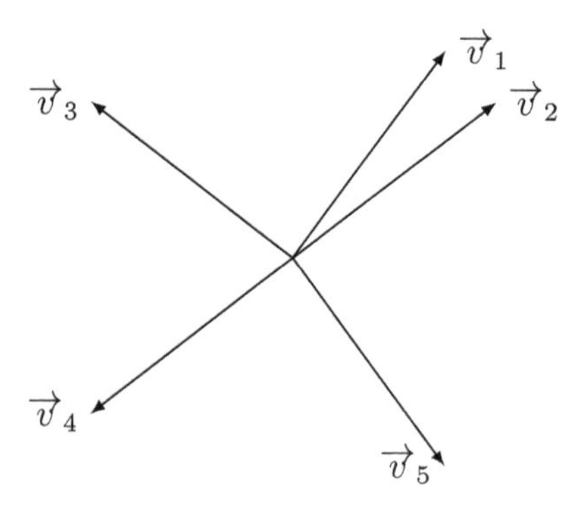

Put in order of *decreasing* value:

$$\vec{v}_1 \cdot \vec{v}_2 \qquad \vec{v}_1 \cdot \vec{v}_3 \qquad \vec{v}_1 \cdot \vec{v}_4 \qquad \vec{v}_1 \cdot \vec{v}_5$$

Cosine similarity played an important role in early search engines.

Activity 1.32: Search Engines

An early approach to matching search queries ("What sort of bears live in the arctic?") to relevant documents relied on comparing document vectors (see Activity 1.4): the search query itself would be treated as a document vector, and the search engine would look for a document vector that most closely matched the search query.

A1.32.1 Suppose documents in a library correspond to document vectors $\vec{v}_1 = \langle 153, 28, 97, 54\rangle$, $\vec{v}_2 = \langle 385, 58, 107, 23\rangle$, and $v_3 = \langle 153, 11, 132, 27\rangle$.

a) Which documents are most similar to each other?

b) Suppose a search query corresponds to the document vector $\langle 1, 0, 1, 1\rangle$. Rank the three documents in order of relevance to the search query.

A1.32.2 Suppose we choose the following six keywords: the, arctic, bear, is, platypus, Australia.

a) Using the keywords from Activity A1.4.1, find the document vector corresponding to the search query, "What sort of bears live in the arctic?"

b) Rank the three document vectors from Activity A1.4.1 in terms of their cosine similarity to the search query.

c) Based on the cosine similarity, which of the three is most relevant to the query?

d) Based on the actual documents, which of the three documents is most similar to the query?

A1.32.3 Cosine similarity typically fails to produce good results, in the sense that the documents retrieved are often unrelated to the search query. The following activity suggests a reason why. Assume the same keywords and search query as in Activity A1.32.2, and consider two documents, one with document vector $\vec{v}_1 = \langle 125, 5, 3, 75, 0, 0\rangle$, and $\vec{v}_2 = \langle 250, 0, 4, 100, 0, 5\rangle$.

a) Based on the *meaning* of the components of the document vector, which document is more relevant to the search query?

b) Based on the cosine similarity, which document is more similar to the search query?

c) Why does this show that the keywords must be chosen carefully? Alternatively, that "corrected count" must exist for certain keywords?

1.7 Unit and Orthogonal Vectors

Since a vector has both a direction and magnitude, we can consider the simplest possible vector pointing in a certain direction:

Definition 1.17 (Unit Vector). *A unit vector is a vector with a length of* 1.

This poses a new problem: Given an arbitrary vector, how can we find a unit vector that points in the same direction?

Activity 1.33: Unit Vectors

In the following, let $\overrightarrow{v} = \langle v_1, v_2, \ldots, v_n \rangle$ be a vector in $\mathbb{R}^n$.

A1.33.1 Suppose $k > 0$ is a real number. Show $||k\overrightarrow{v}|| = k||\overrightarrow{v}||$.

A1.33.2 Let θ be the angle between $\overrightarrow{u}$ and $\overrightarrow{v}$.

a) Suppose $k > 0$. Show the angle between $\overrightarrow{u}$ and $\overrightarrow{v}$ is the same as the angle between $k\overrightarrow{u}$ and $\overrightarrow{v}$.
b) What happens to the angle between $\overrightarrow{u}$ and $\overrightarrow{v}$ if $k < 0$?

A1.33.3 Let $k = ||\overrightarrow{v}||$. Find $||\frac{1}{k}\overrightarrow{v}||$.

Activity A1.33.2 proves:

Theorem 1.10. *For any vector $\overrightarrow{v}$ and any scalar k, the direction of $k\overrightarrow{v}$ is*

- *The same as $\overrightarrow{v}$, if $k > 0$,*
- *The opposite of $\overrightarrow{v}$, if $k < 0$.*

The fact that the direction of $\overrightarrow{v}$ doesn't change if we multiply it by a scalar $k > 0$ means we can scale a vector to any desired length while preserving its direction. If we **normalize a vector**, we scale it so that it is a unit vector.

In Activity A1.33.3 provides a general strategy for normalizing vectors.

Example 1.13. *Find a unit vector that has the same direction as $\overrightarrow{v} = \langle 3, -5, 1, -2 \rangle$.*

Solution. *We find*

$$\begin{aligned} ||\overrightarrow{v}|| &= \sqrt{3^2 + (-5)^2 + 1^2 + (-2)^2} \\ &= \sqrt{39} \end{aligned}$$

So a unit vector that has the same direction will be

$$\frac{1}{\sqrt{39}}\langle 3, -5, 1, -2 \rangle = \left\langle \frac{3}{\sqrt{39}}, -\frac{5}{\sqrt{39}}, \frac{1}{\sqrt{39}}, -\frac{2}{\sqrt{39}} \right\rangle$$

Normalizing a vector leads to an important simplification when using cosine similarity.

Activity 1.34: More About the Dot Product

A1.34.1 Suppose we compute the cosine similarity between two vectors.

a) Explain why the cosine similarity of a vector $\vec{u}$ with itself should be greater than the cosine similarity of $\vec{u}$ with any other vector.

b) Let $\vec{\hat{u}}$ be the unit vector with the same direction as $\vec{u}$. Find $\vec{\hat{u}} \cdot \vec{\hat{u}}$.

c) Let $\vec{\hat{v}}$ be any other unit vector. What can you say about the value of $\vec{\hat{u}} \cdot \vec{\hat{v}}$?

Activity 1.34 leads to:

Theorem 1.11. *The greater the dot product of two normalized vectors, the more similar the vectors.*

In other applications, it's desirable to have **orthogonal vectors**: vectors that are perpendicular to each other.

Activity 1.35: Orthogonal Vectors

A1.35.1 Suppose $\vec{u}$, $\vec{v}$ are perpendicular.

a) What is the angle θ between the two vectors?

b) Prove: If $\vec{u}$ is perpendicular to $\vec{v}$, then $\vec{u} \cdot \vec{v} = 0$.

A1.35.2 Suppose $\vec{u}$, $\vec{v}$ are nonzero vectors where $\vec{u} \cdot \vec{v} = 0$. Explain why the two vectors must be perpendicular.

A1.35.3 Suppose P, Q, R are the *points* $(3, 1)$, $(5, 8)$, and $(6, 5)$; and O is the origin.

a) Find the *vectors* $\overrightarrow{OP}$, $\overrightarrow{PQ}$, and $\overrightarrow{QR}$.

b) Show that $\overrightarrow{OP}$ and $\overrightarrow{QR}$ are orthogonal.

A1.35.4 Let $\vec{u} = \langle 2, 1, 5 \rangle$.

a) Find a nonzero vector $\vec{v}$ that is orthogonal to $\vec{u}$.

b) Find a nonzero vector $\vec{w}$ that is orthogonal to $\vec{u}$, but points in direction a *different* from that of $\vec{v}$.

Activities A1.35.1 and A1.35.2 lead to the following result:

Theorem 1.12. *If* $\mathbf{u}, \mathbf{v}$ *are orthogonal vectors,* $\mathbf{u} \cdot \mathbf{v} = 0$. *Conversely, if* $\mathbf{u} \cdot \mathbf{v} = 0$, *then* $\mathbf{u}$, $\mathbf{v}$ *are orthogonal.*

2

Systems of Linear Equations

In this chapter, we'll consider the problem of solving systems of linear equations. This is a critical part of linear algebra, and we offer the following general strategy:

Strategy. *Every problem in linear algebra begins with a system of linear equations.*

2.1 Standard Form

When trying to work with a collection of objects, it's easiest if they're all in some standardized format. You've seen such standardized formats before: it's convenient to rewrite a given quadratic equation in the form $ax^2 + bx + c = 0$; it's convenient to write the equation of a line in the form $y = mx + b$; and so on. In this activity, we'll develop a standard form for a system of linear equations.

Activity 2.1: Standard Form

We'll say that an equation has been reduced to simplest form if it has as few terms as possible.

A2.1.1 The equation $3x - 5y = 8$ has three terms: $3x$, $5y$, and 8. If possible, rewrite the equation so it has fewer terms; if not possible, explain why not.

A2.1.2 The equation $2x + 7z = 1 + 2y + 5x$ has five terms. If possible, rewrite the equation so it has fewer terms; if not possible, explain why not.

A2.1.3 The equation $5x - 2y + z + 12 = 0$ has five terms. If possible, rewrite the equation so it has fewer terms; if not possible, explain why not.

Activity 2.1 motivates the following definition:

Definition 2.1 (Standard Form). *A linear equation is in* ***standard form*** *when*

- *All of the variables are on one side of the =,*
- *The constant is on the other side of the =.*

As Activity 2.1 shows, equations in standard form will have fewer terms than equations not in standard form.

In Activity A1.3.2, we represented linear equations as vectors. If we have a system of equations, we can represent the entire system as a collection of vectors and, as long as *all* equations are in standard form, we'll be able to work with the vectors as easily as we could have worked with the equations.

For example, the system of equations

$$\begin{aligned} 3x + 5y &= 11 \\ 2x - 7y &= 4 \end{aligned}$$

could be represented as the vectors $\langle 3, 5, 11 \rangle$ and $\langle 2, -7, 4 \rangle$, where the first component is the coefficient of x; the second component is the coefficient of y; and the last component is the constant.

Actually, if we write our two equations as two vectors, we might forget that they are in fact related. To reinforce their relationship, we'll represent them as a single object by throwing them inside a set of parentheses. We might represent the evolution of our notation as

$$\left\{ \begin{array}{rcl} 3x + 5y & = & 11 \\ 2x - 7y & = & 4 \end{array} \right. \quad \rightarrow \quad \begin{array}{c} \langle 3, 5, 11 \rangle \\ \langle 2, -7, 4 \rangle \end{array} \quad \rightarrow \quad \begin{pmatrix} 3 & 5 & 11 \\ 2 & -7 & 4 \end{pmatrix}$$

As a general rule, mathematicians are terrible at coming up with new words, so we appropriate existing words that convey the essential meaning of what we want to express.

In geology, when precious objects like gemstones or fossils are embedded in a rock, the rock is called a *matrix*. The term is also used in biology for the structure that holds other precious objects—the internal organs—in place, and in construction for the mortar that holds bricks in place. Thus in 1850, the mathematician James Joseph Sylvester (1814–1897) appropriated the term for the structure that keeps the coefficients and constants of a linear equation in place.

Since the matrix shown includes both the coefficients of our equation and the constants, we can describe it as the coefficient matrix augmented by a column of constants or, more simply, the **augmented coefficient matrix**. It's traditional to use a vertical bar separating the coefficients from the constants, but this is a convenience, nothing more:

$$\begin{pmatrix} 3 & 5 & 11 \\ 2 & -7 & 4 \end{pmatrix} \text{ is the same as } \left(\begin{array}{cc|c} 3 & 5 & 11 \\ 2 & -7 & 4 \end{array} \right)$$

Around 200 B.C., the Chinese began writing systems of linear equations by recording the coefficients and constants, an approach they called *fang cheng shu* ("rectangular tabulation method"). The idea would be reinvented in 1693, by Gottfried Wilhelm Leibniz (1646–1716).[1]

In general, to write the augmented matrix for a system of equations:

- Choose an order for the variables,
- Rewrite all equations in standard form,
- Using negative, 0, and 1 coefficients as needed, make sure every equation includes all variables,
- Each row of the matrix corresponds to the coefficients and constant term of one equation.

Example 2.1. *Write the augmented coefficient matrix for the system*

$$\begin{cases} 3x &= 6 - 7y \\ 2x - 7z &= y \end{cases}$$

Solution. *We see that the variables are x, y, and z. We'll choose the order x, y, and z, and rewrite our system of equations so that all the variables are on one side and the constant is on the other, then (using coefficients of 1 and 0) ensure that all equations include all variables:*

$$\begin{cases} 3x &= 6 - 7y \\ 2x - 7z &= y \end{cases} \rightarrow \begin{cases} 3x + 7y &= 6 \\ 2x - 7z - y &= 0 \end{cases} \rightarrow \begin{cases} 3x + 7y + 0z &= 6 \\ 2x - 1y - 7z &= 0 \end{cases}$$

The coefficients and constants become the entries of the augmented coefficient matrix:

$$\left(\begin{array}{ccc|c} 3 & 7 & 0 & 6 \\ 2 & -1 & -7 & 0 \end{array}\right)$$

Because a system of linear equations can have any number of equations and any number of variables, it follows that an augmented coefficient matrix can be of any size. We define:

Definition 2.2 (Dimensions of a Matrix). *A $m \times n$ matrix has m horizontal rows and n vertical columns.*

For obvious reasons:

Definition 2.3 (Square Matrix). *A $n \times n$ matrix is called a* ***square matrix****.*

Although Leibniz was not the first to use matrices to represent systems linear equations, he was the first to use **index notation** to refer to the entries of the matrix:

Definition 2.4 (Index Notation). *The matrix A has entries a_{ij}, where a_{ij} is the entry in the ith horizontal row, jth vertical column.*

---

[1] Leibniz was fascinated by Chinese culture and philosophy, so it's not improbable that he learned of the Chinese method.

For clarity, we should write a_{ij} as $a_{i,j}$. But we won't.[2]

Example 2.2. *In the matrix*

$$A = \begin{pmatrix} 1 & 2 & -3 & 5 \\ 3 & 0 & 1 & -8 \end{pmatrix}$$

find the dimensions; then find a_{12}, a_{21}, *and* a_{32}.

Solution. *This matrix has 2 horizontal rows and 4 vertical columns, so it is a* 2×4 *matrix.*

a_{12} *is the entry in the first horizontal row, second vertical column, so* $a_{12} = 2$.

a_{21} *is the entry in the second horizontal row, first vertical column, so* $a_{21} = 3$.

a_{32} *is the entry in the third horizontal row, second vertical column. However, there are only two rows, so* a_{32} *is undefined.*

Activity 2.2: The Coefficient Matrix

A2.2.1 Write each system as an augmented coefficient matrix.

a) $\begin{cases} 3x + 5y & = & 2 \\ x - 7y & = & 4 \end{cases}$

b) $\begin{cases} 3x & = & 4 - 2y \\ y & = & 3x - 4 \end{cases}$

c) $\begin{cases} 3x + 4y & = & 8 \\ 2x - 5z & = & 5 \\ x + 2y & = & 3z \end{cases}$

A2.2.2 Write each augmented coefficient matrix as a system of equations.

a) $\begin{pmatrix} 3 & 5 & 2 \\ 1 & 2 & 0 \\ -2 & 3 & 8 \end{pmatrix}$

b) $\begin{pmatrix} 2 & 3 & 5 & 8 \\ 1 & -4 & 0 & 0 \end{pmatrix}$

c) $\begin{pmatrix} 3 & 0 & 0 & 5 \\ 0 & 4 & 0 & 9 \\ 0 & 0 & 7 & 6 \end{pmatrix}$

[2]Because in most cases, it won't matter: we will talk generally about the entry a_{ij}, and no greater clarity is obtained by writing this as $a_{i,j}$.

A2.2.3 Suppose an augmented coefficient matrix has n rows and m columns. How many equations are in the corresponding system of equations, and how many variables?

2.2 Solving Systems

Suppose we want to solve the system of equations represented by an augmented coefficient matrix. How can we solve the corresponding system of equations by manipulating the matrices?

Activity 2.3: Elementary Row Operations

We say that two systems of equations are **equivalent** if they have the same solutions.

A2.3.1 In the following, two systems of equations are given. First, describe how the first system of equations was altered to produce the second. Then determine whether the two systems of equations are equivalent by deciding whether the alteration is an allowable transformation of the system of equations.

a) $\left\{\begin{array}{rcl} 3x+6y &=& 2 \\ 2x+3y &=& 7 \end{array}\right.$ and $\left\{\begin{array}{rcl} 2x+3y &=& 7 \\ 3x+6y &=& 2 \end{array}\right.$

b) $\left\{\begin{array}{rcl} 2x+3y &=& 4 \\ 5x+6y &=& 7 \end{array}\right.$ and $\left\{\begin{array}{rcl} 3x+2y &=& 4 \\ 6x+5y &=& 7 \end{array}\right.$

c) $\left\{\begin{array}{rcl} 8x+9y &=& 4 \\ 2x-5y &=& 7 \end{array}\right.$ and $\left\{\begin{array}{rcl} (8+2)x+(9+2)y &=& 4+2 \\ 2x-5y &=& 7 \end{array}\right.$

d) $\left\{\begin{array}{rcl} 8x+9y &=& 4 \\ 2x+5y &=& 7 \end{array}\right.$ and $\left\{\begin{array}{rcl} (8+2)x+(9+5)y &=& 4+7 \\ 2x+5y &=& 7 \end{array}\right.$

e) $\left\{\begin{array}{rcl} 8x+9y &=& 4 \\ 2x+5y &=& 7 \end{array}\right.$ and $\left\{\begin{array}{rcl} 8x+9y &=& 4 \\ 3\cdot 2x+3\cdot 5y &=& 3\cdot 7 \end{array}\right.$

f) $\left\{\begin{array}{rcl} 8x+9y &=& 4 \\ 2x+5y &=& 7 \end{array}\right.$ and $\left\{\begin{array}{rcl} 8x+9y &=& 4 \\ 8\cdot 2x+9\cdot 5y &=& 4\cdot 7 \end{array}\right.$

A2.3.2 For each pair of systems of equations in Activity A2.3.1, write down the augmented matrices, and indicate which augmented matrices correspond to equivalent systems.

A2.3.3 Consider the system of equations

$$\begin{cases} 3x + 6y &= 2 \\ 2x + 3y &= 7 \end{cases}$$

This system can be solved by

$$\begin{cases} 3x + 6y &= 2 \\ 2x + 3y &= 7 \end{cases} \longrightarrow \begin{cases} 3x + 6y &= 2 \\ 6x + 9y &= 21 \end{cases} \longrightarrow \begin{cases} -6x - 12y &= -4 \\ 6x + 9y &= 21 \end{cases}$$

$$\longrightarrow \begin{cases} -6x - 12y &= -4 \\ -3y &= 17 \end{cases} \longrightarrow \begin{cases} 3x + 6y &= 2 \\ -3y &= 17 \end{cases}$$

$$\longrightarrow \begin{cases} 3x + 6y &= 2 \\ -6y &= 34 \end{cases} \longrightarrow \begin{cases} 3x &= 36 \\ -6y &= 34 \end{cases} \longrightarrow \begin{cases} x &= 12 \\ y &= -\frac{17}{3} \end{cases}$$

a) Explain why all systems in the sequence are equivalent by specifically identifying what was done to each system to obtain the next.

b) Replace **each** system of with an augmented matrix. (This will give you a sequence of augmented matrices.)

A2.3.4 Write down a sequence of augmented coefficient matrices, corresponding to equivalent systems of equations, that lead to a solution to

$$\begin{cases} 2x + 7y &= 12 \\ 3x + 5y &= 8 \end{cases}$$

A2.3.5 Suppose M is an augmented coefficient matrix that represents a system of linear equations. Which of the following operations on M will produce an augmented coefficient matrix for an equivalent system of linear equations?

a) Switching two rows of M.

b) Switching two columns of M.

c) Multiplying every entry in a row of M by a constant c.

d) Adding a constant c to every entry in a row of M.

e) Multiplying a row of M, and adding the entries componentwise to another row of M.

f) Multiplying one row of M by p, a second row by q, and then adding the two rows componentwise to get a new row.

Based on the results of Activity 2.3, we define:

Definition 2.5 (Elementary Row Operations). *Given a matrix, an* ***elementary row operation*** *is one of the following:*

- *Switching two rows of the matrix,*
- *Multiplying the entries of one row of a matrix by a constant c,*
- *Adding the entries of one row to another row componentwise,*
- *Multiplying one row by a constant, then adding the results to another row componentwise.*

Suppose we have a matrix M, which represents a system of linear equations. Any of the elementary row operations will produce a matrix N, which represents a different system of linear equations. However, because the elementary row operations correspond to operations we can perform on the equations in a system of equations, the new system will have the same solutions as the old.

It will be helpful to record the elementary row operations, so we use the following notation. If M is our original matrix, and N is the matrix produced by applying an elementary row operation, then:

- $M \xrightarrow{R_i \leftrightarrow R_j} N$ indicates that rows i, j of M have been switched, to produce matrix N,
- $M \xrightarrow{cR_i \to R_i} N$ means the entries of row i of M have been multiplied by c,
- $M \xrightarrow{cR_i + R_j \to R_j} N$ means that c times the entries of row i have been added componentwise to the entries of row j.

Note that in the third operation, only row j changes.

Example 2.3. *Let*

$$A = \left(\begin{array}{ccc|c} 2 & 1 & 3 & 5 \\ 1 & 4 & 2 & 8 \\ 5 & 3 & 1 & 6 \end{array}\right)$$

Apply the following row operations to A:

- $\xrightarrow{2R_2 \to R_2}$
- $\xrightarrow{R_1 \leftrightarrow R_2}$
- $\xrightarrow{3R_2 + R_3 \to R_3}$

Solution. $\xrightarrow{2R_2 \to R_2}$ *will multiply the second row of* A *by* 2*, so the entries in the second row will become* $2 \cdot 1 = 2$, $4 \cdot 2 = 8$, $2 \cdot 2 = 4$, *and* $8 \cdot 2 = 16$. *The other two rows remain the same, so we have*

$$\left(\begin{array}{ccc|c} 2 & 1 & 3 & 5 \\ 1 & 4 & 2 & 8 \\ 5 & 3 & 1 & 6 \end{array}\right) \xrightarrow{2R_2 \to R_2} \left(\begin{array}{ccc|c} 2 & 1 & 3 & 5 \\ 2 & 8 & 4 & 16 \\ 5 & 3 & 1 & 6 \end{array}\right)$$

$\xrightarrow{R_1 \leftrightarrow R_2}$ *switches the first and second rows of* A*; the third row remains unchanged:*

$$\left(\begin{array}{ccc|c} 2 & 1 & 3 & 5 \\ 1 & 4 & 2 & 8 \\ 5 & 3 & 1 & 6 \end{array}\right) \xrightarrow{R_1 \leftrightarrow R_2} \left(\begin{array}{ccc|c} 1 & 4 & 2 & 8 \\ 2 & 1 & 3 & 5 \\ 5 & 3 & 1 & 6 \end{array}\right)$$

Finally $\xrightarrow{3R_2 + R_3 \to R_3}$ *multiplies the second row of* A *by* 3*, and then adds the row componentwise to the third row. Thus the entries in the third row will be* $5 + 3 \cdot 1 = 8$, $3 + 3 \cdot 4 = 15$, $1 + 3 \cdot 2 = 7$, *and* $6 + 3 \cdot 8 = 30$, *and we have*

$$\left(\begin{array}{ccc|c} 2 & 1 & 3 & 5 \\ 1 & 4 & 2 & 8 \\ 5 & 3 & 1 & 6 \end{array}\right) \xrightarrow{3R_2 + R_3 \to R_3} \left(\begin{array}{ccc|c} 2 & 1 & 3 & 5 \\ 1 & 4 & 2 & 8 \\ 8 & 15 & 7 & 30 \end{array}\right)$$

While we can use elementary row operations to transform a system of equations into another system of equations, it's helpful to have a goal in mind.

Activity 2.4: Row Echelon Form

A2.4.1 Suppose A, B, C are the following augmented matrices

$$A = \left(\begin{array}{ccc|c} 2 & 2 & 3 & 12 \\ 6 & 12 & 14 & 30 \\ 10 & 6 & 12 & 18 \end{array}\right), B = \left(\begin{array}{ccc|c} 4 & 4 & 6 & 24 \\ 0 & 12 & 10 & -12 \\ 0 & 0 & 2 & -276 \end{array}\right),$$

$$C = \left(\begin{array}{ccc|c} 48 & 0 & 0 & 4752 \\ 0 & 12 & 0 & 1368 \\ 0 & 0 & 2 & -276 \end{array}\right)$$

a) Write down the system of equations corresponding to A; to B; and to C. (Use x, y, z for your variables.)

b) Of the three systems of equations, which would be the easiest to solve for the variables? Why?

c) Solve one of the systems of equations. (Pick whichever one you want to solve.)

d) Show that all three systems have the same solution.

A2.4.2 Suppose D and E are the following augmented coefficient matrices.

$$D = \left(\begin{array}{ccc|c} 2 & 2 & -4 & 20 \\ 3 & 5 & 5 & -1 \\ 5 & 2 & 5 & 2 \end{array}\right), E = \left(\begin{array}{ccc|c} 2 & 2 & -4 & 20 \\ 0 & 2 & 11 & -31 \\ 0 & 0 & 21 & -63 \end{array}\right)$$

a) Write down the system of equations corresponding to D and the system of equations corresponding to E. Use x, y, z as your variables.
b) Of the two systems, which one will be easier to solve for x, y, z? Why?
c) Solve either system.
d) Show that the other system has the same solution.

A2.4.3 Suppose you're given an augmented coefficient matrix M. Through a sequence of allowable row operations, you can transform this into an augmented coefficient matrix N. Based on your answers to Activities A2.4.1 and A2.4.2, what form should N have to make it as easy as possible to determine the solutions to the original system of equations?

A2.4.4 If possible, find the solution to the system represented by each augmented coefficient matrix. If not possible, explain why not.

a) $\left(\begin{array}{cc|c} 2 & 5 & 8 \\ 0 & -1 & 3 \end{array}\right)$

b) $\left(\begin{array}{ccc|c} 5 & 0 & 0 & 3 \\ 0 & 2 & 0 & 9 \\ 0 & 0 & 3 & 6 \end{array}\right)$

c) $\left(\begin{array}{cc|c} 3 & 1 & 2 \\ 0 & 3 & 5 \\ 0 & 0 & 1 \end{array}\right)$

Discussing the work in Activity 2.4 is easier if we introduce the following definitions:

Definition 2.6 (Pivot). *In a matrix, the* ***pivot*** *(or* ***row pivot*** *if more specificity is required) is the first nonzero entry of a row.*

Definition 2.7 (Row Echelon Forms). *A matrix is in* ***row echelon form*** *if the entries below every pivot are 0. The matrix is in* ***reduced row echelon form*** *if the entries below and above every pivot are 0.*

As you saw in Activity 2.9, if an augmented coefficient matrix was in reduced row echelon form, it was very easy to solve for the variables.

Example 2.4. *In the matrices below, identify the pivots. Which matrices are in row echelon form? Which are in reduced row echelon form?*

$$A = \begin{pmatrix} 2 & 5 & 3 \\ 0 & 1 & 1 \\ 0 & 0 & 0 \end{pmatrix} \qquad B = \begin{pmatrix} 1 & 5 & 4 \\ 0 & 0 & 1 \\ 0 & 3 & 1 \end{pmatrix} \qquad C = \begin{pmatrix} 8 & 0 & 0 & 0 \\ 0 & 0 & -4 & 0 \\ 0 & 0 & 0 & 3 \end{pmatrix}$$

Solution. *In A, the pivots will be $a_{11} = 2$, $a_{22} = 1$. The third row has no nonzero entries, so it has no pivots. Since there are 0s below the pivots, the matrix is in row echelon form. However, since the second row pivot has an entry above it, the matrix is not in reduced row echelon form.*

In B, the pivots are $b_{11} = 1$, $b_{23} = 1$, and $b_{32} = 3$. However, the second row pivot has a nonzero entry below it, so the matrix is not in row echelon form (and so it cannot be in reduced row echelon form).

In C, the pivots are $c_{11} = 8$, $c_{23} = -4$, $c_{34} = 3$. Since the entries (if any) above and below each pivot are 0s, then this matrix is in reduced row echelon form.

What about matrices in row echelon form? Consider the augmented coefficient matrix

$$\left(\begin{array}{ccc|c} 2 & 2 & -4 & 20 \\ 0 & 2 & 11 & -31 \\ 0 & 0 & 21 & -63 \end{array}\right)$$

This corresponds to the system

$$\begin{aligned} 2x + 2y - 4z &= 20 \\ 2y + 11z &= -31 \\ 21z &= -63 \end{aligned}$$

While it isn't as easy to solve this system as it would be if the augmented matrix were in reduced row echelon form, it's still very easy to solve it using **back substitution**. In this case, consider the last equation

$$21z = -63$$

Since this equation has only one variable, it's easy to solve, and we find $z = -3$.

Now consider the equation above it

$$2y + 11z = -31$$

While this has two variables y and z, we've already determined $z = -3$, so we can substitute and solve for y:

$$\begin{aligned} 2y + 11(-3) &= -31 \\ 2y - 33 &= -31 \\ 2y &= 2 \\ y &= 1 \end{aligned}$$

Finally, consider the first equation:

$$2x + 2y - 4z = 20$$

As before, this seems to consist of three variables, we already know two of them, so we can substitute and solve for the third:

$$\begin{aligned} 2x + 2(1) - 4(-3) &= 20 \\ 2x + 2 + 12 &= 20 \\ 2x &= 6 \\ x &= 3 \end{aligned}$$

It follows that we don't need to get an augmented coefficient matrix into reduced row echelon form, but row echelon form is enough. But how?

One approach is known as **Gaussian elimination**:

- Multiply the working row by a constant to make the pivot equal to 1,
- Subtract a multiple of the working row (now with pivot equal to 1) to the following rows to eliminate the entries below the pivot.

Example 2.5. *Apply Gaussian elimination to eliminate the entries below the first row pivot in the matrix*

$$\left(\begin{array}{ccc|c} 3 & 2 & 5 & 11 \\ 2 & 7 & 4 & 5 \end{array}\right)$$

Solution. *The first row pivot is 3, so we can multiply the first row by $\frac{1}{3}$ to make the pivot equal to 1:*

$$\left(\begin{array}{ccc|c} 3 & 2 & 5 & 11 \\ 2 & 7 & 4 & 5 \end{array}\right) \xrightarrow{\frac{1}{3}R_1 \to R_2} \left(\begin{array}{ccc|c} 1 & \frac{2}{3} & \frac{5}{3} & \frac{11}{3} \\ 2 & 7 & 4 & 5 \end{array}\right)$$

Next, we'll multiply the first row by -2 and add it to the second row:

$$\xrightarrow{-2R_1 + R_2 \to R_2} \left(\begin{array}{ccc|c} 1 & \frac{2}{3} & \frac{5}{3} & \frac{11}{3} \\ 0 & \frac{17}{3} & \frac{2}{3} & -\frac{7}{3} \end{array}\right)$$

Notice that even for this modest matrix, Gaussian elimination immediately produces fractions. In fact, if our coefficient matrix had more rows, our next step would be to divide the second row by $\frac{17}{3}$ to make the first entry equal to 1, which would give us

$$\left(\begin{array}{ccc|c} 1 & \frac{2}{3} & \frac{5}{3} & \frac{11}{3} \\ 0 & \frac{17}{3} & \frac{2}{3} & -\frac{7}{3} \end{array}\right) \xrightarrow{\frac{3}{17}R_2 \to R_2} \left(\begin{array}{ccc|c} 1 & \frac{2}{3} & \frac{5}{3} & \frac{11}{3} \\ 0 & 1 & \frac{2}{17} & -\frac{7}{17} \end{array}\right)$$

and our fractions become rapidly more daunting.

The Chinese, who first used matrices to represent linear equations, also developed an approach to solving systems of linear equations that doesn't rely on fractions. The Chinese simply referred to the process as *fang cheng shu* ("rectangular tabulation method").

Activity 2.5: Row Echelon Form by *Fang Cheng Shu*

The following concerns the augmented coefficient matrix:

$$A = \left(\begin{array}{ccc|c} 3 & 4 & 3 & 11 \\ 2 & 7 & 2 & 5 \\ 5 & 1 & 2 & 9 \end{array}\right)$$

A2.5.1 To what system of equations does the matrix correspond?

A2.5.2 We introduce an approach first described in the *Nine Chapters of Mathematical Art* (ca. 200 BC).

a) Find c so

$$A \xrightarrow{cR_2 \to R_2} B$$

produces matrix B where the first entry of the second row is a multiple of the pivot.

b) Find d so that the row operation

$$B \xrightarrow{dR_1 + R_2 \to R_2} C$$

produces matrix C where the first entry of the second row is 0.

c) Find e so that the row operation

$$C \xrightarrow{eR_3 \to R_3} D$$

produces matrix D where the leading coefficient of the third row is a multiple of the pivot.

d) Find f so that the row operation

$$D \xrightarrow{fR_1 + R_3 \to R_3} E$$

produces matrix E where the first entry of the third row is 0.

A2.5.3 Complete the row reduction of A. (Start with the matrix E.)

Activity 2.5 introduces a method first described in the *Nine Chapters of Mathematical Art* (ca. 200 BC). We can express the approach as an algorithm:

- Identify the pivot of the working row,
- Multiply all *succeeding* rows by the pivot,
- Subtract multiples of the working row to eliminate the entries below the pivot.

Optionally, if all entries in a row have a common factor, that common factor can be removed to prevent the entries from becoming too large.

The Chinese probably used it for the same reason we should: not because it avoids fractions, but because fractions require special handling.[3]

Example 2.6. *Reduce to row echelon form:*

$$\left(\begin{array}{ccc|c} 3 & 2 & 5 & 1 \\ 2 & 7 & 4 & 5 \\ 5 & 4 & 7 & 2 \end{array}\right)$$

Then solve.

Solution. *The first row pivot is* 3, *so we multiply the second and third rows by* 3*:*

$$\left(\begin{array}{ccc|c} 3 & 2 & 5 & 1 \\ 2 & 7 & 4 & 5 \\ 5 & 4 & 7 & 2 \end{array}\right) \xrightarrow[3R_3 \to R_3]{3R_2 \to R_2} \left(\begin{array}{ccc|c} 3 & 2 & 5 & 1 \\ 6 & 21 & 12 & 15 \\ 15 & 12 & 21 & 6 \end{array}\right)$$

We can eliminate the entries below the first row pivot by multiplying the first row by -2 *and adding it to the second row, and then multiplying the first row by* -5 *and adding it to the third row:*

$$\xrightarrow{-2R_1+R_2 \to R_2} \left(\begin{array}{ccc|c} 3 & 2 & 5 & 1 \\ 0 & 17 & 2 & 13 \\ 15 & 12 & 21 & 6 \end{array}\right) \xrightarrow{-5R_1+R_3 \to R_3} \left(\begin{array}{ccc|c} 3 & 2 & 5 & 1 \\ 0 & 17 & 2 & 13 \\ 0 & 2 & -4 & 1 \end{array}\right)$$

Now we move on to the second row, which has pivot 17*. Multiplying the rows below by* 17*, then adding* 2 *times the second row to the third gives us:*

$$\xrightarrow{17R_3 \to R_3} \left(\begin{array}{ccc|c} 3 & 2 & 5 & 1 \\ 0 & 17 & 2 & 13 \\ 0 & 34 & -68 & 17 \end{array}\right) \xrightarrow{-2R_2+R_3 \to R_3} \left(\begin{array}{ccc|c} 3 & 2 & 5 & 1 \\ 0 & 17 & 2 & 13 \\ 0 & 0 & -72 & -9 \end{array}\right)$$

Since the matrix is in row echelon form, we can use back substitution to solve. The third equation gives us

$$\begin{aligned} -72x_3 &= -9 \\ x_3 &= \frac{9}{72} \end{aligned}$$

or $x_3 = \frac{1}{8}$.

[3]The Chinese computing device was a "counting board," essentially a large grid of squares where the coefficients and constants of an equation could be recorded. In computer terms, each square was a memory register, and integer arithmetic could be done very quickly and efficiently. However, trying to record fractions was problematic.

Notice that we can't avoid fractions forever! However, at this point, the roundoff error and increased computational requirements of floating point arithmetic are minimized.

The next equation gives

$$\begin{aligned} 17x_2 + 2x_3 &= 13 \\ 17x_2 + 2\left(\frac{1}{8}\right) &= 13 \\ x_2 &= \frac{3}{4} \end{aligned}$$

The first equation gives us

$$\begin{aligned} 3x_1 + 2x_2 + 5x_3 &= 1 \\ 3x_1 + 2\left(\frac{3}{4}\right) + 5\left(\frac{1}{8}\right) &= 1 \\ x_1 &= -\frac{3}{8} \end{aligned}$$

and so our solution is

$$x_1 = -\frac{3}{8}, x_2 = \frac{3}{4}, x_3 = \frac{1}{8}$$

It will be convenient to express **solutions in vector form**:

$$\langle x_1, x_2, x_3 \rangle = \left\langle -\frac{3}{8}, \frac{3}{4}, \frac{1}{8} \right\rangle$$

What if we wanted to produce the reduced row echelon form of our matrix? The ancient Chinese did not concern themselves with producing the reduced row echelon form of an augmented coefficient matrix. However, it's easy enough to modify the *fang cheng* method to produce one.

Activity 2.6: Reduced Row Echelon Form by *Fang Cheng Shu*

For the following, consider the augmented matrix whose row echelon form is

$$A = \left(\begin{array}{ccc|c} 2 & 3 & 1 & -4 \\ 0 & 5 & 2 & 2 \\ 0 & 0 & 3 & 1 \end{array}\right)$$

A2.6.1 Consider the third row.

a) Identify the third row pivot.

b) Find a so

$$A \xrightarrow{aR_2 \to R_2} B$$

produces a matrix where the entry in the second row that is above the third row pivot will be a multiple of the third row pivot.

c) Find b so

$$B \xrightarrow{bR_3+R_2 \to R_2} C$$

produces a matrix where the second row has a 0 above the third row pivot.

d) Find c, d so

$$C \xrightarrow{cR_1 \to R_1} D \xrightarrow{dR_3+R_1 \to R_1} E$$

produces a matrix where the first row has a 0 above the third row pivot.

A2.6.2 Find a reduced row echelon form of A, starting from the matrix E.

A2.6.3 What is the solution to the system of equations that A represents?

Example 2.7. *Find the reduced row echelon form of* $\left(\begin{array}{ccc|c} 3 & 2 & 5 & 1 \\ 0 & 17 & 2 & 13 \\ 0 & 0 & -72 & -9 \end{array}\right)$ *(the matrix whose row echelon form we found in Example 2.6).*

Solution. *We'll want to clear the entries above the third row pivot. While we could use the current pivot (−72), you might observe that all the terms of the row are multiples of −9, so we can reduce the row terms by this factor:*

$$\left(\begin{array}{ccc|c} 3 & 2 & 5 & 1 \\ 0 & 17 & 2 & 13 \\ 0 & 0 & -72 & -9 \end{array}\right) \xrightarrow{-\frac{1}{9}R_3 \to R_3} \left(\begin{array}{ccc|c} 3 & 2 & 5 & 1 \\ 0 & 17 & 2 & 13 \\ 0 & 0 & 8 & 1 \end{array}\right)$$

Now multiply the second row and first rows by 8, the third row pivot:

$$\xrightarrow[\substack{8R_2 \to R_2 \\ 8R_1 \to R_1}]{} \left(\begin{array}{ccc|c} 24 & 16 & 40 & 8 \\ 0 & 136 & 16 & 104 \\ 0 & 0 & 8 & 1 \end{array}\right)$$

We'll add −2 times the third row from the second, and −5 times the third row from the first:

$$\xrightarrow{-2R_3+R_2 \to R_2} \left(\begin{array}{ccc|c} 24 & 16 & 40 & 8 \\ 0 & 136 & 0 & 102 \\ 0 & 0 & 8 & 1 \end{array}\right) \xrightarrow{-5R_3+R_1 \to R_3} \left(\begin{array}{ccc|c} 24 & 16 & 0 & 3 \\ 0 & 136 & 0 & 102 \\ 0 & 0 & 8 & 1 \end{array}\right)$$

While it's not necessary, we can simplify the second row by removing a common factor of 34:

$$\xrightarrow{\frac{1}{34}R_2 \to R_2} \begin{pmatrix} 24 & 16 & 0 & 3 \\ 0 & 4 & 0 & 3 \\ 0 & 0 & 8 & 1 \end{pmatrix}$$

For illustrative purposes, we'll ignore the fact that the second row pivot (4) is a divisor of the entry above it (16). Instead, we'll approach the algorithm as an algorithm, and multiply the first row by 4, the second row pivot:

$$\xrightarrow{4R_1 \to R_1} \begin{pmatrix} 96 & 64 & 0 & 12 \\ 0 & 4 & 0 & 3 \\ 0 & 0 & 8 & 1 \end{pmatrix}$$

Finally we'll subtract 16 times the second row from the first row:

$$\xrightarrow{R_1 - 16R_2 \to R_1} \begin{pmatrix} 96 & 0 & 0 & -36 \\ 0 & 4 & 0 & 3 \\ 0 & 0 & 8 & 1 \end{pmatrix}$$

The third row gives us the equation $8x_3 = 1$*, so* $x_3 = \frac{1}{8}$.

The second row gives us the equation $4x_2 = 3$*, so* $x_2 = \frac{3}{4}$.

The first row gives us the equation $96x_1 = -36$*, so* $x_1 = -\frac{3}{8}$.

2.3 Coefficient Matrices

Before we move on, let's see why we say the matrix representing our system of equations is called an *augmented* matrix.

Activity 2.7: Coefficient Matrices

A2.7.1 Row reduce the augmented coefficient matrices given. Record the elementary row operations used.

a) $\left(\begin{array}{cc|c} 2 & 7 & 3 \\ 3 & -5 & 1 \end{array}\right)$

b) $\left(\begin{array}{cc|c} 2 & 7 & 5 \\ 3 & -5 & 9 \end{array}\right)$

c) $\left(\begin{array}{cc|c} 2 & 7 & 11 \\ 3 & -5 & 17 \end{array}\right)$

d) What do you notice about the row operations required?

Activity 2.7 points to an important idea: The row operations necessary to transform an augmented matrix into row echelon form depend only on the coefficients, and not on the column of constants. Thus, we might consider the matrix of coefficients separately from the column of constants.

For example, the augmented coefficient matrix corresponding to the system of equations

$$\begin{aligned} 3x_1 + 2x_2 - 5x_3 &= 8 \\ 2x_1 - 4x_2 + 8x_3 &= 7 \\ 5x_1 - x_2 - x_3 &= 1 \end{aligned}$$

has coefficient matrix $A = \begin{pmatrix} 3 & 2 & -5 \\ 2 & -4 & 8 \\ 5 & -1 & -1 \end{pmatrix}$ and column of constants $\mathbf{b} = \begin{pmatrix} 8 \\ 7 \\ 1 \end{pmatrix}$
Because the column of constants is a 3-tuple of real numbers, it's also a vector, so we can label it using vector notation, either $\mathbf{b}$ or even $\vec{b}$. Moreover, since we're used to seeing vectors written horizontally, as $\langle v_1, v_2, \ldots, v_n \rangle$, but $\mathbf{b}$ is a vector written vertically, we'll call it a **column vector**. Putting this all together, we can express our augmented coefficient matrix as a **block matrix**

$$\left(\begin{array}{ccc|c} 3 & 2 & -5 & 8 \\ 2 & -4 & 8 & 7 \\ 5 & -1 & -1 & 1 \end{array}\right) = (\ A \mid \mathbf{b}\ )$$

Thus we see that the augmented coefficient matrix is the coefficient matrix A, augmented by the column of constants $\mathbf{b}$.

One particularly important case occurs when the column of constants is the zero vector. In that case, we say that we have a **homogeneous system of equations.**

Activity 2.8: Homogeneous and Inhomogeneous Systems

A2.8.1 Suppose E_1, E_2, ...,E_n are elementary row operations, where

$$\begin{pmatrix} 5 & 3 \\ 2 & 7 \end{pmatrix} \xrightarrow{E_1} A_1 \xrightarrow{E_2} A_2 \xrightarrow{\cdots} A_n \xrightarrow{E_{n+1}} \begin{pmatrix} 1 & 0 \\ 0 & 1 \end{pmatrix}$$

a) Find the elementary row operations E_1, E_2,..., E_n.

b) Apply the row operations you found in A2.8.1a to the vector $\begin{pmatrix} 4 \\ 13 \end{pmatrix}$. In other words, find x_1, x_2, where

$$\begin{pmatrix} 4 \\ 13 \end{pmatrix} \xrightarrow{E_1} B_1 \xrightarrow{E_2} B_2 \ldots \xrightarrow{E_n} \begin{pmatrix} x_1 \\ x_2 \end{pmatrix}$$

c) What does the final matrix from Activity A2.8.1b tell you about the solution to the system represented by the augmented coefficient matrix $\left(\begin{array}{cc|c} 5 & 3 & 4 \\ 2 & 7 & 13 \end{array}\right)$?

d) Solve the system represented by the coefficient matrix $\left(\begin{array}{cc|c} 5 & 3 & -3 \\ 2 & 7 & 6 \end{array}\right)$.

e) Solve the system represented by the coefficient matrix $\left(\begin{array}{cc|c} 5 & 3 & a \\ 2 & 7 & b \end{array}\right)$.

A2.8.2 Let M be any matrix, and suppose there is a sequence of row operations that reduces

$$M \xrightarrow{E_1} M_1 \xrightarrow{E_2} \ldots \xrightarrow{E_n} \begin{pmatrix} 1 & 0 & 0 \\ 0 & 1 & 0 \\ 0 & 0 & 1 \end{pmatrix}$$

a) Let $\mathbf{b}$ be any column vector in $\mathbb{R}^3$. Explain why there is always a solution to the system of linear equations represented by $\left(M \mid \mathbf{b} \right)$.

b) Remember $\mathbf{0}$ is the vector consisting of all zeros. What is the solution to the system of linear equations represented by $\left(M \mid \mathbf{0} \right)$?

Activity A2.8.2 leads to the following result:

Theorem 2.1. *A homogeneous system of equations always has a solution.*

However, the guaranteed solution is not very interesting, since it consists of setting all variables equal to 0. We often call such a noninteresting solution a **trivial solution**; thus, a trivial solution to $a^5 + b^5 = c^5$ is $a = 0$, $b = 0$, $c = 0$.

A more interesting question is: Are there *non*trivial solutions? We'll try to answer that question next.

2.4 Free and Basic Variables

The row echelon form of a matrix allows us to distinguish between two types of variables:

- Variables that correspond to pivots,
- Variables that do not correspond to pivots.

Is there an important difference between the two types of variables?

Activity 2.9: Free and Basic Variables

A2.9.1 Let

$$C = \left(\begin{array}{cccc|c} 1 & 2 & 5 & 8 & 0 \\ 0 & 2 & 7 & 4 & 0 \\ 0 & 0 & 0 & 3 & 0 \end{array} \right)$$

be the row echelon form of the augmented coefficient matrix of a system of equations in x_1, x_2, x_3, x_4.

a) To which variables do the pivots correspond?

b) Consider the variable(s) that correspond to a pivot. Choose distinct, nonzero values for these variables. Then, if possible, solve for the remaining variables. If it's not possible to solve for the remaining variables, explain why.

c) Consider the variable(s) that *don't* correspond to a pivot. Choose distinct, nonzero values for these variables. Then, if possible, solve for the remaining variables. If it's not possible to solve for the remaining variables, explain why.

d) Find a nontrivial solution to the system of equations.

A2.9.2 Let

$$D = \left(\begin{array}{cccc|c} 0 & 3 & 5 & 8 & 0 \\ 0 & 0 & 7 & 4 & 0 \end{array} \right)$$

a) To which variables do the pivots correspond?

b) Consider the variable(s) that correspond to a pivot. Choose distinct, nonzero values for these variables. Then, if possible, solve for the remaining variables. If it's not possible to solve for the remaining variables, explain why.

c) Consider the variable(s) that *don't* correspond to a pivot. Choose distinct, nonzero values for these variables. Then, if possible, solve for the remaining variables. If it's not possible to solve for the remaining variables, explain why.

d) Find a nontrivial solution to the system of equations.

In Activity 2.9, you found that if you assigned arbitrary values to the variables corresponding to pivots, you could *not*, in general, solve for the remaining variables. On the other hand, you *could* assign arbitrary values to the nonpivot variables and still solve the system of equations. This leads to the following definition:

Definition 2.8 (Free and Basic Variables). *A variable in a system of equations is* ***free*** *if, in the row echelon form of the corresponding matrix, it does not correspond to a pivot; otherwise, it is* **basic**.

The free variables are named such because we are free to choose any value whatsoever for these variables, and you found in Activity 2.9 that you could choose values for the free variables and solve for the values of the basic variables.

However, this only gives one solution. To find all solutions, we rely on a **parameterization**: a formula that gives the values of the basic variable in terms of the values of the free variables.

Example 2.8. *Suppose the row echelon form for a system of linear equations in x_1, x_2, x_3 is*

$$\left(\begin{array}{ccc|c} 2 & -1 & 3 & 0 \\ 0 & 1 & -4 & 0 \end{array}\right)$$

Identify the free and basic variables; then parameterize the solutions to the system. Finally, find three solutions to the system.

Solution. *The first row pivot (2) corresponds to the variable x_1, and the second row pivot (1) corresponds to the variable x_2, so these are basic variables. The variable x_3 does not correspond to a pivot, so it is a free variable. This means that we should be able to express both x_1 and x_2 in terms of x_3.*

Note that the last row of the matrix corresponds to the equation

$$x_2 - 4x_3 = 0$$

Since x_2 is a basic variable and x_3 is a free variable, we can solve for x_2 in terms of x_3:

$$x_2 = 4x_3$$

The first row of the matrix corresponds to the equation

$$2x_1 - x_2 + 3x_3 = 0$$

Since x_1 is a basic variable, we can solve for it in terms of the free variable x_3. First, we solve for x_1:

$$2x_1 = x_2 - 3x_3$$

Next, we know that $x_2 = 4x_3$, so we can eliminate x_2 from our equation:

$$\begin{aligned} 2x_1 &= 4x_3 - 3x_3 \\ 2x_1 &= x_3 \\ x_1 &= \frac{1}{2}x_3 \end{aligned}$$

This allows us to express our basic variables x_1 and x_2 entirely in terms of our free variable x_3:

$$x_1 = \frac{1}{3}x_3 \qquad\qquad x_2 = 4x_3$$

where x_3 can have any value.

If we choose a value for x_3, we obtain a solution. For example, we might let $x_3 = 1$, in which case $x_1 = \frac{1}{3}, x_2 = 4, x_3 = 1$. Or we might let $x_3 = 2$, giving us $x_1 = \frac{2}{3}, x_2 = 8, x_3 = 2$. For a third solution, we let $x_3 = 3$ and obtain $x_1 = 1, x_2 = 12, x_3 = 3$.

We can express our solution in vector form:

$$\langle x_1, x_2, x_3 \rangle = \left\langle \frac{1}{3}x_3, 4x_3, x_3 \right\rangle$$

It will be convenient to go two steps further. First, in the vector on the right, all components have a factor of x_3, so we express it as the result of a scalar multiplication of some vector by x_3:

$$\left\langle \frac{1}{3}x_3, 4x_3, x_3 \right\rangle = x_3 \left\langle \frac{1}{3}, 4, 1 \right\rangle$$

We call this the **vector form** of the solution.

Second, to distinguish between the *variable* x_3, and our choice of a value for x_3, we'll introduce a **parameter** variable, say s, which allows us to write our solution

$$\left\langle \frac{1}{3}x_3, 4x_3, x_3 \right\rangle = s \left\langle \frac{1}{3}, 4, 1 \right\rangle$$

where we may choose any value we want for s. We call this the **parameterized form** of the solution.

Since there are an infinite number of possible choices we could make for s, each of which would give us a solution to our system of equations, we say that our system has an infinite number of solutions.

Example 2.9. *Suppose the row echelon form for a system of linear equations in x_1, x_2, x_3, x_4 is*

$$\left(\begin{array}{cccc|c} 1 & -3 & 7 & 8 & 0 \\ 0 & 1 & 3 & -2 & 0 \end{array} \right)$$

Parameterize the solutions. Then find three different solutions.

Solution. *We note that the free variables are x_3 and x_4, while the basic variables are x_1 and x_2, so we should be able to express the basic variables in terms of the free variables.*

The last equation gives us

$$x_2 + 3x_3 - 2x_4 = 0$$

Since x_2 is a basic variable and x_3, x_4 are free, we should express x_2 in terms of x_3 and x_4:

$$x_2 = -3x_3 + 2x_4$$

The first equation gives us

$$x_1 - 3x_2 + 7x_3 + 8x_4 = 0$$

Since x_1 is a basic variable, we should express it in terms of the free variables x_3 and x_4:

$$x_1 = 3x_2 - 7x_3 - 8x_4$$

Using $x_2 = -3x_3 + 2x_4$:

$$\begin{aligned} x_1 &= 3(-3x_3 + 2x_4) - 7x_3 - 8x_4 \\ x_1 &= -16x_3 - 2x_4 \end{aligned}$$

Again, we can express our solution in vector form:

$$\langle x_1, x_2, x_3, x_4 \rangle = \langle -16x_3 - 2x_4, -3x_3 + 2x_4, x_3, x_4 \rangle$$

Since we have two free variables, x_3 and x_4, let's separate the solution vector into two vectors, where the first only contains x_3 terms, and the second only contains x_4 terms:

$$\begin{aligned} \langle x_1, x_2, x_3, x_4 \rangle &= \langle -16x_3, -3x_3, x_3, 0 \rangle + \langle -2x_4, 2x_4, 0, x_4 \rangle \\ &= x_3 \langle -16, -3, 1, 0 \rangle + x_4 \langle -2, 2, 0, 1 \rangle \end{aligned}$$

or, introducing parameters s and t,

$$= s\langle -16, -3, 1, 0 \rangle + t\langle -2, 2, 0, 1 \rangle$$

We can choose values of s and t to find different solutions.

If we let $s = 1$, $t = 0$, we obtain:

$$\langle x_1, x_2, x_3, x_4 \rangle = \langle -16, -3, 1, 0 \rangle$$

If we let $s = 0$, $t = 1$, we obtain:

$$\langle x_1, x_2, x_3, x_4 \rangle = \langle -2, 2, 0, 1 \rangle$$

If we let $s = 1$, $t = 1$, we obtain:

$$\begin{aligned} \langle x_1, x_2, x_3, x_4 \rangle &= \langle -16, -3, 1, 0 \rangle + \langle -2, 2, 0, 1 \rangle \\ &= \langle -18, -1, 1, 1 \rangle \end{aligned}$$

When parameterizing solutions this way, we often end with fractional coefficients, so arbitrary choices for the parameters will give us fractional solutions. However, many problems in mathematics require we find *integer* solutions.

Activity 2.10: Integer Solutions

A2.10.1 Suppose the row echelon form of an augmented coefficient matrix for a system of equations in x_1, x_2, x_3 is

$$\left(\begin{array}{ccc|c} 2 & 3 & 5 & 0 \\ 0 & 7 & 2 & 0 \end{array}\right)$$

a) Which are the free variables? Which are the basic variables?
b) Express the basic variable(s) in terms of the free variable(s). To make the following parts easier, **DON'T** reduce or otherwise simplify fractional coefficients.
c) Your expression for the basic variable(s) should include several fractional coefficients. Why does this occur? (In other words, what did you do that produced fractional coefficients?)
d) Remember you can choose any value you want for the free variables. What values could you choose for the free variables to obtain *integer* solutions for all variables?

A2.10.2 Suppose the row echelon form of an augmented coefficient matrix for a system of equations in x_1, x_2, x_3 is

$$\left(\begin{array}{cccc|c} 5 & 1 & 3 & -5 & 0 \\ 0 & 0 & 2 & 1 & 0 \end{array}\right)$$

a) Express the basic variable(s) in terms of the free variable(s).
b) Remember you can choose any value you want for the free variables. What values must you choose for the free variables to obtain *integer* solutions for all variables?

A2.10.3 Suppose the row echelon form of an augmented coefficient matrix for a system of equations in x_1, x_2, x_3, x_4 is

$$\left(\begin{array}{cccc|c} 0 & 4 & 2 & -5 & 0 \\ 0 & 0 & 0 & 3 & 0 \end{array}\right)$$

a) Solve for the basic variable(s) in terms of the free variable(s).
b) What values should you choose for the free variables to obtain *integer* solutions for all variables?

A2.10.4 Suppose the row echelon form of an augmented coefficient matrix for a system of equations in x_1, x_2, x_3, x_4 is

$$\left(\begin{array}{cccc|c} a & 5 & 2 & 5 & 0 \\ 0 & b & 2 & 3 & 0 \end{array}\right)$$

where a and b are nonzero integers. Based on your observations in Activity A2.10.1 through A2.10.3, how should you choose the free variables to obtain integer solutions for all variables?

A2.10.5 Suppose the row echelon form of an augmented coefficient matrix for a system of equations in $x1$, x_2, x_3 is

$$\left(\begin{array}{ccc|c} 2 & 5 & 1 & 0 \\ 0 & 3 & -1 & 0 \end{array}\right)$$

a) Based on your conclusion in Activity A2.10.4, what is the least values should you choose for the free variable to obtain integer solutions for all variables?

b) Can you find a smaller choice of the free variable lead to a smaller integer solution for all variables?

Activity 2.10 provides a way to generate integer solutions, though as Activity A2.10.5 shows, we won't necessarily be able to find all integer solutions this way.

One of the most important differences between elementary and advanced mathematics is that, for the most part, elementary mathematics is concerned with questions like "*How* can we solve ...," but higher mathematics is concerned with questions like "*When* can we solve ...."

Activity 2.11: Rows of 0s

A2.11.1 If possible, solve the systems corresponding to the given augmented coefficient matrices. Parameterize your solution(s) as appropriate.

a) $\left(\begin{array}{cc|c} 2 & -3 & 11 \\ -5 & 1 & 8 \end{array}\right)$ and $\left(\begin{array}{cc|c} 2 & -3 & 0 \\ -5 & 1 & 0 \end{array}\right)$

b) $\left(\begin{array}{cc|c} 1 & 5 & 2 \\ 2 & 10 & 4 \end{array}\right)$ and $\left(\begin{array}{cc|c} 1 & 5 & 0 \\ 2 & 10 & 0 \end{array}\right)$

c) $\left(\begin{array}{cc|c} 1 & 5 & 3 \\ 2 & 7 & 4 \\ 5 & -2 & 6 \end{array}\right)$ and $\left(\begin{array}{cc|c} 1 & 5 & 0 \\ 2 & 7 & 0 \\ 5 & -2 & 0 \end{array}\right)$

d) $\left(\begin{array}{ccc|c} 2 & 3 & 5 & 3 \\ 5 & 0 & 6 & 3 \end{array}\right)$ and $\left(\begin{array}{ccc|c} 2 & 3 & 5 & 0 \\ 5 & 0 & 6 & 0 \end{array}\right)$

e) $\left(\begin{array}{ccc|c} 2 & -3 & 5 & 4 \\ -3 & 7 & 1 & 5 \\ 1 & 3 & -1 & 8 \end{array}\right)$ and $\left(\begin{array}{ccc|c} 2 & -3 & 5 & 0 \\ -3 & 7 & 1 & 0 \\ 1 & 3 & -1 & 0 \end{array}\right)$

f) $\left(\begin{array}{cccc|c} -3 & 5 & 2 & 7 & 3 \\ 6 & -10 & -4 & -14 & 9 \end{array}\right)$ and $\left(\begin{array}{cccc|c} -3 & 5 & 2 & 7 & 0 \\ 6 & -10 & -4 & -14 & 0 \end{array}\right)$

A2.11.2 Suppose in the row reduction of a $n \times n$ *coefficient* matrix, you end with **NO** rows of 0s. What does this say about the number of solutions to:

a) The corresponding homogeneous system?
b) The corresponding nonhomogeneous system?

A2.11.3 Suppose in the row reduction of a $n \times m$ *coefficient* matrix, where $m > n$, you end with **NO** rows of 0s. What does this say about the number of solutions to:

a) The corresponding homogeneous system?
b) The corresponding nonhomogeneous system?

A2.11.4 Suppose in the row reduction of a $n \times m$ *coefficient* matrix, where $m < n$, you end with **NO** rows of 0s. What does this tell you?

A2.11.5 Suppose in the row reduction of a $n \times m$ *coefficient* matrix, where $m > n$, you end with one (or more) rows of 0s. What does this say about the number of solutions to:

a) The corresponding homogeneous system?
b) The corresponding nonhomogeneous system?
c) Would your answer change if $n = m$?

The number of rows of 0s in the row echelon form of a coefficient matrix seems to shed light on whether our system of equations has no solution; a unique solution; or an infinite number of solutions. Let's introduce the definition:

Definition 2.9 (Rank). *The rank of a matrix is the number of nonzero rows in the row echelon form of the matrix.*

Before we proceed, we introduce a standard riddle in mathematics: How many numbers are there? Three: 0, 1, and infinity. What this means in the context of "When" can we solve an equation is the following:

- There might be no solution at all (0),
- The solution might be unique (1),
- There might be infinitely many solutions.

We'll explore that next.

Activity 2.12: Rank

A2.12.1 A useful thing to do in higher mathematics is to find bounds on a quantity: the least and greatest possible values. In the following, assume A is a $n \times m$ coefficient matrix.

a) Could the rank of A be greater than n? Why/why not?

b) Could the rank of A be greater than m? Why/why not?

c) Determine the greatest possible value for the rank of a 3×5 matrix. Then write down a 3×5 matrix with this rank.

d) Determine the greatest possible value for the rank of a 5×3 matrix. Then write down a 5×3 matrix with this rank.

A2.12.2 Let A be a $n \times m$ coefficient matrix corresponding to a homogeneous system of equations, and suppose the rank of A is k.

a) How many variables are in the system of equations? How many of these are free variables, and how many are basic?

b) Could the system have no solution? Why/why not?

c) Suppose the system has a unique solution. What does this tell you about the rank?

d) Suppose the system has an infinite number of solutions. What does this tell you about the rank?

A2.12.3 Another useful strategy in mathematics is to consider **trichotomy**: Given any quantities a, b, we might consider the three situations where $a > b$; $a = b$; and $a < b$. Let A be a $n \times m$ coefficient matrix corresponding to a homogeneous system of equations, and suppose the rank of A is k.

a) Can $k > m$? Why/why not?

b) Suppose $k = m$. What does this tell you about the number of solutions to the system of equations?

c) Suppose $k < m$. What does this tell you about the number of solutions to the system of equations?

A2.12.4 Suppose A is a $n \times m$ coefficient matrix for a *non*homogeneous system of equations. What must be true about the rank of A for there to be a unique solution?

Activity A2.12.2 tell us that if our homogeneous system of equations has a unique solution, the rank must equal the number of columns (which corresponds to the number of variables). Conversely, Activity A2.12.3 tells us that if the rank is equal to the number of columns, the system has a unique equation. This proves:

Theorem 2.2 (Rank). *Let A be a $n \times m$ coefficient matrix. The corresponding system of homogeneous equations will have a unique solution if and only if the rank is equal to m.*

2.5 Computational Considerations

In practice, most "interesting" systems of equations have thousands of variables, so row reduction is something that will rarely be done by hand; instead, it is the job of the scientist or engineer to *frame* the question as a system of linear equations, then let a computing device solve it.

At the same time, the mathematician and the computer scientist should concern themselves with the problem of *how* the machine computes an answer. Indeed, one of the reasons it's important to understand how computations are done "by hand" is that it provides insight on how best to program a computer to solve a problem.

Activity 2.13: Roundoff Errors

A full introduction to the issue of roundoff error is properly the subject of computer science. This activity serves to introduce the basic problem.

When we round a number to a certain number of decimal places, we accept that the rounded value differs from the exact value by some amount. Equivalently, there is some value, called **machine-ϵ**, that is the smallest recognized difference between two numbers. For example, if we round to three decimal places, the smallest recognized difference between two numbers is 0.001. Regardless of what computing device we use, there is some nonzero machine-ϵ; the only difference is how large it is.

A2.13.1 Suppose our computing devices has $\epsilon = 0.01$ (equivalent to rounding to two decimal places).

a) Find the *exact* value of $(7 \div 3) \times (7 \div 6)$. Then round your answer to two decimal places. This corresponds to the answer the computer *should* give you.

b) Find $(7 \div 3) \times (7 \div 6)$, but this time round every intermediate computation to two decimal places before proceeding. This corresponds to the answer the computer *does* give you.

c) The difference between the value the computer *should* give and the answer it *does* give is the round-off error. Will decreasing ϵ eliminate the round-off error? Provide evidence of your claim. (Suggestion: See if smaller values of ϵ eliminate the roundoff error.)

A2.13.2 Consider the system

$$
\begin{aligned}
0.01x_1 + x_2 &= 1 \\
x_1 - x_2 &= 0
\end{aligned}
$$

a) Row reduce and find the exact solution, then round your answer to two decimal places. For purposes of this problem, do **NOT** switch rows.
b) Solve the system a second time, but this time round the results of every intermediate computation to two decimal places. For purposes of this problem, do **NOT** switch rows. **Note**: If your row reduction results in a fraction, find the decimal equivalent of the fraction, rounded to two decimal places, before proceeding.
c) How well does the calculated solution match the exact solution?

A2.13.3 Consider the system

$$
\begin{aligned}
0.01x_1 + 0.1x_2 &= 1 \\
x_1 - 0.1x_2 &= 0
\end{aligned}
$$

a) Row reduce and find the exact solution, then round your answer to two decimal places. For purposes of this problem, do **NOT** switch rows.
b) Solve the system a second time, but this time round the results of every intermediate computation to two decimal places. For purposes of this problem, do **NOT** switch rows. **Note**: If your row reduction results in a fraction, find the decimal equivalent of the fraction, rounded to two decimal places, before proceeding.
c) How well does the calculated solution match the exact solution?

A2.13.4 The preceding leads to the **partial pivoting** algorithm:

- Switch rows of the matrix so the leading term of the first row is greater (in absolute value) than all terms below it. This becomes our row pivot.
- Use row reduction as before, to eliminate all entries below the pivot.
- Again, switch rows of the matrix so the leading term of the second row is greater (in absolute value) than all the terms below it. This will become our next pivot.
- Repeat until the matrix is in row echelon form.

a) Use partial pivoting to solve

$$
\begin{aligned}
0.01x_1 + x_2 &= 1 \\
x_1 - x_2 &= 0
\end{aligned}
$$

Remember to round the results of every intermediate computation to two decimal places.

b) How well does the calculated solution match the exact solution?

A2.13.5 For each of the following systems, solve the system three times: First by using correct row reduction; second, row reduction without row switching, but rounding the results of every intermediate calculation to two decimal places; and finally, row reduce using partial pivoting (again rounding the results of every intermediate calculation to two decimal places).

a) $\left(\begin{array}{cc|c} 0.01 & 7 & 100 \\ 1 & 100 & 0 \end{array}\right)$

b) $\left(\begin{array}{ccc|c} 0.01 & 1 & 0.1 & 5 \\ 2 & 207 & 0.03 & 7 \\ 3 & 15 & 0.07 & 5 \end{array}\right)$

2.6 Applications of Linear Algebra

As long as we can frame a problem in the form of a system of linear equations, we can use the tools of linear algebra to solve it. In the following activities, we'll take a look at several seemingly unrelated problems, all of which can be reduced to systems of linear equations and solved.

Activity 2.14: Finding Orthogonal Vectors

A2.14.1 Let $\vec{v} = \langle v_1, v_2 \rangle$.

a) Set up an equation or system of equations to find $\vec{x} = \langle x_1, x_2 \rangle$, where $\vec{x}$ is orthogonal (perpendicular) to the given vector $\vec{v}$.

b) How many solutions are there to this system?

c) Explain, from a *geometric* viewpoint, why your answer is not surprising.

A2.14.2 Let $\vec{v}_1 = \langle v_{11}, v_{12} \rangle$ and $\vec{v}_2 = \langle v_{21}, v_{22} \rangle$.

a) Set up an equation or system of equations to find $\vec{x} = \langle x_1, x_2 \rangle$, where $\vec{x}$ is orthogonal (perpendicular) to $\vec{v}_1$ **AND** $\vec{v}_2$.

b) Find $\langle x_1, x_2 \rangle$.

c) Explain the *geometric* significance of your answer.

A2.14.3 Let $\vec{v}_1 = \langle v_{11}, v_{12}, v_{13} \rangle$.

a) Set up an equation or system of equations to find $\vec{x} = \langle x_1, x_2, x_3 \rangle$, a vector perpendicular to $\vec{v}_1$.

b) Solve the system.

c) What is the *geometric* significance of the solution?

A2.14.4 Let $\overrightarrow{v}_1 = \langle v_{11}, v_{12}, v_{13} \rangle$, $\overrightarrow{v}_2 = \langle v_{21}, v_{22}, v_{23} \rangle$.

a) Set up an equation or system of equations to find $\overrightarrow{x} = \langle x_1, x_2, x_3 \rangle$, a vector perpendicular to $\overrightarrow{v}_1$ **AND** $\overrightarrow{v}_2$.

b) Assume the components of $\overrightarrow{v}_1$ and $\overrightarrow{v}_2$ are integers. Find the simplest *integer* solution $\langle x_1, x_2, x_3 \rangle$.

A2.14.5 In each of the following, a set of vectors is given. If possible, find a vector (preferably with integer coordinates) that is perpendicular to *all* of the vectors in the given set; if there is more than one such vector, provide a parameterized solution. If no vectors exist, explain why not, basing your explanation on linear algebra considerations.

a) The vectors $\langle 1, 4, 8 \rangle$, $\langle -1, -7, 4 \rangle$.

b) The vectors $\langle 2, 1, 4 \rangle$, $\langle 1, -1, 3 \rangle$, and $\langle 3, -1, -7 \rangle$.

c) The vectors $\langle 1, 4, 7 \rangle$, $\langle 5, 3, -4 \rangle$, and $\langle 7, 11, 10 \rangle$.

d) The vectors $\langle 1, 1, 0, -3 \rangle$, $\langle 2, 5, -1, 4 \rangle$.

e) The vectors $\langle 3, 1, 4, -1 \rangle$, $\langle 1, -1, 2, 5 \rangle$, and $\langle 3, 1, 4, 8 \rangle$.

We can also work backward: suppose we have a problem that corresponds to a system of linear equations. Then the problem can be solved using the tools and techniques of linear algebra.

Activity 2.15: Bezout's Algorithm

Modern information security is based around a cryptographic system called RSA (after the initials of its three inventors). Implementation of RSA requires, among other things, finding *integer* solutions x, y to the **linear Diophantine equation**

$$ax + by = m$$

for integers a, b, m.

A2.15.1 Suppose we want to find integers x and y that satisfy $17x - 41y = 1$.

a) Show that $x = 2y + \dfrac{7y+1}{17}$.

b) Let $z = \dfrac{7y+1}{17}$. Show that $y = 2z + \dfrac{3z-1}{7}$.

c) Let $w = \dfrac{3z-1}{7}$. Show that $z = 2w + \dfrac{w+1}{3}$.

d) Let $t = \frac{w+1}{3}$. Show that $w = 3t - 1$.

e) Explain why, if t is an integer, the other variables will also be integers.

f) Set up a system of linear equations with *integer* coefficients relating x, y, z, w, t.

g) Find an integer solution to $17x - 41y = 1$.

A2.15.2 Use the approach of Activity A2.15.1 to find an integer solution to $23x - 57y = 1$.

A2.15.3 The **Euclidean algorithm** to find the GCD of two integers a, b begins by dividing the larger by the smaller number to form a quotient and remainder. The smaller number is then divided by the remainder to form a new quotient and remainder, and the process is repeated; the last nonzero remainder is the GCD. We illustrate the process for the GCD of 17 and 41:

$$\begin{aligned} 41 \div 17 &= 2, \text{ remainder } 7 \\ 17 \div 7 &= 2, \text{ remainder } 3 \\ 7 \div 3 &= 2, \text{ remainder } 1 \\ 3 \div 1 &= 3, \text{ remainder } 0 \end{aligned}$$

a) Compare the equations you found in A2.15.1 with the terms of the Euclidean algorithm. What do you notice? (**Note**: It may be helpful to make the row pivots positive.)

b) Use the Euclidean algorithm to find the GCD of 23 and 57, then compare your results with the equations you found in A2.15.2

A2.15.4 Find an integer solution to $37x - 47y = 1$.

A2.15.5 Find an integer solution to $179x - 299y = 1$.

A classic brain teaser is known as the **hundred fowls problem**. The oldest version appears in *The Mathematical Classic of Zhang Qiujian* (5th cent.):

> A rooster costs 5 qian; a hen costs 3 qian; and 3 chicks cost 1 qian. If you buy 100 birds for 100 qian, how many roosters, hens, and chicks did you buy?

This requires finding an integer solution to a system of two equations in three unknowns.

Activity 2.16: The Hundred Fowls Problem

In the following, qian, shillings, dirham, and pana are all units of currency.

A2.16.1 Suppose a rooster costs 5 qian; a hen costs 3 qian; and 3 chicks cost 1 qian; and suppose you buy 100 birds for 100 qian.

a) Let x be the number of roosters, y the number of hens, and z the number of chicks. Set up a system of linear equations corresponding to the hundred fowls problem. (Suggestion: As needed, rewrite an equation so the coefficients are integers.)

b) Row reduce the system; identify the free and basic variables; and parameterize the solutions.

c) Find a solution to the problem (remember the number of each type of bird must be a whole number). If there is more than one solution, find all of them.

A2.16.2 Around 800, the English mathematician Alcuin of York posed the following problem: A man bought 100 animals for 100 shillings. Horses costs 3 shillings, cows cost 1 shilling, and 24 sheep cost a shilling. How many horses, cows, and sheep did the man buy? If there is more than one solution, find all of them.

A2.16.3 In the 10th century, the Egyptian mathematician Abu Kamil posed the following problem: A man buys 100 birds for 100 dirham. Geese cost 4 dirhams each; chickens 1 dirham; pigeons at 1 dirham for two; and starlings at 1 dirham for 10. How many of each type did the man buy? If there is more than one solution, find all of them.

A2.16.4 Also in the 10th century, the Indian mathematician Mahavira posed the problem: 100 birds are bought for 100 pana. 5 Pigeons cost 3 pana; 7 cranes cost 5 pana; 9 swans cost 7 pana; and 3 peacocks cost 9 pana. How many birds of each type were bought? If there is more than one solution, find all of them.

Another use is casting shadows.

Activity 2.17: Shadows

Realistic images can be generated in computer graphics using a technique known as **ray tracing**, which relies on the following assumptions:

- Light travels in straight lines,
- Opaque objects block the path of light beams.

In the figure, suppose light travels in parallel straight lines in the direction given by $\vec{v}$. Let $\overline{XY}$ be some opaque object that casts shadow $\overline{X'Y'}$ onto a surface $\overleftrightarrow{PQ}$.

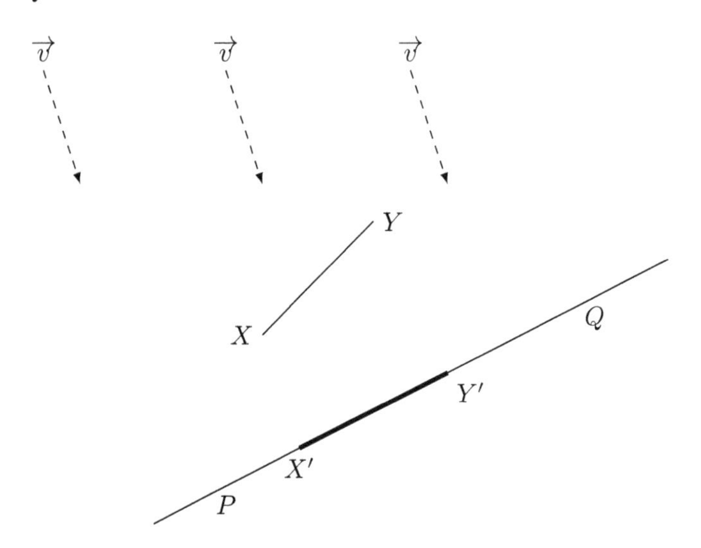

A2.17.1 Suppose P, Q, X, Y are points in $\mathbb{R}^2$.

a) Explain why there exist values x_1, x_2 where $x_1\overrightarrow{PQ} + x_2\vec{v} = \overrightarrow{XY}$.

b) What is the significance of $x_1\overrightarrow{PQ}$?

c) Suppose $P = (0,0)$, $Q = (14,7)$, $X = (3,5)$, $Y = (6,8)$, and $\vec{v} = \langle 1,-3\rangle$. Find $\overrightarrow{X'Y'}$, the vector corresponding to the shadow cast by $\overline{XY}$ on the line $\overleftrightarrow{PQ}$.

A2.17.2 Suppose $\overline{XY}$ is a line segment in $\mathbb{R}^3$, which casts a shadow $\overline{X'Y'}$ onto plane PQR when illuminated by light traveling in directions parallel to some vector $\vec{v}$.

a) Explain why there must be values x_1, x_2, x_3 where $x_1\overrightarrow{PQ} + x_2\overrightarrow{QR} + x_3\vec{v} = \overrightarrow{XY}$.

b) What is the significance of $x_1\overrightarrow{PQ} + x_2\overrightarrow{QR}$?

c) Suppose $P = (0,0,0)$, $Q = (1,1,4)$, $R = (2,-3,5)$, $X = (3,-4,1)$, $Y = (5,-7,-4)$, $\vec{v} = \langle 1,5,4\rangle$. Find $\overrightarrow{XY}$, the vector corresponding to the shadow cast by $\overline{XY}$ on the plane PQR.

3

Transformations

3.1 Geometric Transformations

An important concept in mathematics is that of a **transformation**, where we take some object and alter it to produce a new object. The most familiar type of transformation is a geometric transformation. We can look at these transformations in two ways:

- Geometrically: We reflect a point across an axis, or rotate it about a center of rotation, or translate it some distance horizontally and vertically,
- Algebraically: We take the coordinates of the point (x, y) and alter them to produce a new set of coordinates for the point (x', y').

In either case, we say that the new point is the **image of the original under the transformation**. Additionally, we might say the original point is the **preimage of the transformed point**.

One way to view the transformation is as a set of functions, where the new coordinates (x', y') are functions of the old coordinates,

$$x' = f(x, y)$$
$$y' = g(x, y)$$

where $f(x, y)$ and $g(x, y)$ are some formulas that involve x and y.

Example 3.1. *Consider a translation that shifts every point 3 units to the right. Express the new coordinates (x', y') in terms of the old coordinates (x, y).*

Solution. *Let's draw a picture showing how a generic point with coordinates (x, y) is affected by this translation:*

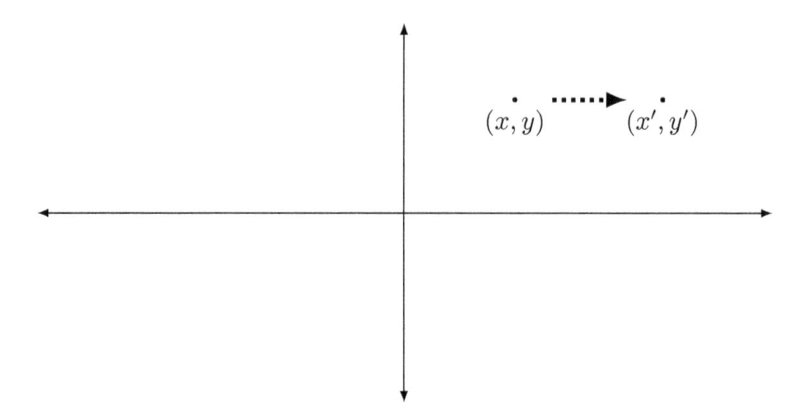

Since the point has been shifted 3 units to the right, the new x-coordinate must be 3 more than the original:

$$x' = x + 3$$

But since the point has not been shifted up or down, the new y-coordinate is the same as the original:

$$y' = y$$

Activity 3.1: Geometric Transformations

A3.1.1 For each transformation given, find formulas that compute the new coordinates (x', y') from the original coordinate (x, y).

a) M_x, the reflection of a point across the x-axis.
b) M_y, the reflection of a point across the y-axis.
c) R_{90°, the rotation of a point 90° counterclockwise about the origin.
d) R_{-90°, the rotation of a point 90° clockwise about the origin.
e) R_{360°, the rotation of a point 360° clockwise about the origin.
f) Z_k, where the distance of a point from the origin is increased by a factor of k.
g) I, where nothing whatsoever is done to the point.

A3.1.2 For each transformation given, find formulas that compute the new coordinates (x', y') from the original coordinate (x, y).

a) H_a, which shifts a point a units horizontally.
b) V_b, which shifts a point b units vertically.

A3.1.3 For each of the following, a geometric transformation is applied to a point (x, y). Express the new coordinates x' and y' as formulas of the original coordinates x and y.

a) Find X_a, which scales the horizontal coordinates by a factor of a but leaves the vertical coordinates the same.

b) Find Y_b, which scales the vertical coordinates by a factor of b but leaves the horizontal coordinates the same.

We can also consider rotations by angles not a multiple of 90°.

Activity 3.2: More Rotations

You may wish to review polar coordinates (Activity 9.10).

A3.2.1 Consider the transformation R_ϕ, which rotates the point with rectangular coordinates (x, y) counterclockwise around the origin through an angle of ϕ.

a) Find $[r, \theta]$, the polar coordinates of the point with rectangular coordinates (x, y).

b) Find (x', y'), the rectangular coordinates of the point with polar coordinates $[r, \theta + \phi]$.

c) Express (x', y') in terms of x, y, and ϕ only.

3.2 Vector Transformations

Suppose X' and Y' are the images of X and Y under a transformation. Then we say that $\overrightarrow{X'Y'}$ is the image of $\overrightarrow{XY}$ under the transformation.

Activity 3.3: Transformations of Vectors

In the following, let X and Y be distinct points in $\mathbb{R}^2$, and let X' and Y' be the images of the points after some transformation has been applied. Compare the magnitude and direction of $\overrightarrow{XY}$ to the magnitude and direction of $\overrightarrow{X'Y'}$ after the given transformation has been applied.

A3.3.1 The transformation R_θ, corresponding to the rotation of the points around the origin through an angle of θ.

A3.3.2 The transformation Z_k, corresponding to increasing the distance of a point to the origin by a factor of k.

A3.3.3 The transformation M_y, corresponding to reflecting a point across the y-axis.

A3.3.4 The transformation H_a, corresponding to shifting each point horizontally by a units.

Activity 3.3 shows that rotations, reflections, and scalings of points will affect vectors in the same way that they affect the points. However, a translation will have no effect on a vector. This merits further investigation.

Activity 3.4: More Vector Transformations

In the following, remember that the geometric description of a vector should refer to its direction and magnitude.

Suppose M is the transformation given by

$$x' = 5x + 8y + h$$
$$y' = 3x - 7\frac{x}{y} + k$$

and N is the transformation given by

$$x' = 5x + 8y$$
$$y' = 3x - 7\frac{x}{y}$$

A3.4.1 Let $P = (5, 8)$ and $Q = (-3, 6)$.

a) Let M be applied to points P, Q to produce points P', Q'. Find $\overrightarrow{P'Q'}$.

b) Let N be applied to points P, Q to produce points P'', Q''. Find $\overrightarrow{P''Q''}$.

c) What is the relationship between $\overrightarrow{P'Q'}$ and $\overrightarrow{P''Q''}$?

d) Suppose K is the transformation given by

$$x' = 5x + 8y - \sqrt{7}$$
$$y' = 3x - 7\frac{x}{y} + 11$$

If K is applied to P, Q to produce P''', Q''', what would you expect $\overrightarrow{P'''Q'''}$ to be? Verify your claim.

Activity 3.4 leads to a useful simplification: If we consider our transformation as acting on a *vector* instead of a point, then we may ignore any added or subtracted constants. This will be true regardless of the form of the transformation.

3.3 The Transformation Matrix

While a transformation of a vector could utilize any formula of the original vector components, we'll start with the simplest possible formulas: namely, linear functions of the components. We define:

Definition 3.1 (Transformation Matrix). *Let T be a transformation that takes the vector $\langle x_1, x_2, \ldots, x_n \rangle$ and produces image $\langle y_1, y_2, \ldots, y_m \rangle$ according to formulas*

$$\begin{aligned}
y_1 &= a_{11}x_1 + a_{12}x_2 + \ldots + a_{1n}x_n \\
y_2 &= a_{21}x_1 + a_{22}x_2 + \ldots + a_{2n}x_n \\
&\vdots \\
y_m &= a_{m1}x_1 + a_{m2}x_2 + \ldots + a_{mn}x_n
\end{aligned}$$

Then the transformation can be represented by the ***transformation matrix*** *T, where*

$$T = \begin{pmatrix} a_{11} & a_{12} & \cdots & a_{1n} \\ a_{21} & a_{22} & \cdots & a_{2n} \\ \vdots & \vdots & \ddots & \vdots \\ a_{m1} & a_{m2} & \cdots & a_{mn} \end{pmatrix}$$

where the matrix entries are the coefficients of the formulas used to compute the y_is from the x_is.

Example 3.2. *If possible, find the transformation matrix for the transformation*

$$\begin{aligned}
y_1 &= 2x_1 + 5x_2 \\
y_2 &= 3x_1 - 7x_2
\end{aligned}$$

Solution. *The transformation matrix consists of the coefficients of the input values used to compute the output values, so it will be*

$$T = \begin{pmatrix} 2 & 5 \\ 3 & -7 \end{pmatrix}$$

If our transformation that can be described using a matrix T, we could indicate the result of the transformation on a vector $\vec{v}$ using function notation; $T(\vec{v})$ or, if we write our vectors as $\mathbf{v}$, $T(\mathbf{v})$. However, we often use **operator notation**, which is function notation without the parentheses: $T\vec{v}$ or $T\mathbf{v}$.

"How you speak influences how you think," so we can read this in several different ways:

- *T applied to $\vec{v}$* is $T\vec{v}$,
- *T acts on $\vec{v}$* to produce $T\vec{v}$,
- *T takes the vector $\vec{v}$* and gives $T\vec{v}$.
- $T\vec{v}$ is the *image of the vector $\vec{v}$* under the transformation T.

Note that there's no requirement that the original vector and the resultant vector are in the same space. In fact, we've already worked with transformations from a lower dimensional space into a higher dimensional space. Also, we don't always distinguish between vectors $\langle x_1, x_2, \ldots, x_n \rangle$ and points $(x_1, x_2, \ldots, x_n)$.

Activity 3.5: Embeddings

A lower dimensional space can be **embedded** in a higher dimensional space: for example, two-dimensional planes "live" in three-dimensional space.

A3.5.1 Suppose the vector equation of the line $\overleftrightarrow{KL}$ is

$$\overrightarrow{OX} = t\langle -3, 5 \rangle$$

a) Explain why any point on the line can be described in terms of how far it is from the origin along a single straight line.

b) Why does this mean that the points on a line in $\mathbb{R}^2$ correspond to vectors in $\mathbb{R}^1$?

c) Find the transformation matrix that takes the vector $\vec{x}$ in $\mathbb{R}^1$ to $\overrightarrow{OX}$ in $\mathbb{R}^2$.

A3.5.2 Suppose the vector equation of the plane PQR is

$$\overrightarrow{OX} = s\langle 1, 1, 2 \rangle + t\langle -2, 4, -1 \rangle$$

a) Explain why we can interpret the values s, t as the vector $\langle s, t \rangle$ in $\mathbb{R}^2$.

b) Find the transformation matrix that acts on vector $\langle s, t \rangle$ in $\mathbb{R}^2$ to produce $\overrightarrow{OX}$ in $\mathbb{R}^3$, where X is a point on the plane PQR.

A3.5.3 Suppose the vector equation of a plane is

$$\overrightarrow{OX} = s\langle 1, 5, 3, 4\rangle + t\langle 2, 7, 5, 8\rangle$$

a) The transformation takes vectors in $\mathbb{R}^m$ and produces vectors in $\mathbb{R}^n$. What are m and n?

b) Find the corresponding transformation matrix.

What about taking a point in a higher dimensional space and transforming into an image in a lower dimensional space? Ironically, while the *process* is more familiar, the *mathematics* is more challenging.

Activity 3.6: More Shadows

In computer graphics, highly realistic images are obtained by **ray tracing**, which is based on the assumption that light travels in straight lines from its source until its interaction with an object change its trajectory or other feature of the light. The simplest case occurs when the object is opaque: If the straight line between a light source P and a point Q passes through an opaque object, then point Q is in the object's shadow.

A3.6.1 Suppose P is a point source of light located at $(0, h)$, and X is a point at (x, y). Extend PX until it meets the x-axis at X'; we say that X' is the **projection of X onto the x-axis.**

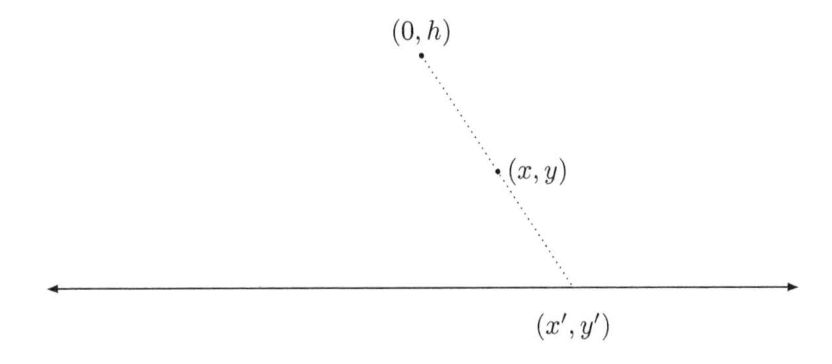

a) Explain why this transformation can be viewed as a transformation from $\mathbb{R}^2$ to $\mathbb{R}^1$. In particular, why do we not need a formula for y'?

b) Find a formula giving x' from the original coordinates (x, y).

c) If possible, find the matrix corresponding to this transformation. If not possible, explain why.

A3.6.2 Consider a point X in $\mathbb{R}^3$, and a light source P at $(0, 0, h)$.

a) Find the point X' in $\mathbb{R}^2$ corresponding to the shadow of X. Suggestion: Write the vector equation of the line $\overleftrightarrow{PX}$, then find the point on the line that is on the xy-plane.

b) Find a formula giving the coordinates (x', y') from the original coordinates (x, y, z).

c) If possible, find the matrix corresponding to this transformation. If not possible, explain why.

A3.6.3 If a light source is “infinitely far” away, the shadow is directly beneath the object.

a) Suppose an object's coordinates are (x, y, z). Find a formula giving the coordinates (x', y') of its shadow.

b) If possible, find the matrix corresponding to this transformation. If not possible, explain why.

3.4 Domain, Codomain, and Range

Since a transformation takes vector and produces a vector, we can think of it as operating on an input to produce an output. This leads to the following definition:

Definition 3.2 (Domain and Range). *The* ***domain*** *of a transformation is the set of all possible input vectors. The* ***range*** *of a transformation is the set of all possible output vectors.*

In practice, the domain is easy to find, but the range is much harder. Thus it's often useful to consider:

Definition 3.3 (Codomain). *The* ***codomain*** *is a set that contains the range.*

Example 3.3. *Let the transformation matrix T be*

$$T = \begin{pmatrix} 0 & -1 \\ 1 & 0 \end{pmatrix}$$

Write the corresponding transformation functions. Then find the domain and codomain of the transformation.

Solution. *Since the matrix has two rows, and each row corresponds to an equation, we have two outputs, which we'll call x' and y'. Since the matrix has two columns, there are two input values, which we'll call x and y. The* *transformation matrix gives us the coefficients of the formulas that determine the output values, so we have:*

$$x' = 0x + (-1)y$$
$$y' = 1x + 0y$$

Our domain must supply values x, y, so the domain consists of vectors in $\mathbb{R}^2$.

The transformation itself produces two values x', y', so the codomain consists of the vectors in $\mathbb{R}^2$.

If the transformation T takes vectors in $\mathbb{F}^m$ and produces vectors in $\mathbb{F}^n$, we write $T : \mathbb{F}^m \to \mathbb{F}^n$. Thus in the previous example, we can express the domain and codomain of T concisely by writing $T : \mathbb{R}^2 \to \mathbb{R}^2$.

Activity 3.7: Domain and Codomain

A3.7.1 Consider the transformation

$$x' = 3x + 4y - 2z$$
$$y' = 2x - 3y + z$$

a) Find the matrix T that corresponds to this transformation.

b) Find the domain and codomain of the transformation.

A3.7.2 Consider the transformation

$$x' = 3x - 2y$$
$$y' = 5x - 4y$$
$$z' = 2x - y$$

a) Find the matrix V that corresponds to this transformation.

b) Find the domain and codomain of the transformation.

A3.7.3 For each transformation matrix given, write down the explicit formulas for computing the outputs from the inputs. Then find the domain and codomain.

a) $\begin{pmatrix} 3 & 5 \\ 2 & -7 \end{pmatrix}$

b) $\begin{pmatrix} 1 & 3 & 5 \\ -2 & 4 & 3 \end{pmatrix}$

c) $\begin{pmatrix} 1 & 2 \\ 2 & -3 \\ -1 & 4 \end{pmatrix}$

A3.7.4 Suppose

$$C = \begin{pmatrix} 1 & 3 & 5 \\ 1 & 2 & -4 \\ 0 & 0 & 0 \end{pmatrix} \qquad D = \begin{pmatrix} 1 & 3 & 5 \\ 1 & 2 & -4 \end{pmatrix}$$

a) Find the domain and codomain of C.

b) Find the domain and codomain of D.

c) Are C and D the same transformation? Defend your conclusion.

Finding the range of a function is generally difficult or even impossible, since we must find *every* possible output value, without including *any* impossible value. But remember: Every problem in linear algebra begins with a system of linear equations. Thus, if we want to find the range, we should ask "What system of linear equations can be used to find the range?"

Activity 3.8: Finding the Range, Part One

A3.8.1 Consider the transformation

$$\begin{aligned} x' &= 2x + 5y \\ y' &= 3x - 7y \end{aligned}$$

a) Find the domain and codomain.

b) If $(5, 7)$ in the range, find the value(s) in the domain that will produce it. If not, explain how you know that *no* value in the domain will produce it.

c) If $(-3, 5)$ in the range, find the value(s) in the domain that will produce it. If not, explain how you know that *no* value in the domain will produce it.

A3.8.2 Consider the transformation

$$\begin{aligned} x' &= 3x + 5y + 4z - 2w \\ y' &= 2x + y - 3z + 5w \end{aligned}$$

a) Find the domain and codomain.

b) If $(3, -2)$ is in the range, find the value(s) in the domain that will produce it. If not, explain how you know that *no* value in the domain will produce it.

c) If $(-8, 5)$ is in the range, find the value(s) in the domain that will produce it. If not, explain how you know that *no* value in the domain will produce it.

A3.8.3 Consider the transformation

$$\begin{aligned} x' &= 3x + 2y \\ y' &= x - y \\ z' &= 2x + 3y \end{aligned}$$

a) Find the domain and codomain.
b) If $(12, 7, 5)$ is in the range, find the value(s) in the domain that will produce it. If not, explain how you know that *no* value in the domain will produce it.
c) If $(4, -8, 6)$ is in the range, find the value(s) in the domain that will produce it. If not, explain how you know that *no* value in the domain will produce it.

A3.8.4 Given transformation T, how could you determine whether a point is in the range?

Activity 3.8 shows that when a transformation is given by a matrix, we can determine if a particular element of the codomain is in the range by solving a system of linear equations. This suggests that we can find all of the range by seeing if a generic element of the codomain is in the range.

Example 3.4. *Find the range of the transformation*

$$\begin{aligned} x' &= 3x + 5y \\ y' &= 2x + 4y \\ z' &= x + y \end{aligned}$$

Solution. *We want to find whether (x', y', z') is in the range, so we row reduce:*

$$\left(\begin{array}{cc|c} 3 & 5 & x' \\ 2 & 4 & y' \\ 1 & 1 & z' \end{array}\right) \xrightarrow[3R_3 \to R_3]{3R_2 \to R_2} \left(\begin{array}{cc|c} 3 & 5 & x' \\ 6 & 12 & 3y' \\ 3 & 3 & 3z' \end{array}\right)$$

$$\xrightarrow[R_3 - R_1 \to R_3]{R_2 - 2R_1 \to R_2} \left(\begin{array}{cc|c} 3 & 5 & x' \\ 0 & 2 & 3y' - 2x' \\ 0 & -2 & 3z' - x' \end{array}\right)$$

$$\xrightarrow{R_3 + R_2 \to R_3} \left(\begin{array}{cc|c} 3 & 5 & x' \\ 0 & 2 & 3y' - 2x' \\ 0 & 0 & 3z' - 3x' + 3y' \end{array}\right)$$

Note that the last line of the final matrix corresponds to the equation

$$3z' - 3x' + 3y' = 0$$

This means that if (x', y', z') is in the domain, we require that $3z' - 3x' + 3y' = 0$, so the range is all values (x', y', z') where $3z' - 3x' + 3y' = 0$.

It will often be convenient to express the range in vector form. If $3z' - 3x' + 3y' = 0$, we can row reduce the augmented coefficient matrix

$$\left(\begin{array}{ccc|c} -3 & 3 & 3 & 0 \end{array} \right) \xrightarrow{\frac{1}{3}R_1 \rightarrow R_1} \left(\begin{array}{ccc|c} -1 & 1 & 1 & 0 \end{array} \right)$$

which has y', z' as free variables. This gives us parameterized solution

$$x' = s + t \qquad y' = s \qquad z' = t$$

which we can express in vector form as

$$\langle x', y', z' \rangle = s\langle 1, 1, 0 \rangle + t\langle 1, 0, 1 \rangle$$

Example 3.5. *Find the domain, codomain, and range of the transformation*

$$\begin{aligned} x' &= x + 2y - z + 3w \\ y' &= x + y + 3z + 2w \\ z' &= x + 3y - 5z + 4w \\ w' &= 2x + y + 10z + 3w \end{aligned}$$

Solution. *The transformation requires four input values, x, y, z, w, so the domain consists of vectors in $\mathbb{R}^4$. It produces four output values, $x', y'z', w'$, so the codomain consists of vectors in $\mathbb{R}^4$.*

We want to find whether x', y', z', w' is in the range, so we row reduce:

$$\left(\begin{array}{cccc|c} 1 & 2 & -1 & 3 & x' \\ 1 & 1 & 3 & 2 & y' \\ 1 & 3 & -5 & 4 & z' \\ 2 & 1 & 10 & 3 & w' \end{array} \right) \xrightarrow[-2R_1 + R_4 \rightarrow R_4]{\substack{-R_1 + R_2 \rightarrow R_2 \\ -R_1 + R_3 \rightarrow R_3}} \left(\begin{array}{cccc|c} 1 & 2 & -1 & 3 & x' \\ 0 & -1 & 4 & -1 & y' - x' \\ 0 & 1 & -4 & 1 & z' - x' \\ 0 & -3 & 12 & -3 & w' - 2x' \end{array} \right)$$

$$\xrightarrow[-3R_2 + R_4 \rightarrow R_4]{R_2 + R_3 \rightarrow R_3} \left(\begin{array}{cccc|c} 1 & 2 & -1 & 3 & x' \\ 0 & -1 & 4 & -1 & y' - x' \\ 0 & 0 & 0 & 0 & y' - 2x' + z' \\ 0 & 0 & 0 & 0 & w' + x' - 3y' \end{array} \right)$$

Notice the last two rows give us the system

$$\begin{aligned} -2x' + y' + z' &= 0 \\ x' - 3y' + w' &= 0 \end{aligned}$$

Consequently any $\langle x', y', z', w' \rangle$ in the range must satisfy this system of equations.

Since this is a homogeneous system, we can work with just the coefficient matrix. Row reducing:

$$\begin{pmatrix} -2 & 1 & 1 & 0 \\ 1 & -3 & 0 & 1 \end{pmatrix} \xrightarrow{R_1 \leftrightarrow R_2} \begin{pmatrix} 1 & -3 & 0 & 1 \\ -2 & 1 & 1 & 0 \end{pmatrix} \xrightarrow{2R_1 + R_2 \rightarrow R_2} \begin{pmatrix} 1 & -3 & 0 & 1 \\ 0 & -5 & 1 & 2 \end{pmatrix}$$

This gives us z', w' as free variables. To avoid fractions, we'll let $z' = 5t$ and $w' = 5s$. Rewriting our second equation:

$$\begin{aligned} -5y' + z' + 2w' &= 0 \\ -5y' + 5t + 2(5s) &= 0 \\ y' &= t + 2s \end{aligned}$$

Rewriting our first equation:

$$\begin{aligned} x' - 3y' + w' &= 0 \\ x' - 3(t + 2s) + (5s) &= 0 \\ x' &= 3t + s \end{aligned}$$

So

$$\langle x', y', z', w' \rangle = \langle 3t + s, t + 2s, 5t, 5s \rangle$$

Separating these into s and t components:

$$\begin{aligned} \langle x', y', z', w' \rangle &= \langle s, 2s, 0, 5s \rangle + \langle 3t, t, 5t, 0 \rangle \\ \langle x', y', z', w' \rangle &= s\langle 1, 2, 0, 5 \rangle + t\langle 3, 1, 5, 0 \rangle \end{aligned}$$

Activity 3.9: Finding the Range, Part Two

A3.9.1 Find the domain, codomain, and range of the transformations corresponding to the given matrix.

a) $\begin{pmatrix} 2 & -5 \\ 3 & -2 \end{pmatrix}$

b) $\begin{pmatrix} 2 & 3 \\ 3 & -2 \\ 1 & 3 \end{pmatrix}$

c) $\begin{pmatrix} 1 & 3 & 5 \\ 2 & 1 & 3 \end{pmatrix}$

A3.9.2 In some cases you found the range and the codomain were the same; in others, the range only formed part of the codomain. What seems to be the relationship between the domain and codomain when the range is only part of the codomain?

3.5 Discrete Time Models

Thus far we've been focusing on geometric transformations. But if we view transformations as giving new values from old values, we can expand the scope of linear algebra. In this section, we'll introduce one of the most important applications of linear algebra: the **discrete time model**. As its name suggests, the discrete time model assumes time, which we usually regard as continuous, to be broken into discrete intervals.

For example, imagine watching the customers at a coffee shop. If security cameras take pictures of the line at thirty second intervals, then we only know the number of customers in line at certain discrete points in time.[1] What we'd like to be able to do is to model (predict) the number of people in line in the *next* frame, based on the number of people in line in the *current* frame.

Activity 3.10: The Rabbit Problem

In 1202, Leonardo of Pisa (1170–1250, also known as Fibonacci) posed the following problem: Suppose you begin the year with a pair of (immature) rabbits. How many pairs would you have after a year, assuming that rabbits breed according to the following rules:

- Rabbits that were immature at the start of the month are mature by the start of the next month,
- Rabbits that were mature at the start of the month produce a pair of immature rabbits at the start of the next month,
- No rabbits die.

To solve this problem, let x_n be the number of pairs of immature rabbits you have at the start of month n, and y_n be the number of mature rabbits you have at the start of month n; we can then represent our rabbit population as the vector $\langle x_n, y_n \rangle$.

A3.10.1 Find x_1 and y_1.

A3.10.2 "It is easier to know where you came from than to know where you're going." Why must $x_2 = 0$?

A3.10.3 Find y_2.

A3.10.4 Find x_3 and y_3.

[1] Even if the security camera shoots a video, video frames themselves represent instants in time separated by some fraction of a second.

A3.10.5 Write down explicit formulas for finding x_{n+1}, y_{n+1} from x_n, y_n.

A3.10.6 If possible, find F, the matrix corresponding to the transformation $F\langle x_n, y_n\rangle = \langle x_{n+1}, y_{n+1}\rangle$.

We can introduce any number of alternate assumptions.

Example 3.6. *Suppose Leonardo of Pisa assumed that mature rabbits produce two pairs of immature rabbits. What are the resulting formulas for x_{n+1}, y_{n+1}?*

Solution. *The number of mature rabbits at $t = n + 1$ will be the number of mature rabbits at $t = n$, plus the number of immature rabbits at $t = n$ who became mature, so $y_{n+1} = x_n + y_n$ still holds.*

Since each pair of mature rabbits at $t = n$ produces two pairs of immature rabbits at $t = n + 1$, we have $x_{n+1} = 2y_n$. So

$$\begin{aligned} x_{n+1} &= 2y_n \\ y_{n+1} &= x_n + y_n \end{aligned}$$

which corresponds to the transformation matrix $\begin{pmatrix} 0 & 2 \\ 1 & 1 \end{pmatrix}$.

Leonardo's rabbit problem is the oldest example of what is now known as a **Leslie model**, named after Patrick Holt Leslie (1900-1972). In 1945, Leslie considered models for populations of animals that had distinct life stages: for example, egg, duckling, duck.

Activity 3.11: Leslie Models

A3.11.1 Suppose we make the following assumptions about the growth rate of a population of ducks:

- Every egg at time n becomes a duckling at time $n + 1$,
- Every duckling at time n becomes a duck at time $n + 1$,
- Every duck at time n produces 2 eggs at time $n + 1$,
- No ducks die.

Let e_n, c_n, and d_n be the number of eggs, ducklings, and ducks at time n.

a) Write down the equations to find $e_{n+1}, c_{n+1}, d_{n+1}$ from the values of e_n, c_n, and d_n.
b) Find the corresponding transition matrix.

A3.11.2 Suppose we make the following assumptions about the growth rate of a butterfly population with life cycle stages

$$\text{Egg} \rightarrow \text{Larva} \rightarrow \text{Pupae} \rightarrow \text{Adult}$$

- 10% of the eggs at time n become larvae at time $n+1$,
- 20% of the larvae at time n become pupae at time $n+1$,
- 30% of the pupae at time n become adults at time $n+1$,
- Each adult at time n lays 200 eggs, and then dies.

Let e_n, l_n, p_n, a_n be the number of eggs, larvae, pupae, and adults at time n.

a) Write down equations to find e_{n+1}, l_{n+1}, p_{n+1}, a_{n+1} from e_n, l_n, p_n, a_n.

b) Find the corresponding transition matrix.

A3.11.3 Consider an animal that passes through three distinct life stages (for example, egg-tadpole-frog), where x_n, y_n, z_n are the number of individuals in each of the three stages at time n. Let T be the transition matrix corresponding to the equations to find $x_{n+1}, y_{n+1}, z_{n+1}$ from x_n, y_n, z_n.

a) What assumptions are being made about the growth rate of the population, if the transition matrix is

$$A = \begin{pmatrix} 0 & 0 & 10 \\ 0.1 & 0 & 0 \\ 0 & 0.3 & 0.7 \end{pmatrix}$$

b) Describe the growth of the animal population corresponding to the transition matrix

$$B = \begin{pmatrix} 0.1 & 0 & 10 \\ 0.1 & 0.2 & 0 \\ 0 & 0.3 & 0.7 \end{pmatrix}$$

A3.11.4 Consider the matrices

$$A = \begin{pmatrix} 0 & 0 & 50 \\ 0.1 & 0 & 0 \\ 0 & 2 & 0.3 \end{pmatrix}, B = \begin{pmatrix} 0 & 0 & 50 \\ 0 & 0.2 & 0.3 \\ 0.1 & 0 & 0 \end{pmatrix},$$

$$C = \begin{pmatrix} 0 & 0 & 50 \\ 0.1 & 0 & 0 \\ 0 & 0.2 & 0.3 \end{pmatrix}, D = \begin{pmatrix} 0 & 0.3 & 50 \\ 0.1 & 0 & 0 \\ 0 & 0.2 & 0.3 \end{pmatrix}$$

a) Which (if any) of these could be transition matrices for the population of a species with three distinct life stages (assume these to be egg, hatchling, adult)? What are the corresponding assumptions about the population growth rate?

b) Which (if any) of these **CANNOT** be transition matrices for the population of a species with three distinct life stages? Defend your answer, based on what the transition matrix would imply about the growth rate of the species.

The Leslie model determines the actual number of individuals at each of the life stages. We might also look at the *fraction* of the total population in each life cycle. More generally, we might consider a population that can be in any one of several **states**. This leads to as **stochastic matrix**, also known as a **Markov matrix**, after the Russian mathematician Andrey Markov (1856–1922), who first investigated them while considering the distribution of letters in a written text.

Activity 3.12: Stochastic Matrices

A3.12.1 Imagine a park with three locations: a picnic area; a lake; and a playground.

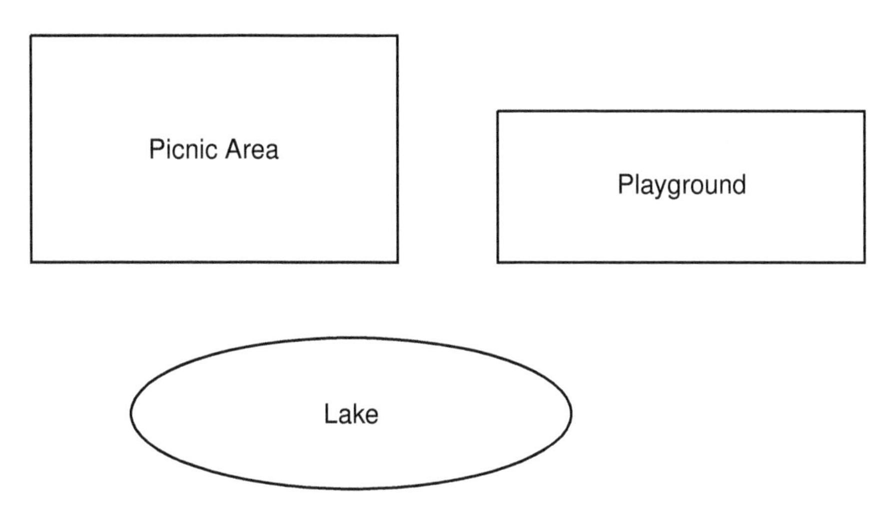

Let p_n, l_n, and g_n be the number of persons in the picnic area, lake, and playground at time n, and assume people move under the following assumptions:

- Of those at the picnic area at time n, one-half will be at the lake at time $n+1$; one-third will be at the playground at time $n+1$; and the remainder will still be at the picnic area at time $n+1$.
- Of those at the playground at time n, one-fourth will be at the lake at time $n+1$; one-fifth will be at the picnic area at time $n+1$; and the rest will still be at the playground at time $n+1$.
- Of those at the lake at time n, one-half will be at the picnic area at time $n+1$, and one-half will be at the playground at time $n+1$.

a) Write the equations to compute $p_{n+1}, l_{n+1}, g_{n+1}$ from p_n, l_n, g_n. Suggestion: It's easier to determine where you're *from* than where you're going.

b) Find the corresponding transition matrix T.

c) Suppose 1000 persons start in the picnic area at time $t = 0$. How many will be in each location at time $t = 1$?

d) How many will be in each location at time $t = 2$?

A3.12.2 Suppose at a different time of year, the transition matrix for the park is

$$T = \begin{pmatrix} 0.2 & 0.5 & 0.1 \\ 0 & 0.2 & 0.3 \\ 0.8 & 0.3 & 0.6 \end{pmatrix}$$

Describe how parkgoers are moving around the park. (In other words: For each location, describe where the persons in that location at time n can be found at time $n + 1$)

A3.12.3 Suppose one of the areas is closed, so no one can enter it. How would this be reflected in the transition matrix? (In other words, how could you tell, from the transition matrix alone, that an area was closed?)

A3.12.4 Could

$$M = \begin{pmatrix} 0.2 & 0.1 & 0.5 \\ 0.3 & 0.4 & 0.4 \\ 0.5 & 0.5 & 0.3 \end{pmatrix}$$

describe how parkgoers move around the park? Why/why not?

A3.12.5 Could

$$N = \begin{pmatrix} 0.2 & 0.1 & 0.5 \\ 0.3 & 0.4 & 0.4 \\ 0.3 & 0.5 & 0.1 \end{pmatrix}$$

describe how parkgoers move around the park? Why/why not?

A3.12.6 Suppose P is a matrix that describes how parkgoers move around the park. What must be true about the entries of P? Why? (Identify as many things as you can that must be true about the entries)

As a general rule, the transformation $T\vec{v} \neq \vec{v}$. But it's possible that there might be *some* vector for which $T\vec{v} = \vec{v}$. We define:

Definition 3.4 (Steady State Vector). *Let T be a transformation that acts on vectors in $\mathbb{F}^n$ to produce vectors in $\mathbb{F}^n$. The* ***steady state vector*** *is a vector $\vec{v}$ where $T\vec{v} = \vec{v}$.*

Activity 3.13: Steady State Vectors

A3.13.1 Let T be the transformation

$$x' = 3x + 2y$$
$$y' = 5x - 4y$$

Let $\langle x, y\rangle$ be the steady state vector.

a) Set up the system of equations needed to find x and y.
b) Explain why this system is always has at least one solution. What is the guaranteed solution?
c) Does this system have any other solutions?

A3.13.2 Suppose a park has three areas: a picnic area; a lake; and a playground. Let $\langle p_n, l_n, g_n\rangle$ be a vector whose components give the number of parkgoers in the picnic area, lake, and playground at $t = n$, and let

$$T = \begin{pmatrix} 0.2 & 0.1 & 0.5 \\ 0.3 & 0.2 & 0.2 \\ 0.5 & 0.7 & 0.3 \end{pmatrix}$$

be the transition matrix where $T\langle p_n, l_n, g_n\rangle = \langle p_{n+1}, l_{n+1}, g_{n+1}\rangle$.

a) Interpret the transition matrix in terms of how people move about in a park with three areas. In particular: of the people in each of the three areas, what fraction move to which area, and what (if any) fraction remain in the same location?
b) Find the steady state vector $\langle p, l, g\rangle$.
c) Does the fact that $T\langle p, l, g\rangle = \langle p, l, g\rangle$ mean that no one is moving around the park? Why/why not?

As Activity 3.13 suggests, finding the steady state vector is not a particularly challenging problem, nor is it conceptually difficult. Thus it is with some chagrin that mathematicians realized they lost billions of dollars by failing to identify one particularly important use.

Activity 3.14: How to Lose a Billion Dollars

Google's ranking of web pages is based on the following assumptions about how surfers move around a set of connected web pages:

- Surfers on a page follow one of the outgoing links,
- Outgoing traffic splits equally among all outgoing links.

There is an additional assumption, called the **teleport probability** (see Activity A3.14.2). The algorithm is known as Page rank (after the inventor, Lawrence Page).

A3.14.1 Imagine the connected set of pages shown:

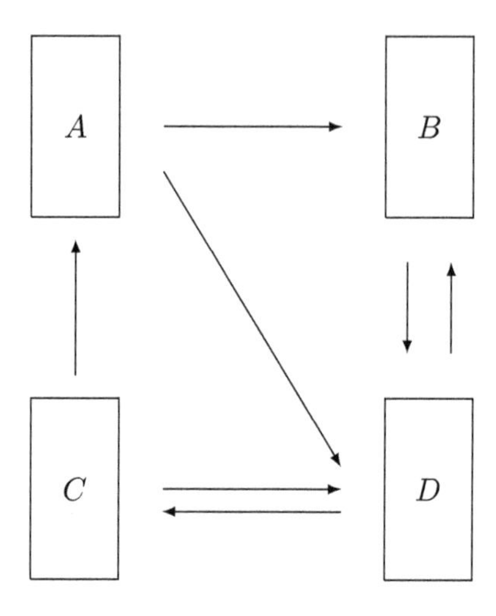

Let a_n, b_n, c_n, d_n be the number of surfers on each web page at time n.

a) Write explicit formulas for determining a_{n+1}, b_{n+1}, c_{n+1}, d_{n+1} from a_n, b_n, c_n, d_n.
b) Find the corresponding transformation matrix T.
c) Find the steady state vector.
d) Google's ranking of a web page is based on the magnitude of the corresponding component of the steady state vector. What would be the ranking of the four web pages above?

A3.14.2 One problem is that a web page might have no outgoing links:

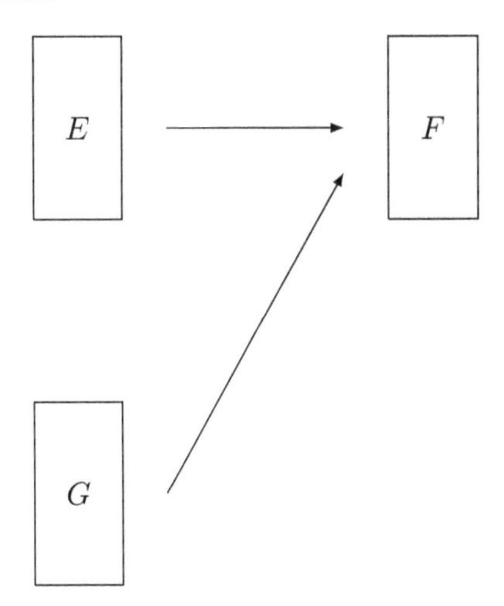

a) Write the transformation matrix V corresponding to the web traffic between the web pages.
b) Find the steady state vector.
c) Since all surfers will eventually end on web page F and stay there, Google incorporates a **teleport probability**, which is the fraction of surfers on a web page who will randomly select one of the web pages and go there, without following a link. Using a teleport probability of 0.15 (so 15% of the surfers on each web page will randomly select one of the three web pages, including possibly the page they are currently on), find a corresponding transition matrix W.
d) Find the steady state vector for the new transition matrix.

The importance of linear algebra may be gauged by the following observation: Here we are, less than halfway through elementary linear algebra. But with the knowledge that you now have, you could have invented Google. What can we do with the *rest* of elementary linear algebra? And what might we do with more advanced linear algebra?

3.6 Linear Transformations

The advantage to thinking of transformations as acting on vectors as opposed to points is that we've established an arithmetic of vectors: we know how to add two vectors, or to multiply a vector by a scalar. This raises a new question: What happens if we apply a transformation to a linear combination of vectors?

We might hope that the transformation of a linear combination of vectors has a simple form. This leads to one of the most important ideas in linear algebra (and quite possibly in all of mathematics):

Definition 3.5 (Linear Transformation). *A transformation T is linear if, for any vectors* $\mathbf{x}$, $\mathbf{y}$ *in its domain, and any scalars a, b,*

$$T(a\mathbf{x} + b\mathbf{y}) = aT(\mathbf{x}) + b(T\mathbf{y})$$

provided both sides are defined.

Note that we used parentheses in $T(a\mathbf{x} + b\mathbf{y})$ for clarity and, while we didn't need to, we included them in $aT(\mathbf{x})$ and $bT(\mathbf{y})$ for consistency.

Although we define linear transformations in terms of vectors and scalars, it's important to understand that by "vector" we mean anything that serves as an input for the transformation, and by "scalar" we mean anything that the vector can be meaningfully multiplied by to obtain a vector of the same type.

Example 3.7. *Determine whether $f(x) = 5x$ is a linear transformation, where x is a real number.*

Solution. *In this case, our input is a real number, so our scalar is anything you can meaningfully multiply a real number by and obtain a real number: in other words, another real number.*

To decide whether we have a linear transformation, we must verify or refute that $f(ax + by) = af(x) + bf(y)$. "You can always write down one side of an equality," so let's write down $f(ax+by)$ and see what it gives us. "Definitions are the whole of mathematics," so our definition $f(x) = 5x$ gives us

$$\begin{aligned} f(ax + by) &= 5(ax + by) \\ &= 5ax + 5by \end{aligned}$$

Next, we'll write down $af(x) + bf(y)$ and see where this takes us:

$$af(x) + bf(y) = a(5x) + b(5y)$$

Rearranging:

$$= 5ax + 5by$$

which is the same as $f(ax + by)$, so f is a linear transformation.

Before we go any further, it will be convenient to establish some important features about linear transformations.

Activity 3.15: Functions

We say f is a *function* if every input x gives us a unique output $f(y)$. Conversely, if f could give two or more different outputs, then f is not a function.

A3.15.1 Suppose T is a linear transformation. Let c be any real number, and assume all vectors are in the domain or codomain, as necessary.

a) Prove $T(c\mathbf{0}) = cT(\mathbf{0})$.
b) Prove $T(c\mathbf{0}) = T(\mathbf{0})$.
c) While c could be any real number, why are these statements uninteresting if $c = 1$? (Recall Shannon's definition of information as "answers you didn't already know.")
d) Suppose $c \neq 1$. What do these two statements say about $T(\mathbf{0})$? Prove it.
e) Suppose $T(\mathbf{x}) = \mathbf{y}$, and $T(\mathbf{x}) = \mathbf{z}$. Show that $\mathbf{y} = \mathbf{z}$. Suggestion: To show two things are equal, find their difference.
f) Are linear transformations functions? Why/why not?

Activity 3.15 proves two very important properties of linear transformations. First:

Theorem 3.1 (Transformation of Zero). *For any linear transformation T, $T(\mathbf{0}) = \mathbf{0}$ for the appropriate zero vectors.*

Moreover:

Theorem 3.2. *A linear transformation is a function.*

Activity 3.16: Linear Transformations and Matrices

A3.16.1 Determine whether the given function is a linear transformation.

a) $f(x) = 3x$
b) $g(x) = x^2$
c) $h(x) = 5x + 7$

A3.16.2 Suppose T is a linear transformation, where $T(\mathbf{u}) = \mathbf{x}$, and $T(\mathbf{v}) = \mathbf{y}$. Find the following.

a) $T(3\mathbf{u})$.
b) $T(-2\mathbf{v})$.
c) $T(5\mathbf{u} - 4\mathbf{v})$.

d) $T(\mathbf{0})$.

A3.16.3 Suppose A is a linear transformation, and

$$A(\mathbf{e}_1) = \langle 1, 3, -2 \rangle \qquad A(\mathbf{e}_2) = \langle 4, -1, -5 \rangle \qquad A(\mathbf{e}_3) = \langle -3, 2, 5 \rangle$$

where $\mathbf{e}_1$, $\mathbf{e}_2$, $\mathbf{e}_3$ are the elementary vectors in $\mathbb{R}^3$.

a) Express $\langle 2, 5, -4 \rangle$ as a linear combination of the elementary vectors $\mathbf{e}_1$, $\mathbf{e}_2$, $\mathbf{e}_3$.
b) Find $A(\langle 2, 5, -4 \rangle)$.
c) Find $A(\langle 0, 0, 0 \rangle)$.
d) Find $A(\langle x_1, x_2, x_3 \rangle)$.
e) Find the transformation matrix corresponding to A.

A3.16.4 Let T be a linear transformation with domain $\mathbb{R}^m$ and codomain $\mathbb{R}^n$, and suppose $T(\mathbf{e}_i) = \langle t_{1i}, t_{2i}, \ldots, t_{ni} \rangle$, where $\mathbf{e}_i$ are the elementary vectors in $\mathbb{R}^m$. Find the matrix corresponding to the transformation T.

In Activity A3.16.4, you found you could recover the transformation matrix from what the transformation did to the elementary vectors. Consequently:

Theorem 3.3. *Every linear transformation can be represented by a transformation matrix.*

In fact, based on Activity A3.16.4, the transformation matrix has a particularly simple form:

Theorem 3.4 (Transformation Matrix). *Let T be a linear transformation with domain $\mathbb{R}^m$ and codomain $\mathbb{R}^n$. Let $T(\mathbf{e}_i) = \mathbf{t}_i$, where $\mathbf{e}_i$ are the elementary vectors in $\mathbb{R}^m$. Then the transformation T can be represented by the matrix whose ith column is $\mathbf{t}_i$.*

Example 3.8. *Suppose $T(\langle 1, 0, 0 \rangle) = \langle 1, 5, 3, 8 \rangle$, $T(\langle 0, 1, 0 \rangle) = \langle 0, 1, -3, 4 \rangle$, and $T\langle 0, 0, 1 \rangle) = \langle 4, 1, 0, 1 \rangle$. Find the domain and codomain of T, then find the transformation matrix.*

Solution. *We see that T acts on vectors in $\mathbb{R}^3$, which will be the domain; and produces vectors in $\mathbb{R}^4$, which will be the codomain.*

The transformation matrix will be the matrix whose columns are the images of the elementary vectors:

$$T = \begin{pmatrix} 1 & 0 & 4 \\ 5 & 1 & 1 \\ 3 & -3 & 0 \\ 8 & 4 & 1 \end{pmatrix}$$

A good habit to develop as a mathematician is to switch questions around. Thus, we showed that every linear transformation can be represented as a matrix. Is it also true that every matrix corresponds to a linear transformation?

Activity 3.17: Matrices and Linear Transformations

A3.17.1 Let

$$A = \begin{pmatrix} a_{11} & a_{12} & a_{13} \\ a_{21} & a_{22} & a_{23} \end{pmatrix}$$

be the matrix for a transformation.

a) Find the domain and codomain of the transformation.
b) Suppose $\overrightarrow{u}$ is a vector in the domain, and c is a scalar. Find $A(c\overrightarrow{u})$ and $c(A\overrightarrow{u})$. Does $A(c\overrightarrow{u}) = cA\overrightarrow{u}$?
c) Suppose $\overrightarrow{v}$ is another vector in the domain, and d another scalar. Find $A(d\overrightarrow{v})$ and $d(A\overrightarrow{v})$.
d) Find $A(c\overrightarrow{u} + d\overrightarrow{v})$
e) Is A a linear transformation?

The results of Activity 3.17 can be generalized, leading to:

Theorem 3.5. *Every matrix corresponds to a linear transformation.*

3.7 Transformation Arithmetic

We can make transformations more useful by considering *transformation arithmetic*: If A, B are transformations, what would be the meaning of something like $A+B$ or AB? To begin with, we'll introduce an important transformation:

Definition 3.6 (Identity Transformation). *The **identity transformation** I is the transformation where $I\mathbf{v} = \mathbf{v}$ for all vectors I can be applied to.*

The matrix corresponding to the identity transformation is called the **identity matrix**.

Example 3.9. *Find the identity transformation $I : \langle x, y \rangle \to \langle x', y' \rangle$ and the corresponding identity matrix.*

Solution. *This transformation takes as input a vector $\langle x, y \rangle$ in $\mathbb{R}^2$ and produces as output $\langle x', y' \rangle$ in $\mathbb{R}^2$. Since it's the identity transformation, $\langle x', y' \rangle = \langle x, y \rangle$, so we have*

$$x' = 1x + 0y$$
$$y' = 0x + 1y$$

The corresponding matrix will be $\begin{pmatrix} 1 & 0 \\ 0 & 1 \end{pmatrix}$

Activity 3.18: The Identity Matrix

A3.18.1 If possible, write down the identity transformation for the given domain and codomain (as a system of equations), and the identity matrix (as a matrix); if not possible, explain why not.

a) Domain $\mathbb{R}^3$; codomain $\mathbb{R}^3$.

b) Domain $\mathbb{R}^5$; codomain $\mathbb{R}^5$.

c) Domain $\mathbb{R}^3$; codomain $\mathbb{R}^4$.

d) Domain $\mathbb{R}^4$; codomain $\mathbb{R}^2$.

A3.18.2 Is

$$\begin{pmatrix} 1 & 0 & 0 & 0 \\ 0 & 1 & 0 & 0 \\ 0 & 0 & 1 & 0 \end{pmatrix}$$

an identity matrix? Why/why not?

A3.18.3 Suppose the domain of the identity transformation is $\mathbb{R}^n$.

a) What is the codomain?

b) What does this say about the dimensions of the identity matrix?

Next, what happens if we apply more than one transformation to a vector? This is known as a **composition of transformations**. For example, we might rotate a vector through some angle, and then reflect it across the x-axis. If we apply a transformation A, then apply transformation B, we *should* represent the sequence of transformations as $B(A\mathbf{v})$, where the parentheses indicates that we must first apply A to $\mathbf{v}$, and then apply B to the resultant vector. However, in practice we either omit the parentheses and write $BA\mathbf{v}$, or write place parentheses around the transformations and write $(BA)\mathbf{v}$. Thus:

Definition 3.7 (Composition). *Let A, B be transformations. The transformation BA is the transformation corresponding to the application of A to a vector, followed by the application of B to the resultant.*

A special case of composition occurs when we apply the same transformation repeatedly: for example, we might rotate through an angle of 90°, and then rotate again through another angle of 90°. We'll express this situation using exponents: A^n is the result of applying the transformation A n times in succession.

Activity 3.19: Composition of Transformations

In the following, assume that the transformation acts on vectors in $\mathbb{R}^2$ and produces vectors in $\mathbb{R}^2$.

A3.19.1 For each of the following, *describe* each composition as a *single* geometric transformation. Then find the transformation matrix for the transformation.

a) $R_{90^\circ}R_{90^\circ}$
b) M_xM_x
c) M_xI
d) IR_{90°

A3.19.2 Find the transformation matrix for the following.

a) $R_{90^\circ}^5$
b) $(M_x)^{1000}$

A3.19.3 The juxtaposition AB looks like a multiplication, so it's natural to wonder if AB and BA are the same thing. Let $\overrightarrow{v} = \langle x, y \rangle$.

a) Find $R_{90^\circ}M_x\overrightarrow{v}$.
b) Find $M_xR_{90^\circ}\overrightarrow{v}$.
c) Does $M_xR_{90^\circ}\overrightarrow{v} = R_{90^\circ}M_x\overrightarrow{v}$?
d) If we claim $AB = BA$, we mean that this is always true, regardless of A or B. Is it true that $AB = BA$?
e) If we claim $AB \neq BA$, we mean that this is always true, regardless of A or B. Is it true that $AB \neq BA$?

A3.19.4 Remember R_{180° is the transformation that rotates a point 180° counterclockwise around the origin.

a) Suppose $X^2 = R_{180^\circ}$. What is the *geometric* transformation corresponding to X?
b) Find X. (There are two solutions; find both.)

Activity A3.19.3 reveals an important feature about transformations: In general, transformations are *not* commutative, and the transformation AB is generally different from the transformation BA.

Next, suppose we apply a transformation A to some vector, giving us $A\mathbf{v}$. We might want to return to our original vector. What we want is some transformation B where $BA\mathbf{v} = \mathbf{v}$ for all vectors A can apply to. This leads to:

Definition 3.8 (Inverse Transformations). *Suppose A is a transformation. The* ***inverse transformation****, written A^{-1}, is the transformation that satisfies $A^{-1}A\mathbf{v} = \mathbf{v}$ for the appropriate identity transformation.*

Remember there are only so many symbols! Thus, while $x^{-1} = \frac{1}{x}$ if x is a real number, A^{-1} has a completely different meaning when A is a transformation or a matrix.

Activity 3.20: Inverse Transformations

In the following, let M_x be the reflection of a point across the x-axis; M_y the reflection of a point across the y-axis; R_θ the rotation of a point around the origin through an angle of θ; and I the "do nothing" transformation. Assume the points are in $\mathbb{R}^2$.

A3.20.1 For each of the following geometric transformations, describe the *geometric* transformation that would "undo" the transformation. Then find the matrix for the inverse transformation.

a) The transformation M_x, corresponding to a reflection across the x-axis.
b) The transformation $R_{90°}$, corresponding to a rotation around the origin by $90°$ counterclockwise.
c) The transformation Z_k, which increases the distance of a point to the origin by a factor of k.
d) The transformation I, corresponding to the identity transformation.
e) The transformation X_k, which produces a horizontal stretch by a factor of k.

A3.20.2 Explain why it will not be possible to find an inverse for the described transformation.

a) The transformation D, which maps the point (x, y) to the point on the x-axis "directly above or below" it.
b) The transformation O, which sends a point to the origin.

A3.20.3 Another way to define the inverse transformation A^{-1} is that it is the transformation that satisfies $A^{-1}A = I$. Explain why this is equivalent to defining it as the transformation that satisfies $A^{-1}A\vec{v} = \vec{v}$.

In Activities 3.19 and 3.20, you began with transformations A, B which could be described using a matrix, and obtained a transformation AB, A^n, and (if it existed) A^{-1}, which could also be described using a matrix. Thus, you began and ended with linear transformations. As a mathematician, the next question to ask would be, "Is this always true?" In other words, if A, B

are linear transformations, will their product, power, and inverse also be linear transformations?

Activity 3.21: Preserving Linearity

In the following, let S be a linear transformation that can be represented by a $n \times m$ matrix. Assume $\vec{u}$, $\vec{v}$ are vectors with real components that S can act on, and that a, b are real numbers. Also assume that T is a linear transformation that can act on $S\vec{u}$ and $S\vec{v}$.

A3.21.1 Find the domain and codomain of S.

A3.21.2 Explain why you know $S(a\vec{u} + b\vec{v}) = aS\vec{u} + bS\vec{v}$.

A3.21.3 What is the domain of T?

A3.21.4 Explain why T might **NOT** be able to act on the vectors $\vec{u}$ or $\vec{v}$.

A3.21.5 We've assumed TS exists. Does this mean ST also exists? Why/why not?

A3.21.6 Prove/disprove: TS is a linear transformation.

A3.21.7 Remember A^n corresponds to applying the transformation A n times in succession.

a) What must be true about the domain and codomain of A in order for A^n to be defined?

b) Prove/disprove: Provided it exists, A^n a linear transformation. Suggestion: Use an induction proof.

A3.21.8 Prove: I (the identity transformation of the appropriate size) is a linear transformation.

A3.21.9 Remember S^{-1} is the transformation that undoes S, so if $S\vec{u} = \vec{x}$, then $S^{-1}\vec{x} = \vec{u}$.

a) Find the domain and codomain of S^{-1}.

b) Let $\vec{x} = S\vec{u}$, $\vec{y} = S\vec{v}$. Express $S(a\vec{u} + b\vec{v})$ in terms of $a, b, \vec{x}, \vec{y}$.

c) Find $S^{-1}(a\vec{x} + b\vec{y})$.

d) Prove/disprove: S^{-1} is a linear transformation.

Activity 3.21 leads to the following:

Theorem 3.6. *Provided they exist, the product, power, and inverse of linear transformations is a linear transformation.*

3.8 Cryptography

One important use of transformations is cryptography, where our goal is to transform a *plaintext message* into a *ciphertext message*. Most encryption techniques begin by breaking the message, which we may assume to consist of a sequence of numbers, into blocks of a specified length: for example, a message might be broken into blocks of four consecutive numbers. These blocks can be treated as vectors, and a transformation can be applied to them.

The simplest possibility is to rearrange the characters. For example, we might reverse the order: SECRET becomes TERCES. This produces a **transposition cipher**.

Activity 3.22: Transposition Ciphers

A **permutation** consists of the rearrangement of an ordered list: for example, *abcde* might be permuted to *dbace*. Since the list elements are ordered, the list can be treated as a vector.

A3.22.1 Find a transformation matrix that yields the given permutations.

a) $A(\langle x, y, z\rangle) = \langle y, z, x\rangle$.
b) $B(\langle x, y, z\rangle) = \langle z, x, y\rangle$.
c) $C(\langle a, b, c, d, e\rangle) = \langle e, a, b, c, d\rangle$.
d) $D(\langle a, b, c, d, e\rangle) = \langle b, a, c, d, e\rangle$.
e) $E(\langle a, b, c, d, e\rangle) = \langle a, b, c, e, d\rangle$.

A3.22.2 The **order of a permutation matrix** T is the smallest n for which $T^n\vec{v} = \vec{v}$. Consider the matrices from Activity A3.22.1

a) Find the order of the matrices.
b) Find the order of AB.
c) In general, composition of transformations is **NOT** commutative. However, $DE = ED$. Why?
d) Find the order of DE. Suggestion: Because $DE = ED$, then $(DE)^n = D^nE^n$.
e) What is the relationship between the order of D, E, and DE? Why does this happen?
f) Suppose T is a permutation matrix that acts on vectors with 7 components. If possible, find a permutation matrix T with order 12.

Another alternative is to replace each message component with something else. This produces a **substitution cipher**. For example, we might replace each letter of a message with its place in the alphabet, so A becomes 1, B becomes 2, and so on. MESSAGE becomes 13-5-19-19-1-7-5. Of course so simple a replacement would be easy to break. However, we can use the tools of linear algebra to improve our encryption.

Activity 3.23: The Hill Cipher

The **Hill cipher**, invented by Lester Hill in 1929, was the first truly mathematical code. It is based on matrix transformations as follows:

- Choose H, a $n \times n$ matrix (this is sometimes called the **Hill matrix**). H must satisfy certain properties, which we won't worry about here.
- Break the message into blocks of length n,
- Encrypt the block $\mathbf{p}$ as $H\mathbf{p}$, where $\mathbf{p}$ is a block of length n.

A3.23.1 Let $M = \begin{pmatrix} 3 & 7 \\ 5 & 12 \end{pmatrix}$.

a) Using M as the Hill matrix, encrypt the plaintext: 8-7-1-3-9-4-5-7.

b) Suppose M produces ciphertext $\langle 99, 168 \rangle$. Find the original plaintext.

A3.23.2 One problem with simple substitution ciphers is assigning the same ciphertext value to the same symbol: for example, if the letter E is always encrypted as 5, then the fact that E is the most common letter in an English text will mean that 5 is the most common letter in the ciphertext.

a) Using the transformation above, encrypt the vector $\langle 5, 8 \rangle$.

b) Using the transformation above, encrypt the vector $\langle 5, 1 \rangle$.

c) Using the transformation above, encrypt the vector $\langle 1, 5 \rangle$.

d) Suppose 5 is the most common value in the plaintext. How does the Hill cipher prevent the encrypted value of 5 from being the most common value in the ciphertext?

A3.23.3 To determine how good a cryptographic system is, cryptographers assume the adversary can mount some sort of **cryptographic attack** on the system. One of the most important is a *known plaintext attack*: the adversary has access to some unencrypted plaintexts, and the corresponding encrypted ciphertexts. (This can happen, for example, if someone makes a credit card purchase and doesn't destroy the paper receipt.)

a) Suppose you know that someone is using a 2×2 matrix for a Hill cipher. Explain why knowing the plaintext message $\langle 5, 11 \rangle$ is encrypted as the ciphertext message $\langle 87, 136 \rangle$ is *not* enough to determine the Hill matrix.

b) Suppose you know the plaintext $\langle 5, 11 \rangle$ becomes the ciphertext $\langle 87, 136 \rangle$, and also that the plaintext $\langle 2, 3 \rangle$ becomes the ciphertext $\langle 25, 39 \rangle$. If possible, find H; if not possible, explain why not.

c) Suppose a system uses a $n \times n$ Hill matrix. How many plaintext/ciphertext pairs is necessary to recover the matrix?

A3.23.4 Another type of cryptographic attack is called a *chosen plaintext attack*: in this case, the adversary can *choose* a plaintext and obtain the encrypted values. (This can happen, for example, in systems intended for public use: thus, in order to encrypt a credit card, you must supply a credit card number, so an attacker might make up a credit card number and record its encrypted version.) Suppose someone is using a $n \times n$ Hill matrix. What sequence of plaintexts could you submit that would allow you to determine the entries of the Hill matrix easily?

The actual Hill cipher requires some familiarity with mod-N arithmetic (Activities 9.6, 9.7, 9.8, 9.9).

Activity 3.24: More Hills

The full Hill cipher is based on mod-N arithmetic, where values are reduced mod N. In principle, we could choose any modulus we want. For example:

- If we're encrypting text, it's convenient to work mod 26, with $A = 0$, $B = 1$, and so on.
- If we're encrypting numerical values, it's convenient to work mod 10, giving us the digits 0 through 9.
- If we're encrypting alphanumeric values (things you can type on a keyboard), we could work mod 256, with the different values from 0 to 255 corresponding to different symbols.

In other contexts, it's convenient to work mod $2^{16} = 65536$ or other numbers.

A3.24.1 Suppose you're working mod 100. Reduce:

a) $3195 \bmod 100$.

b) $37 \bmod 100$

c) $39187 \bmod 100$.

d) What is an easy way to find $N \bmod 100$?

A3.24.2 Suppose you're encrypting a 16-digit credit card number with the Hill cipher, using the matrix $H = \begin{pmatrix} 11 & 17 \\ 31 & 6 \end{pmatrix}$ and working mod 100.

a) What is the block size?

b) Since we're working mod 100, we can work with the numbers between 0 and 99. What does this suggest about how to break up a (16-digit) credit card number so its value can be encrypted using the Hill cipher?

c) Encrypt the credit card number: $3895\ 1704\ 2190\ 0038$.

4

Matrix Algebra

In Activities 3.16 and 3.17, we showed that every linear transformation corresponds to a matrix, and every matrix corresponds to a linear transformation. Thus, we can speak of the matrix A and the linear transformation corresponding to A interchangeably. This allows us to say things like "the domain and codomain of A" instead of "the domain and codomain of the linear transformation corresponding to the matrix A," and we can "apply A to a vector," instead of "applying the linear transformation corresponding to A to a vector."

Our ability to speak of matrices and linear transformations interchangeably offers further possibilities. Suppose A, B are transformations. In Activities 3.19 and 3.20, we determined what we would mean by AB, A^n, and A^{-1}, and in Activity 3.21, we showed that if A, B are linear transformations, so are AB, A^n, and A^{-1}. But since linear transformations can be represented by matrices, this means that AB, A^n, and A^{-1} are also matrices of some kind.

This leads to the following idea: Since we already know what expressions like AB or A^{-1} mean as *transformations*, and can express the corresponding transformations using matrices, then we can we define expressions like AB and A^{-1} as *matrices*: namely, the matrices that produce the same transformations. This allows us to define an algebra of matrices. To do so, we'll use the following strategy:

Strategy. *Things that do the same thing are the same thing.*

4.1 Scalar Multiplication

We'll start by making sense of what it would mean to multiply a matrix T by a scalar c. In the operator notation we've been using, Tc seems to indicate the matrix T applied to the *vector* c, so to avoid confusion, we'll express the product of matrix T with scalar c as cT. What should cT be?

Since our guide is the arithmetic of transformations, we should consider the effect of cT on some vector $\mathbf{v}$. Since "things that do the same thing *are* the same thing," we want to find some matrix M where $cT\mathbf{v} = M\mathbf{v}$: in other words, M should do the same thing to a vector that cT does to the vector.

However, there's one problem: we know what T does to a vector, since it's a transformation. But what do we want the transformation cT to do to a vector? At this point, it's important to understand that we can *define* cT to be anything we want. However, "Definitions are the whole of mathematics; all else is commentary." This means we need to be very careful when we introduce a definition, since it will affect everything we do from this point forward.

Let's consider. cT applied to vector $\mathbf{v}$ should be written $(cT)\mathbf{v}$, to indicate that the transformation is cT. Now c, T, and $\mathbf{v}$ are three different types of objects: one is a scalar, one is a matrix, and one is a vector. Consequently, we have no built-in guarantees that the associative property applies. But the associative property is so very useful that we might *insist* that it applies. This suggests the following:

Definition 4.1 (Scalar Multiplication). *Let T be a matrix, and $\mathbf{x}$ any vector in the domain of the transformation T. Let c be an appropriate scalar. Then cT is the matrix satisfying $(cT)\mathbf{x} = c(T\mathbf{x})$.*

Thus, the transformation cT applied to a vector $\mathbf{v}$ will be the scalar c multiplied by the vector $T\mathbf{v}$.

Example 4.1. *Suppose $A = \begin{pmatrix} 2 & 3 \\ 1 & 4 \end{pmatrix}$. Find the domain and codomain of A; then find $3A$.*

Solution. *Since A is a 2×2 matrix, its domain is $\mathbb{R}^2$ and its codomain is $\mathbb{R}^2$.*

If we take some vector $\mathbf{v}$ in $\mathbb{R}^2$, then $(3A)\mathbf{v} = 3(A\mathbf{v})$. Let $\mathbf{v} = \langle x, y \rangle$. We note that

$$\begin{aligned} A\mathbf{v} &= \langle 2x + 3y, x + 4y \rangle \\ 3(A\mathbf{v}) &= \langle 6x + 9y, 3x + 12y \rangle \end{aligned}$$

So $3A$ will be the matrix where $3A : \langle x, y \rangle \to \langle x', y' \rangle$, where $\langle x', y' \rangle = \langle 6x + 9y, 3x + 12y \rangle$. Consequently we must have

$$\begin{aligned} x' &= 6x + 9y \\ y' &= 3x + 4y \end{aligned}$$

so $3A = \begin{pmatrix} 6 & 9 \\ 3 & 12 \end{pmatrix}$.

Activity 4.1: Scalar Multiplication of a Matrix

A4.1.1 Let

$$T = \begin{pmatrix} t_{11} & t_{12} & t_{13} \\ t_{21} & t_{22} & t_{23} \end{pmatrix}$$

a) Find the domain and codomain of the transformation T.

b) Let $\mathbf{x}$ be a generic vector in the domain, and consider the transformation $c(T\mathbf{x})$. Write explicit formulas for obtaining the components of the output from the components of the input.

c) To what transformation matrix M will $c(T\mathbf{x})$ correspond?

Activity 4.1 leads to the following result:

Definition 4.2. *Let M be any matrix, and c be any scalar. Then cM will be the matrix produced by multiplying every entry of M by c.*

Example 4.2. *Let* $A = \begin{pmatrix} 2 & 1 & 5 \\ 3 & 0 & -8 \end{pmatrix}$. *Find $3A$, $-2A$.*

Solution. *$3A$ will be the matrix formed by multiplying every entry of A by 3:*

$$3A = \begin{pmatrix} 3\cdot 2 & 3\cdot 1 & 3\cdot 5 \\ 3\cdot 3 & 3\cdot 0 & 3\cdot -8 \end{pmatrix}$$
$$= \begin{pmatrix} 6 & 3 & 15 \\ 9 & 0 & -24 \end{pmatrix}$$

Similarly, $-2A$ will be the matrix formed by multiplying every entry of A by -2:

$$-2A = \begin{pmatrix} -2\cdot 2 & -2\cdot 1 & -2\cdot 5 \\ -2\cdot 3 & -2\cdot 0 & -2\cdot -8 \end{pmatrix}$$
$$= \begin{pmatrix} -4 & -2 & -10 \\ -6 & 0 & 16 \end{pmatrix}$$

Before we proceed, we introduce an important idea. Notice that we now have two different definitions for scalar multiplication: Definition 4.1 and Definition 4.2. However, any time we provide two definitions for the same thing, there's the possibility our definitions may lead to conflicting result. This isn't just a hypothetical situation: it actually happens. Thus in 1893, the Supreme Court of the United States had to decide whether a tomato was a fruit (which it was, according to the scientific definition) or a vegetable (which it was, according to the dictionary definition). The Court chose to reject the scientific definition.

Might the different definitions lead to conflicting results? One way to answer this question is to determine if either definition includes the other.

Activity 4.2: Equivalent Definitions: Scalar Multiplication

A4.2.1 Let

$$T = \begin{pmatrix} t_{11} & t_{12} & t_{13} \\ t_{21} & t_{22} & t_{23} \end{pmatrix}$$

and suppose $\mathbf{v}$ is a generic vector in the domain of T.

a) Find the domain and codomain of T.
b) Find cT using Definition 4.2.
c) Find $(cT)\mathbf{v}$.
d) Find $c(T\mathbf{v})$.
e) Prove/disprove: $(cT)\mathbf{v} = c(T\mathbf{v})$.

Put together, Activities 4.1 and 4.2 give us two results:

- If we use Definition 4.1 to define the scalar multiplication cA, then we can find cA by multiplying every entry of A by c.
- If we use Definition 4.2 to define the scalar multiplication cA, then $(cA)\mathbf{v} = c(A\mathbf{v})$.

Intuitively, we've "closed a loop:" *either* definition implies the other. We say that the definitions are *equivalent*. Strictly speaking we should choose one and obtain the other as a theorem—but it doesn't matter which one we choose.

4.2 Matrix Addition

What about the addition of two matrices: How might we define $A+B$? Again, we want to define this in terms of a transformation. As before, if we apply the transformation $A+B$ to a vector $\mathbf{v}$, we should write this as $(A+B)\mathbf{v}$. Once again, while this appears to have the form of the distributive property, the fact that A, B are matrices and $\mathbf{v}$ is a vector means that we can't guarantee that $(A+B)\mathbf{v} = A\mathbf{v} + B\mathbf{v}$. But again, the distributive property is so useful that we can insist that this be true, which leads to the definition:

Definition 4.3 (Matrix Addition)**.** *Let A, B be matrices. Then $(A+B)\mathbf{x} = A\mathbf{x} + B\mathbf{x}$.*

There is an unstated requirement in the preceding definition, which we'll examine in the next activity.

Activity 4.3: Addition of Matrices

A4.3.1 Consider our definition $(A+B)\vec{x} = A\vec{x} + B\vec{x}$.

a) Since A and B can both act on $\vec{x}$, what do you know about the domain of A and B?

b) "There are only so many symbols." Explain why the + operator on the left hand side of the equation is different from the + operator on the right hand side.

c) Since $A\vec{x}$ and $B\vec{x}$ are vectors, then $A\vec{x} + B\vec{x}$ can be found by adding two vectors. What is required for this to be possible?

d) What do you know about the codomain of A and B?

e) What does this say about the dimensions of the matrices A and B, if $A + B$ is to have a meaning?

A4.3.2 Suppose

$$A = \begin{pmatrix} a_{11} & a_{12} \\ a_{21} & a_{22} \\ a_{31} & a_{32} \end{pmatrix}, B = \begin{pmatrix} b_{11} & b_{12} \\ b_{21} & b_{22} \\ b_{31} & b_{32} \end{pmatrix}$$

a) Find the domain and codomain of A, and the domain and codomain of B.

b) Write a generic vector $\vec{x}$ in the domain of A.

c) Find the output $A\vec{x} + B\vec{x}$.

d) What single matrix M satisfies $M\vec{x} = A\vec{x} + B\vec{x}$?

e) Since we want $(A + B)\vec{x} = A\vec{x} + B\vec{x}$, then $M = A + B$. What does this suggest about a rule for the matrix sum $A + B$?

Activity 4.3 leads to the following conclusions. First, in order for $A + B$ to be defined, both A and B have to have the same domain and the same codomain. But in order to have the same domain and the same codomain, they have to have the same dimensions. Provided this requirement is met, then the single matrix M that satisfies $M\mathbf{x} = A\mathbf{x} + B\mathbf{x}$ can be found by adding the entries of A and B componentwise. Thus:

Definition 4.4. *Let A, B be two matrices of the same size. Then $A + B$ is the matrix produced by adding the entries of A, B componentwise.*

Example 4.3. *Let matrices A, B, C be given. If possible, find $A+B$; $A+C$; $B+C$.*

$$A = \begin{pmatrix} 1 & -3 & 0 \\ 2 & 5 & 0 \end{pmatrix} \qquad B = \begin{pmatrix} 6 & 4 \\ 3 & 8 \end{pmatrix} \qquad C = \begin{pmatrix} -3 & 5 \\ 2 & 7 \end{pmatrix}$$

Solution. *Since A, B and A, C have different sizes, then $A + B$ and $A + C$ are undefined.*

We can find $B+C$ by adding the entries componentwise:

$$\begin{aligned} B+C &= \begin{pmatrix} 6 & 4 \\ 3 & 8 \end{pmatrix} + \begin{pmatrix} -3 & 5 \\ 2 & 7 \end{pmatrix} \\ &= \begin{pmatrix} 6+(-3) & 4+5 \\ 3+2 & 8+7 \end{pmatrix} \\ &= \begin{pmatrix} 3 & 9 \\ 5 & 15 \end{pmatrix} \end{aligned}$$

As before, we have two definitions, so we should verify that they are equivalent.

Activity 4.4: Equivalent Definitions: Matrix Addition

Suppose we define $A+B$ as the matrix whose entries are the componentwise sums of the entries of A and B.

A4.4.1 Explain why this requires that A, B have the same dimensions.

A4.4.2 Explain why this requires that A, B, and $A+B$ have the same domain and codomain.

A4.4.3 Let $\mathbf{v}$ be a vector in the domain of A. Using only Definition 4.4, show that $(A+B)\mathbf{v} = A\mathbf{v} + B\mathbf{v}$.

As with scalar multiplication, we have two equivalent definitions of matrix addition: Definitions 4.3 and 4.4, and we can choose which one we want to use.

4.3 Matrix Multiplication

We've already introduced composition of transformations, so we have a meaning for AB as the transformation A applied to the result of the transformation B applied to a vector. In addition, Activity 3.19 shows that the order in which transformations are performed makes a difference: If you rotate, then reflect, you will generally get a different result than if you reflect first, then rotate.

We define:

Definition 4.5 (Matrix Multiplication)**.** *Let A, B be transformations. Provided the right hand side is defined, $(AB)\mathbf{x} = A(B\mathbf{x})$.*

Thus we can find the product AB by finding the single matrix M that has the same effect on $\mathbf{x}$ as $A(B\mathbf{x})$.

Activity 4.5: Product of Matrices

Assume M, N act on vectors with real components.

A4.5.1 Suppose N is a $m \times n$ matrix.

a) Find the domain and codomain of N.
b) In order to apply M to $N\vec{x}$, what must be true about the domain of M?
c) Suppose $\mathbb{R}^p$ is the codomain of M. What is the domain of MN? What is the codomain?
d) What does this say about the dimensions of matrices M, N that can be multiplied to obtain MN?
e) What does this say about the dimensions of the product matrix MN?

A4.5.2 Let

$$P = \begin{pmatrix} 2 & 5 \\ 3 & 1 \\ 1 & 5 \end{pmatrix}, Q = \begin{pmatrix} 1 & 4 & 3 \\ 2 & 1 & 1 \\ 2 & -5 & 4 \end{pmatrix}$$

a) Consider the transformation QP. If possible, write explicit formulas to find the components of the final output vector $QP\vec{x}$ from the components of the input vector $\vec{x}$. If not possible, explain why not.
b) Consider the transformation PQ. If possible, write explicit formulas to find the components of the final output vector $QP\vec{x}$ from the components of the input vector $\vec{x}$. If not possible, explain why not.

A4.5.3 Suppose M is a $n \times m$ matrix, and assume MN is defined for some matrix N.

a) Suppose $MN = I$, the identity transformation. What must be true about the dimensions of N?
b) Suppose $MN = I$. Will NM be defined? Why/why not?

A4.5.4 Suppose

$$M = \begin{pmatrix} m_{11} & m_{12} \\ m_{21} & m_{22} \end{pmatrix}, N = \begin{pmatrix} n_{11} & n_{12} \\ n_{21} & n_{22} \end{pmatrix}, \vec{x} = \begin{pmatrix} x_1 \\ x_2 \end{pmatrix}$$

a) Write down explicit formulas for finding the components of $MN\vec{x}$ from the components of $\vec{x}$.
b) What single transformation T does the same thing as composition MN?
c) Write down explicit formulas for finding the components of $NM\vec{x}$ from the components of $\vec{x}$.
d) What single transformation U does the same thing as the composition NM?
e) Does $MN = NM$?

A4.5.5 Suppose

$$A = \begin{pmatrix} a_{11} & a_{12} & a_{13} \\ a_{21} & a_{22} & a_{23} \end{pmatrix}, B = \begin{pmatrix} b_{11} & b_{12} \\ b_{21} & b_{22} \\ b_{31} & b_{32} \end{pmatrix}$$

a) Let $\vec{x}$ be a vector of the appropriate size. Write down explicit formulas for finding the components of $AB\mathbf{x}$ from the components of the input $\vec{x}$.

b) If $M\vec{x} = AB\vec{x}$, what is M?

The fact that MN is a single transformation that corresponds to the transformation M applied to the result of N leads to our definition of the product of two matrices.

To make it easier to express and compute the product, we'll introduce the following definition:

Definition 4.6 (Row and Column Vectors). *Let A be a matrix. The rows of A are* ***row vectors****, while the columns are* ***column vectors****.*

Example 4.4. *Let* $A = \begin{pmatrix} 2 & -1 & -3 \\ 0 & 5 & 4 \end{pmatrix}$. *Find the row and column vectors.*

Solution. *The row vectors are vectors corresponding to the rows of A:*

$$\mathbf{u}_1 = \langle 2, -1, -3 \rangle \qquad \mathbf{u}_2 = \langle 0, 5, 4 \rangle$$

The column vectors are vectors corresponding to the columns of A:

$$\mathbf{v}_1 = \langle 2, 0 \rangle \qquad \mathbf{v}_2 = \langle -1, 5 \rangle \qquad \mathbf{v}_3 = \langle -3, 4 \rangle$$

If we want, we can express our matrix using the row or column vectors:

$$\begin{pmatrix} 2 & -1 & -3 \\ 0 & 5 & 4 \end{pmatrix} = \begin{pmatrix} \mathbf{u}_1 \\ \mathbf{u}_2 \end{pmatrix} = \begin{pmatrix} \mathbf{v}_1 & \mathbf{v}_2 & \mathbf{v}_3 \end{pmatrix}$$

This allows us to find the product of two matrices as:

Definition 4.7. *Let A be a matrix whose rows are the vectors $\mathbf{a}_1, \mathbf{a}_2, \ldots, \mathbf{a}_m$, and B the matrix whose columns are the vectors $\mathbf{b}_1, \mathbf{b}_2, \ldots, \mathbf{b}_n$. The product AB will be the matrix whose ijth entry will be $\mathbf{a}_i \cdot \mathbf{b}_j$.*

In other words, the ijth entry of the product MN will be the dot product of the ith row vector of M with the jth column vector of N.

Example 4.5. *Find the product*

$$\begin{pmatrix} 2 & 1 & 7 \\ 3 & 5 & -1 \end{pmatrix} \begin{pmatrix} 1 & 4 \\ 2 & 7 \\ -3 & 8 \end{pmatrix}$$

Solution. *The rows of the first matrix are the vectors* $\mathbf{a}_1 = \langle 2, 1, 7 \rangle$, $\mathbf{a}_2 = \langle 3, 5, -1 \rangle$. *Meanwhile the columns of second matrix are vectors* $\mathbf{b}_1 = \langle 1, 2, -3 \rangle$ *and* $\mathbf{b}_2 = \langle 4, 7, 8 \rangle$. *Thus we can express our product as:*

$$\begin{pmatrix} \mathbf{a}_1 \\ \mathbf{a}_2 \end{pmatrix} \begin{pmatrix} \mathbf{b}_1 & \mathbf{b}_2 \end{pmatrix} = \begin{pmatrix} \mathbf{a}_1 \cdot \mathbf{b}_1 & \mathbf{a}_1 \cdot \mathbf{b}_2 \\ \mathbf{a}_2 \cdot \mathbf{b}_1 & \mathbf{a}_2 \cdot \mathbf{b}_2 \end{pmatrix}$$

Finding the various dot products gives us:

$$= \begin{pmatrix} -17 & 71 \\ 16 & 39 \end{pmatrix}$$

Activity 4.6: Equivalent Definitions: Matrix Multiplication

For the following, let

$$A = \begin{pmatrix} a & b \\ c & d \end{pmatrix}, B = \begin{pmatrix} f & g & h \\ p & q & r \end{pmatrix}$$

A4.6.1 Find the domain and codomain of A, B, and AB.

A4.6.2 Suppose $\mathbf{v}$ is in the domain of AB.

a) Explain why if $\mathbf{v}$ is a vector in the domain of AB, it will also be in the domain of B.
b) Let $\mathbf{v}$ be a generic vector in the domain of B. Find $B\mathbf{v}$.
c) Find $A(B\mathbf{v})$. Remember expressions in parentheses must be evaluated first.
d) Find $(AB)\mathbf{v}$, where AB is found using Definition 4.7.
e) Do your results support $(AB)\mathbf{v} = A(B\mathbf{v})$?

Activity 4.6 suggests that Definitions 4.5 and 4.7 are equivalent.

One of the important features of matrix arithmetic is that in general, it is not commutative: $AB \neq BA$.

Activity 4.7: The Game of Matrix Products

It helps to have a partner in this game; if you don't have one, generate a random matrix by rolling dice.

In each round:

- Both you and your partner make up (in secret) a 2×2 matrix with nonzero entries, all of which must be different.

- Choose a statement: $AB = BA$, $AB \neq BA$.
- Reveal your matrices. You win 1 point if your statement is correct, and lose 1 point if your statement is incorrect.

For example, you choose the matrix $A = \begin{pmatrix} 2 & 5 \\ 3 & 1 \end{pmatrix}$ (remember the entries must all be different and not equal to 0), and the statement $AB = BA$.

Your partner chooses the matrix $B = \begin{pmatrix} 1 & 2 \\ 5 & 3 \end{pmatrix}$. Then you compute AB:

$$\begin{pmatrix} 2 & 5 \\ 3 & 1 \end{pmatrix} \begin{pmatrix} 1 & 2 \\ 5 & 3 \end{pmatrix} = \begin{pmatrix} 27 & 19 \\ 8 & 9 \end{pmatrix}$$

and BA:

$$\begin{pmatrix} 1 & 2 \\ 5 & 3 \end{pmatrix} \begin{pmatrix} 2 & 5 \\ 3 & 1 \end{pmatrix} = \begin{pmatrix} 8 & 7 \\ 19 & 28 \end{pmatrix}$$

Since you chose $AB = BA$, but $AB \neq BA$, you lose 1 point.

A4.7.1 Play the game several times. Which statement should you choose?

A4.7.2 Suppose you're allowed to repeat entries. Could you choose a matrix A with at least one nonzero entry, and one of the two statements $AB = BA$ or $AB \neq BA$, which would allow you to win, regardless of the matrix your partner chose?

Remember that in a discrete time model, the transformation matrix T corresponds to the formulas used to find the values of the quantities at time $n+1$, given the values of the quantities at time n. If we apply T twice, we can obtain the values at time $n+2$ from the quantities at time n; and if we apply T k times, we obtain the values at time $n+k$ from the quantities at time n.

Activity 4.8: Powers of a Matrix and Fast Powering

A4.8.1 In order for A^n to be defined, what must be true about the dimensions of A?

A4.8.2 The use of exponential notation for A^n is neither arbitrary nor meaningless. In the following, use the definition of A^n to prove that the normal rules of exponents for real numbers apply to the exponentiation of matrices. Assume m, n are whole numbers.

a) Show $A^m A^n = A^{m+n}$.

b) Show $(A^m)^n = A^{mn}$.

c) Find A^0. Suggestion: From $A^m A^n = A^{m+n}$, we have $A^m = A^m A^0$. Then "Things that do the same thing are the same thing."

d) Explain why $A^{-m} \neq \frac{1}{A^m}$.

e) What interpretation could be made of A^{-m}?

A4.8.3 Consider the Fibonacci rabbit problem (Activity 3.10). If x_n, y_n represent the number of immature and mature pairs of rabbits at the end of month n, then $\begin{pmatrix} x_{n+1} \\ y_{n+1} \end{pmatrix} = F \begin{pmatrix} x_n \\ y_n \end{pmatrix}$ where $F = \begin{pmatrix} 0 & 1 \\ 1 & 1 \end{pmatrix}$.

a) Interpret F in terms of an explicit formula for finding x_{n+1}, y_{n+1} from x_n, y_n.

b) Apply F twice to find x_{n+2}, y_{n+2} from x_n, y_n.

c) Explain why your formulas correspond to the matrix F^2.

d) Use matrix multiplication to find F^2, and verify this is the same as the answer you got for F^2 using direct application of the transformation matrix.

A4.8.4 One way to find very high powers of a matrix is to use the **fast powering algorithm**. We illustrate it below. (F is the matrix from the previous problem)

a) Find $(F^2)^2$.

b) Explain why $(F^2)^2 = F^4$.

c) Find F^8.

d) Find F^{12}.

e) The original rabbit problem posed by Leonardo of Pisa was to determine the number of rabbits after $n = 12$ months. If $x_0 = 1, y_0 = 1$, find x_{12}, y_{12}, and solve the problem.

f) How many of each type of rabbit were there after 36 months?

Another use of matrix products involves a mathematical object known as a *graph*.

Activity 4.9: Graphs and Matrices

In discrete mathematics, a **graph** consists of a set of vertices (also known as points or nodes), some of which are joined by edges (also known as lines or links); a graph with vertices a, b, c, d and some edges is shown below.

The **adjacency matrix** for a graph is the matrix A where a_{ij} indicates the *number* of edges between vertex i and vertex j. If we let a correspond to vertex 1, b vertex 2, c vertex 3, and d vertex 4, then we obtain the adjacency matrix shown..

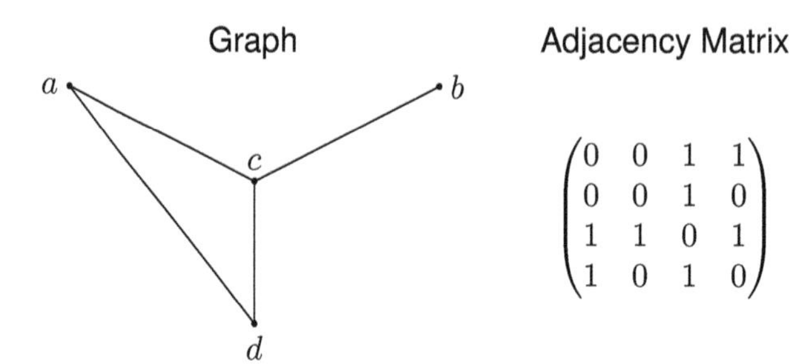

A4.9.1 Find the adjacency matrix for each graph. Assume a is vertex 1, b is vertex 2, etc.

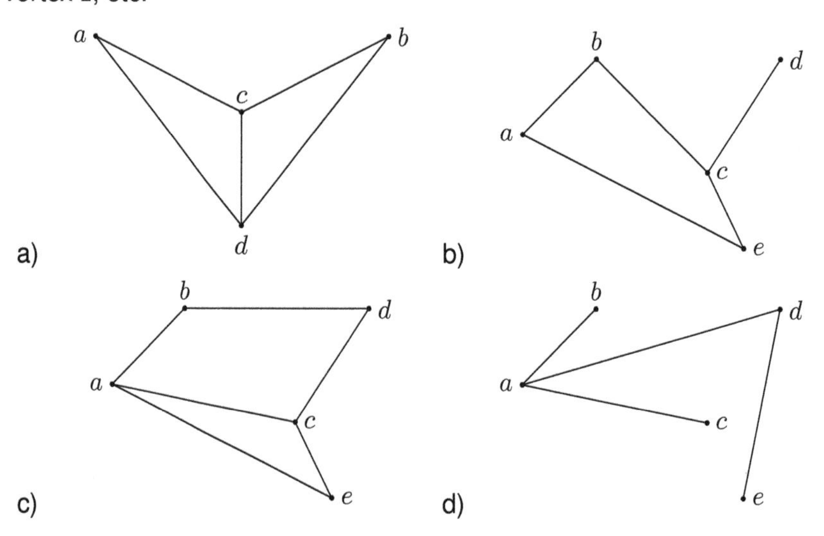

A4.9.2 If possible, draw a graph with the given adjacency matrix; if not possible, explain why not.

a) $\begin{pmatrix} 0 & 1 & 0 & 1 \\ 1 & 0 & 0 & 1 \\ 0 & 0 & 0 & 1 \\ 1 & 1 & 1 & 0 \end{pmatrix}$

b) $\begin{pmatrix} 0 & 0 & 0 \\ 0 & 0 & 0 \\ 0 & 0 & 0 \end{pmatrix}$

c) $\begin{pmatrix} 0 & 1 & 1 & 1 \\ 1 & 0 & 1 & 1 \\ 1 & 1 & 0 & 1 \\ 1 & 1 & 1 & 0 \end{pmatrix}$

d) $\begin{pmatrix} 0 & 1 & 1 & 0 \\ 1 & 0 & 0 & 0 \\ 1 & 0 & 0 & 1 \\ 0 & 0 & 0 & 0 \end{pmatrix}$

e) $\begin{pmatrix} 1 & 0 & 1 \\ 0 & 0 & 1 \\ 1 & 1 & 0 \end{pmatrix}$

A4.9.3 The **fundamental counting principle** states that if there p ways of making a first choice, and q ways of making a second choice, there are pq ways of making the two choices in sequence.

a) Suppose A is the adjacency matrix for a graph. What is the significance of the location of the 1s in the first row?
b) Consider the second column of A. What is the significance of the location of the 1s in the second column?
c) Consider the dot product of the first row of A with the second column of A. What is the significance of the terms in the dot product that *aren't* equal to 0?
d) What does the value of the dot product of the first row of A with the second column of A tell you?
e) What does this suggest about the entries of A^2?

Activity A4.9.3 leads to the following useful result:

Proposition 1. *Let A be the adjacency matrix for a graph. The ijth entry of A^k is the number of paths between vertices i and j that cross k edges.*

We close with several claims about matrix arithmetic.

Activity 4.10: Properties of Matrix Arithmetic

To prove properties of matrix arithmetic, remember that we can define the entries of the resulting matrix in terms of the entries of the starting matrices.

A4.10.1 In this activity, you'll prove that scalar multiplication is associative, so for scalars m, n and matrix A, $m(nA) = (mn)A$. Let the i, j entry of A be a_{ij}.

a) Find the i, j entry of (nA).
b) Find the i, j entry of $m(nA)$. (Remember the operations inside parentheses must be done first)
c) Find the i, j entry of $(mn)A$. (Remember the operations inside parentheses must be done first)

d) Why do your results prove that $m(nA) = (mn)A$?

A4.10.2 Prove: Provided the sum exists, $A + B = B + A$.

A4.10.3 Prove: Provided the sums exist, $(A + B) + C = A + (B + C)$.

A4.10.4 Prove: Matrix multiplication is associative, so $(AB)C = A(BC)$.

A4.10.5 Prove: The distributive property of scalar multiplication over addition holds, so $c(A + B) = cA + cB$, where c is a scalar and A, B are matrices.

A4.10.6 Prove: The distributive property of matrix multiplication over addition holds, so $A(B + C) = AB + AC$.

A4.10.7 Consider the product $(A + B)^2 = (A + B)(A + B)$.

a) Suppose A is a $m \times n$ matrix. What must be true about m and n in order for $(A + B)^2$ to exist?

b) Explain why $(A + B)^2 \neq A^2 + 2AB + B^2$.

c) Assuming it exists, what is $(A + B)^2$?

d) Prove it.

Activity 4.10 proves:

Theorem 4.1. *Let A, B, C be matrices of the appropriate size. Provided both sides are defined, matrices satisfy*

$$\begin{aligned} A + B &= B + A, \quad \textit{commutativity of addition} \\ (A + B) + C &= A + (B + C), \quad \textit{associativity of addition} \\ (AB)C &= A(BC), \quad \textit{associativity of multiplication} \\ A(B + C) &= AB + AC, \quad \textit{distributivity of multiplication over addition} \end{aligned}$$

4.4 Elementary Matrices

One goal of mathematics is to be able to treat seemingly unrelated problems in the same way. We've seen that matrices can be viewed as linear transformations, and by doing so, we've been able to define what it means to multiply two matrices. But we began by viewing matrices as representations of systems of equations, which we could then solve by row reduction. Could we treat row reduction as a linear transformation, and in that way perform row reductions through matrix multiplication?

Activity 4.11: Elementary Matrices

In the following, assume M is a $n \times m$ coefficient matrix.

A4.11.1 Suppose M' is the matrix formed by multiplying the third row of M by a constant c. If possible, find a matrix T where $TM = M'$. If not possible, explain why not.

A4.11.2 Suppose M' is the matrix formed by multiplying the third row of M by a constant c, then adding the result to the second row. If possible, find a matrix T where $TM = M'$. If not possible, explain why not.

A4.11.3 Suppose M' is the matrix formed by multiplying the third row of M by a constant c. If possible, find a matrix T where $MT = M'$. If not possible, explain why not.

A4.11.4 Suppose

$$A = \left(\begin{array}{cc|c} 2 & 5 & 1 \\ 5 & 3 & 2 \end{array}\right)$$

a) Reduce A to reduced row echelon form. Be sure to record the individual steps of the row reduction.

b) Find matrices $T_1, T_2, \ldots, T_n$ where T_i corresponds to the row reductions in the ith step.

c) What is the significance of $T_n T_{n-1} \cdots T_2 T_1$? (Suggestion: What is $T_n T_{n-1} \cdots T_2 T_1 A$?)

Activity 4.11 leads to the following:

Definition 4.8 (Elementary Matrix). *An elementary matrix E is the result of performing an elementary row operation on an identity matrix.*

Moreover:

Theorem 4.2. *Let E be an elementary matrix corresponding to some elementary row operation. If EM exists, it will equal the matrix produced by applying the elementary row operation to M.*

Example 4.6. *Find the 3×3 elementary matrix corresponding to multiplying the first row by 2 and adding it to the third row.*

Solution. *If we perform these operations on the identity matrix,*

$$\begin{pmatrix} 1 & 0 & 0 \\ 0 & 1 & 0 \\ 0 & 0 & 1 \end{pmatrix} \xrightarrow{2R_1 + R_3 \rightarrow R_3} \begin{pmatrix} 1 & 0 & 0 \\ 0 & 1 & 0 \\ 2 & 0 & 1 \end{pmatrix}$$

The significance of the elementary matrix is that it allows us to perform a row operation by a matrix multiplication. Thus the row operation

$$\begin{pmatrix} a & b & c \\ d & e & f \\ g & h & i \end{pmatrix} \xrightarrow{2R_1+R_3 \to R_3} \begin{pmatrix} a & b & c \\ d & e & f \\ 2a+g & 2b+h & 2c+i \end{pmatrix}$$

can be performed by multiplying by the corresponding elementary matrix:

$$\begin{pmatrix} 1 & 0 & 0 \\ 0 & 1 & 0 \\ 2 & 0 & 1 \end{pmatrix} \begin{pmatrix} a & b & c \\ d & e & f \\ g & h & i \end{pmatrix} = \begin{pmatrix} a & b & c \\ d & e & f \\ 2a+g & 2b+h & 2c+i \end{pmatrix}$$

4.5 More Transformations

Remember that a matrix represents a linear transformation.

Activity 4.12: Matrix Multiplication and Transformation

A4.12.1 Suppose A transforms vectors $\langle x_1, x_2 \rangle$ into vectors $\langle y_1, y_2 \rangle$ according to the formulas

$$\begin{aligned} y_1 &= 2x_1 + 3x_2 \\ y_2 &= x_1 - 4x_2 \end{aligned}$$

a) Find the transformation matrix A.

b) We have been (and should be) writing the application of A to a vector as $A(\langle x_1, x_2 \rangle) = \langle y_1, y_2 \rangle$. Explain why we can also write $A \begin{pmatrix} x_1 \\ x_2 \end{pmatrix} = \begin{pmatrix} y_1 \\ y_2 \end{pmatrix}$. In other words, we can write vectors as column vectors.

A4.12.2 Some authors prefer to avoid writing column vectors, and write the vector $\langle x_1, x_2 \rangle$ as the matrix $\begin{pmatrix} x_1 & x_2 \end{pmatrix}$.

a) Consider the transformation

$$\begin{aligned} y_1 &= 2x_1 + 3x_2 \\ y_2 &= x_1 - 4x_2 \end{aligned}$$

Find A' so that $\begin{pmatrix} x_1 & x_2 \end{pmatrix} A' = \begin{pmatrix} y_1 & y_2 \end{pmatrix}$

b) Notice that this is the same transformation as before. What is the relationship between A and A'?

A4.12.3 Suppose we wish to apply the transformation A to several vectors $\vec{v}_1 = \langle x_1, y_1 \rangle$, $\vec{v}_2 = \langle x_2, y_2 \rangle$, ..., $\vec{v}_n = \langle x_n, y_n \rangle$. Rather than finding $A\vec{v}_1$, $A\vec{v}_2$, etc., individually, explain why you can find all the transformed vectors at once by computing AV, where the vectors $\vec{v}_1, \vec{v}_2, \ldots, \vec{v}_n$ are the columns of V.

Activity A4.12.2 suggests a matrix operation which has no analog in elementary arithmetic:

Definition 4.9 (Transpose). *The transpose of A, written A^T, is the matrix M where $m_{ij} = a_{ji}$.*

You can think about the transpose as a "flip" of the matrix elements along its diagonal, where the columns of the original matrix become the rows of the transposed matrix.

Example 4.7. *Find the transpose of* $\begin{pmatrix} 1 & 2 \\ 3 & 4 \end{pmatrix}$

Solution. *The rows of the transpose matrix will be the columns of the original (we'll **bold** the first column of the original and the first row of the transpose to emphasize the relationship):*

$$\begin{pmatrix} \mathbf{1} & 2 \\ \mathbf{3} & 4 \end{pmatrix}^T = \begin{pmatrix} \mathbf{1} & \mathbf{3} \\ 2 & 4 \end{pmatrix}$$

Example 4.8. *Find the transpose of* $\begin{pmatrix} 1 & 5 & 3 \end{pmatrix}$.

Solution. *The columns of the matrix (1, 5, and 3) will become the rows of the transpose, so:*

$$\begin{pmatrix} 1 & 5 & 3 \end{pmatrix}^T = \begin{pmatrix} 1 \\ 5 \\ 3 \end{pmatrix}$$

Notice that the 1×3 matrix $\begin{pmatrix} 1 & 5 & 3 \end{pmatrix}$ looks like the vector $\langle 1, 5, 3 \rangle$. This leads to the following terminology:

Definition 4.10. *Let* $\mathbf{x}$ *be a vector in* $\mathbb{F}^n$.

- *The $1 \times n$ matrix $\mathbf{x}$ whose components are the components of $\mathbf{x}$ is the **row vector** form of $\mathbf{x}$,*
- *The $n \times 1$ matrix $\mathbf{x}$ whose components are the components of $\mathbf{x}$ is the **column vector** form of $\mathbf{x}$.*

The orientation (row or column) of the matrix is generally "understood" from the expression: In $A\mathbf{x}$, $\mathbf{x}$ must be a column vector in order for the multiplication to be performed, while in $\mathbf{x}A$, $\mathbf{x}$ must be a row vector. However, if there is any possibility of uncertainty, we can use the transpose to indicate whether the vector is a row or column vector; our "default" assumption is that $\mathbf{x}$ is a column vector, so we would write $A\mathbf{x}$ and $\mathbf{x}^T A$. Note that it is not generally necessary to write expressions like $\vec{x}^T$, since $\vec{x}$ is clearly a vector and, as such, it can be arranged as a column vector or a row vector as needed.

We can use the row or column vector concept to write any matrix as an ordered set of row or column vectors.

Example 4.9. *Let* $A = \begin{pmatrix} 2 & 3 & 1 \\ 5 & 1 & 0 \end{pmatrix}$. *Write A as a matrix of row vectors; and as a matrix of column vectors. Specify the vectors in both cases.*

Solution. *The rows of A can be read as the vectors* $\mathbf{r}_1 = \langle 2, 3, 1 \rangle$, *and* $\mathbf{r}_2 = \langle 5, 1, 0 \rangle$, *so we can write* $A = \begin{pmatrix} \mathbf{r}_1 \\ \mathbf{r}_2 \end{pmatrix}$.

Similarly, we can express the columns of A as the vectors $\mathbf{c}_1 = \langle 2, 5 \rangle$, $\mathbf{c}_2 = \langle 3, 1 \rangle$, $\mathbf{c}_3 = \langle 1, 0 \rangle$, *and our matrix will be* $A = \begin{pmatrix} \mathbf{c}_1 & \mathbf{c}_2 & \mathbf{c}_3 \end{pmatrix}$

There is a useful relationship between the dot product and the transpose.

Activity 4.13: Properties of the Transpose

A4.13.1 Let $\vec{u} = \langle 1, 2, 3 \rangle$ and $\vec{v} = \langle 2, -1, 5 \rangle$, and let $\mathbf{u}$, $\mathbf{v}$ be column vectors corresponding to $\vec{u}$, $\vec{v}$. If possible, find the products. If not possible, explain why not.

a) $\vec{u} \cdot \vec{v}$.
b) $\mathbf{u}\mathbf{v}$
c) $\mathbf{u}^T\mathbf{v}$
d) $\mathbf{u}\mathbf{v}^T$
e) $\mathbf{u}^T\mathbf{v}^T$

A4.13.2 Let $\vec{u}$, $\vec{v}$ be vectors in $\mathbb{F}^n$, and let $\mathbf{u}$, $\vec{v}$ be the corresponding column vectors. Express $\vec{u} \cdot \vec{v}$ in terms of $\mathbf{u}$, $\mathbf{v}$.

Activity 4.13 leads to the following:

Theorem 4.3. *Suppose $\mathbf{u}$, $\mathbf{v}$ are $n \times 1$ column vectors corresponding to vectors $\vec{u}$, $\vec{v}$ in $\mathbb{F}^n$. Then*

$$\mathbf{u}^T\mathbf{v} = (\vec{u} \cdot \vec{v})$$

where $(\vec{u} \cdot \vec{v})$ is the 1×1 matrix whose entry is $\vec{u} \cdot \vec{v}$.

Let's see where this takes us.

First, let's consider the transpose of a product.

Activity 4.14: The Transpose of a Product, Part One

A4.14.1 Suppose A is a $m \times n$ matrix, and B is a $n \times p$ matrix.

a) Find the dimensions of AB and $(AB)^T$.

b) Determine which (if any) of the following products exist: $A^T B^T$; BA; $B^T A^T$. For the products that exist, what are the dimensions of the product?

c) Which of the four products could $(AB)^T$ be equal to: AB, $A^T B^T$, BA, $B^T A^T$.

A4.14.2 Let

$$A = \begin{pmatrix} 1 & 2 \\ 3 & 5 \end{pmatrix}, B = \begin{pmatrix} -1 & 3 & 5 \\ 2 & -4 & -5 \end{pmatrix}$$

a) Find AB and $(AB)^T$.

b) In Activity A4.14.1c, you concluded that $(AB)^T$ could be one of $A^T B^T$; BA; $B^T A^T$. Verify your conjecture.

A4.14.3 Let

$$M = \begin{pmatrix} 1 & 2 \\ 3 & -1 \end{pmatrix}, N = \begin{pmatrix} 4 & 6 \\ -3 & -1 \end{pmatrix}$$

a) Find MN and $(MN)^T$.

b) Find the products, if they exist: $M^T N^T$; MN; $M^T N^T$.

c) Which is equal to $(MN)^T$?

Activity 4.14 suggests:

Theorem 4.4 (Transpose of a Product). *Provided both sides exist,* $(AB)^T = B^T A^T$.

Let's prove our result.

Activity 4.15: More Transposes

In the following, assume the matrices are of the correct size so that all matrix products exist.

Consider the four matrices: A, B, A^T, B^T.

A4.15.1 Express as a row or column of A or B.

a) The ith row of A^T.

b) The jth column of B^T.

A4.15.2 Express as the dot product of a row or column vector of A with a row or column vector of B.

a) The ijth entry of AB.
b) The ijth entry of $(AB)^T$.
c) The ijth entry of $A^T B^T$.
d) The ijth entry of BA.
e) The ijth entry of $(BA)^T$.
f) The ijth entry of $B^T A^T$.
g) Which matrix will have the same entries as $(AB)^T$?

In some cases, the transpose of a matrix is the same matrix we started with. We define:

Definition 4.11 (Symmetric Matrix). *If $A^T = A$, then A is a symmetric matrix.*

Activity 4.16: Symmetric Matrices

Symmetric matrices play a very important role in some of the more advanced applications of linear algebra. Unfortunately, *most* matrices *aren't* symmetric.

A4.16.1 Which of the following are symmetric matrices?

$$A = \begin{pmatrix} 1 & 5 \\ 2 & -1 \end{pmatrix}, B = \begin{pmatrix} 1 & 2 \\ 2 & 1 \\ 3 & 5 \end{pmatrix}, C = \begin{pmatrix} 1 & 4 & 7 \\ 4 & 2 & 0 \\ 7 & 0 & 3 \end{pmatrix}$$

A4.16.2 Let M be a $n \times m$ matrix. What relationship between m and n is required if M is to be symmetric?

A4.16.3 Prove: For any matrix P, $(P^T)^T = P$.

A4.16.4 Suppose P is a nonsymmetric matrix. Find a matrix Q where QP is a symmetric matrix. Suggestion: The easiest way to make $AB = MN$ is to make $A = M$ and $B = N$.

There is an important relationship between the transpose of a matrix and a matrix corresponding to a rotation.

Activity 4.17: Matrices and Rotations

Let R be a matrix that represents a rotation, and let the columns of R be $\vec{v}_1$, $\vec{v}_2, \ldots, \vec{v}_n$.

A4.17.1 Consider $R\vec{e}_i$, the vector produced when R acts on an elementary vector.

a) Explain why $R\vec{e}_i = \vec{v}_i$, where $\vec{e}_i$ is the elementary vector.

b) Explain why $||\vec{v}_i|| = 1$. Suggestion: Remember that R represents a rotation, and that $\vec{v}_i$ is the image of $\vec{e}_i$ under the rotation.

c) Explain why $\vec{v}_i \cdot \vec{v}_j = 0$ for $i \neq j$.

A4.17.2 Find R^{-1}. Suggestion: What *row* vectors must the $\vec{v}_i$ be multiplied by to get the identity matrix?

Activity 4.17 leads to the following theorem:

Theorem 4.5. *Suppose R represents a rotation. Then:*

- *Each column of R is a unit vector,*
- *For any two column vectors distinct* $\mathbf{v}_i$, $\mathbf{v}_j$, *we have* $\mathbf{v}_i \cdot \mathbf{v}_j = 0$,
- $R^{-1} = R^T$.

The last suggests a useful definition:

Definition 4.12 (Column Orthogonal Matrix)**.** *R is an* ***column orthogonal matrix*** *if its columns are orthogonal unit vectors.*

4.6 Matrix Inverses

We define A^{-1} to be the transformation that "undoes" the transformation A. However, this presupposes that A was already done. This suggests the definition:

Definition 4.13 (Left Inverse)**.** *Let A be a matrix. If* $LA\mathbf{x} = \mathbf{x}$ *for all vectors in the domain of A, then L is a* ***left inverse of*** *A.*

Activity 4.18: Left Inverses

A4.18.1 Let S be a "shadow transformation" that transforms a vector $\overrightarrow{x}$ in $\mathbb{R}^3$ into its shadow $S\overrightarrow{x}$ in $\mathbb{R}^2$. Would you expect S to have a left inverse? Why/why not?

A4.18.2 Suppose

$$A = \begin{pmatrix} 1 & 4 \\ -3 & 2 \\ -1 & 8 \end{pmatrix} \qquad B = \begin{pmatrix} 2 & -1 & 0 \\ 1 & 0 & 5 \end{pmatrix}$$

a) Find the domain and codomain of A; and the domain and codomain of B.
b) Based on Activity A4.18.1, would you expect both A and B to have left inverses? Why/why not?
c) Suppose L is a left inverse of A. What are the dimensions of L?
d) Find LA. Suggestion: Things that *do* the same thing *are* the same thing.
e) Set up and solve (if possible) a system of equations to find L.
f) Suppose M is a left inverse of B. Find MB.
g) Set up and solve (if possible) a system of equations to find M.

"Math ever generalizes," so we could also define:

Definition 4.14 (Right Inverse). *Let A be a matrix, and let $\mathbf{x}$ be any vector in its domain. If $AR\mathbf{x} = \mathbf{x}$ for all vectors in the domain of R, then R is a **right inverse of** A.*

Right inverses are a little peculiar, since the product AR means we are applying the inverse *first*: in effect, we are "undoing" the transformation before it's done.

Activity 4.19: Right Inverses

In the following, let

$$A = \begin{pmatrix} 1 & 4 \\ -3 & 2 \\ -1 & 8 \end{pmatrix} \qquad B = \begin{pmatrix} 2 & -1 & 0 \\ 1 & 0 & 5 \end{pmatrix}$$

A4.19.1 Suppose R is the right inverse of A.

a) Find the domain and codomain of R.
b) Explain why A would be a left inverse of R.

c) Would you expect the left inverse of R to exist? Why/why not?

d) Since A clearly exists, what does this say about R?

A4.19.2 Suppose R is the right inverse of B.

a) Find BR.

b) Set up and solve a system of equations to find the entries of R.

As we saw in Activities 4.18 and 4.19, it's possible for a matrix to have a left inverse but not a right inverse; and it's possible for a matrix to have a right inverse but not a left inverse.

What if a matrix has both?

Activity 4.20: Inverse Matrices

A4.20.1 Let A be a $m \times n$ matrix corresponding to a linear transformation, and suppose A has a left inverse L *and* a right inverse R. Further, suppose $R\vec{x} = \vec{y}$.

a) Find the dimensions of A, R, L.

b) Find $A\vec{y}$.

c) Find $L\vec{x}$.

d) "Things that do the same thing are the same thing." What do your results tell you about R and L?

A4.20.2 Let C be a $n \times n$ square matrix corresponding to a linear transformation, and suppose C has a left inverse D. Let $C\vec{x} = \vec{y}$.

a) What are the dimensions of D?

b) What does the transformation DC do to a vector $\vec{x}$?

c) Find $D\vec{y}$.

d) Find $CD\vec{y}$.

e) "Things that do the same thing are the same thing." What do your results tell you about the transformation CD?

f) Repeat your work, but this time assume C has a *right* inverse E.

Activity 4.20 gives us an important result:

Theorem 4.6 (Right and Left Inverses). *Given matrix A, suppose L and R satisfy $LA = I$, $AR = I$ for appropriately-sized identity matrices. Then $R = L$. Additionally, if A is a square matrix, the existence of one implies the existence of the other.*

Thus, while nonsquare matrices might have a left inverse without a right inverse, or a right inverse without a left inverse, a square matrix either has both—or neither. This means we can simply refer to the *inverse* of a square matrix. We define:

Definition 4.15 (Inverse of a Matrix). *Let A be a square matrix. Provided it exists, the **inverse of** A, written A^{-1}, satisfies $A^{-1}A = AA^{-1} = I$.*

How can we find an inverse matrix? As always, every problem in linear algebra begins with a system of linear equations.

Activity 4.21: Finding the Inverse of a Matrix

A4.21.1 Let $A = \begin{pmatrix} 5 & 3 \\ 3 & 2 \end{pmatrix}$

a) If possible, set up and solve a system of linear equations to find a matrix L where $LA = I$.

b) Find AL. Do your results support the conclusion you reached in Activity A4.20.1d?

A4.21.2 Let $B = \begin{pmatrix} 2 & 3 \\ -4 & -6 \end{pmatrix}$. If possible, set up and solve a system of linear equations to find L where $LB = I$.

As you saw in Activity 4.21, there's no guarantee a matrix will have an inverse, so we introduce the term:

Definition 4.16 (Singular Matrix). *A matrix without an inverse is called a **singular matrix**.*

A matrix that has an inverse is either called nonsingular, or **invertible**.

While we could find the inverse (if it exists) by solving a system of linear equations, let's examine this process closely to see if we can make it more efficient.

Activity 4.22: Double Wide Matrices

A4.22.1 Consider two different systems of equations

$$\begin{cases} 3x + 5y & = & 7 \\ 2x - 3y & = & 10 \end{cases} \quad \text{and} \quad \begin{cases} 3s + 5t & = & 12 \\ 2s - 3t & = & 15 \end{cases}$$

and let $E_1, E_2, \ldots, E_n$ be a sequence of elementary row operations, where

$$\left(\begin{array}{cc|c} 3 & 5 & 7 \\ 2 & -3 & 10 \end{array}\right) \xrightarrow{E_1} \cdots \xrightarrow{E_n} \left(\begin{array}{cc|c} 1 & 0 & b_{11} \\ 0 & 1 & b_{21} \end{array}\right)$$

a) What is the significance of b_{11}, b_{21}?

b) Find b_{12}, b_{22} where

$$\left(\begin{array}{cc|c} 3 & 5 & 12 \\ 2 & -3 & 15 \end{array}\right) \xrightarrow{E_1} \cdots \xrightarrow{E_n} \left(\begin{array}{cc|c} 1 & 0 & b_{12} \\ 0 & 1 & b_{22} \end{array}\right)$$

c) Explain why we can solve both systems at the same time by row reducing the "doubly augmented" coefficient matrix

$$\left(\begin{array}{cc|cc} 3 & 5 & 7 & 12 \\ 2 & -3 & 10 & 15 \end{array}\right) \xrightarrow{E_1} \cdots \xrightarrow{E_n} \left(\begin{array}{cc|cc} 1 & 0 & b_{11} & b_{12} \\ 0 & 1 & b_{21} & b_{22} \end{array}\right)$$

A4.22.2 Consider the equation

$$\begin{pmatrix} 3 & 5 \\ 5 & 8 \end{pmatrix} \begin{pmatrix} x_1 & x_2 \\ x_3 & x_4 \end{pmatrix} = \begin{pmatrix} 1 & 0 \\ 0 & 1 \end{pmatrix}$$

which will find the entries x_1, x_2, x_3, and x_4 of a right inverse of $\begin{pmatrix} 3 & 5 \\ 5 & 8 \end{pmatrix}$.

a) Write down the system of linear equations needed to find x_1, x_2, x_3, x_4.

b) Explain why we can find some of the unknowns x_1, x_2, x_3, x_4 by row reducing the augmented coefficient matrix $\left(\begin{array}{cc|c} 3 & 5 & 1 \\ 5 & 8 & 0 \end{array}\right)$. Which of x_1, x_2, x_3, x_4 can be find by row reducing this matrix?

c) Which of x_1, x_2, x_3, x_4 can be find by row reducing the matrix $\left(\begin{array}{cc|c} 3 & 5 & 0 \\ 5 & 8 & 1 \end{array}\right)$.

d) Suppose E_1, E_2, ..., E_n is a sequence of elementary row operations that allow you to obtain the reduced row echelon form

$$\left(\begin{array}{cc|cc} 3 & 5 & 1 & 0 \\ 5 & 8 & 0 & 1 \end{array}\right) \xrightarrow{E_1} \cdots \xrightarrow{E_n} \left(\begin{array}{cc|cc} 1 & 0 & r_{11} & r_{12} \\ 0 & 1 & r_{21} & r_{22} \end{array}\right)$$

e) What is the significance of the values r_{11}, r_{12}, r_{21}, r_{22}?

Activity 4.22 leads to the following: To find A^{-1}, augmented the matrix A with the identity matrix I of the appropriate size, and perform elementary row operations to reduce

$$(\ A \mid I\ ) \rightarrow (\ I \mid A^{-1}\ )$$

Example 4.10. *Find the inverse of* $A = \begin{pmatrix} 2 & 1 \\ 5 & 7 \end{pmatrix}$.

Solution. *We begin with the multiply augmented matrix*

$$\left(\begin{array}{cc|cc} 2 & 1 & 1 & 0 \\ 5 & 7 & 0 & 1 \end{array}\right)$$

Row reducing:

$$\left(\begin{array}{cc|cc} 2 & 1 & 1 & 0 \\ 5 & 7 & 0 & 1 \end{array}\right) \xrightarrow{2R_2 \to R_2} \left(\begin{array}{cc|cc} 2 & 1 & 1 & 0 \\ 10 & 14 & 0 & 2 \end{array}\right)$$

$$\xrightarrow{-5R_1+R_2 \to R_2} \left(\begin{array}{cc|cc} 2 & 1 & 1 & 0 \\ 0 & 9 & -5 & 2 \end{array}\right) \xrightarrow{9R_1 \to R_1} \left(\begin{array}{cc|cc} 18 & 9 & 9 & 0 \\ 0 & 9 & -5 & 2 \end{array}\right)$$

$$\xrightarrow{-R_2+R_1 \to R_1} \left(\begin{array}{cc|cc} 18 & 0 & 14 & -2 \\ 0 & 9 & -5 & 2 \end{array}\right) \xrightarrow[\frac{1}{9}R_2 \to R_2]{\frac{1}{18}R_1 \to R_1} \left(\begin{array}{cc|cc} 1 & 0 & \frac{7}{9} & -\frac{1}{9} \\ 0 & 1 & -\frac{5}{9} & \frac{2}{9} \end{array}\right)$$

So

$$\begin{pmatrix} 2 & 1 \\ 5 & 7 \end{pmatrix}^{-1} = \begin{pmatrix} \frac{14}{18} & -\frac{2}{18} \\ -\frac{5}{9} & \frac{2}{9} \end{pmatrix}$$

Activity 4.23: More Inverses

A4.23.1 If possible, find the inverses of the given matrices.

a) $\begin{pmatrix} 2 & 7 \\ 3 & 8 \end{pmatrix}$

b) $\begin{pmatrix} 1 & 5 \\ 2 & 3 \end{pmatrix}$

c) $\begin{pmatrix} 0 & 0 \\ 0 & 1 \end{pmatrix}$

d) $\begin{pmatrix} 2 & 6 \\ 1 & 3 \end{pmatrix}$

e) $\begin{pmatrix} 2 & 7 & 1 \\ 3 & 8 & 5 \\ 1 & 4 & 9 \end{pmatrix}$

A4.23.2 Consider the matrices in Activity A4.23.1 that you couldn't find inverses for.

a) Interpret the matrices as geometric transformations.

b) Why is it not surprising that no inverse could be found?

How do inverses and other matrix operations relate?

Activity 4.24: Inverses of Products, Transposes, and Inverses

A4.24.1 Let M_x be the transformation corresponding to reflecting a point across the x-axis, and R_{90° be the transformation corresponding to rotating a point around the origin by an angle of 90° counterclockwise.

a) Describe $M_x R_{90^\circ}$ *geometrically.*

b) Describe M_x^{-1} and $R_{90^\circ}^{-1}$ *geometrically.*

c) Describe $(M_x R_{90^\circ})^{-1}$ *geometrically*, in terms of M_x^{-1} and $R_{90^\circ}^{-1}$. In particular: To "undo" $M_x R_{90^\circ}$, which inverses must be performed, and in what order?

d) What does this suggest about $(AB)^{-1}$ in terms of A^{-1} and B^{-1}?

A4.24.2 Suppose $A^{-1}B = I$.

a) What is the relationship between A^{-1} and B? **Note**: This isn't asking what B *is*; it's asking about the relationship between A^{-1} and B.

b) Since $A^{-1}A = I$, what is the relationship between A and B? Remember: "Things that do the same thing, *are* the same thing."

A4.24.3 Suppose A, B, and AB all have inverses.

a) Suppose $(AB)C = A$. What must C be?

b) Suppose $(AB)(CD) = I$, where C is the matrix you found in Activity A4.24.3. What must D be?

c) What is the relationship between AB and CD?

d) Find AB in terms of A^{-1} and B^{-1}. Then prove your result.

A4.24.4 Suppose N is a $n \times n$ square matrix with inverse N^{-1}.

a) Find two different expressions for $(N^{-1}N)^T$.

b) Find $(N^T)^{-1}$.

Activity A4.24.3 proves:

Theorem 4.7 (Inverse of a Product). *Let A, B be matrices. Provided both sides are defined, $(AB)^{-1} = B^{-1}A^{-1}$.*

We might draw an analogy with wrapping a package: the "transformations" are the steps in the packaging process. To open the packages, we must undo the steps in the reverse of the order they were applied.

Meanwhile Activity A4.24.4 proves:

Theorem 4.8. *Provided the inverse exists, $(A^T)^{-1} = (A^{-1})^T$.*

4.7 Complex Matrices

While we've been focusing on matrices with real entries, our ability to operate with complex numbers (see Section 9.2) means we can also consider matrices with complex entries.

We say that a matrix or vector is *real* if all entries are real; and *complex* if some of the entries are complex numbers. We define the conjugate of a matrix $\overline{A}$ and a vector $\overline{\mathbf{u}}$ to be the matrix or vector of conjugates of the components of the vector or matrix.

Activity 4.25: Complex Matrices

A4.25.1 In the following, assume A, B are matrices, and $\mathbf{u}$, $\mathbf{v}$ are vectors; and all have the appropriate size and/or orientation to make the expression defined.

a) Prove $\overline{AB} = \overline{A}\,\overline{B}$.

b) Prove $\overline{A\mathbf{u}} = \overline{A}\,\overline{\mathbf{u}}$.

A4.25.2 If possible, find real x_i, y_i where the product AB is a real matrix. If not possible, explain why not. Suggestion: It is sufficient to make the complex component of the products equal to 0.

a) $\begin{pmatrix} x_1 + y_1 i & x_2 + y_2 i \end{pmatrix} \begin{pmatrix} 5 - 2i \\ 3 + 4i \end{pmatrix}$

b) $\begin{pmatrix} 2 & 3i \\ i + 1 & 2 \end{pmatrix} \begin{pmatrix} x_1 + y_1 i \\ x_2 + y_2 i \end{pmatrix}$

c) $\begin{pmatrix} 2i & 3 \\ 5 & i \end{pmatrix} \begin{pmatrix} x_{11} + y_{11} i & x_{12} + y_{12} i \\ x_{21} + y_{21} i & x_{22} + y_{22} i \end{pmatrix}$

Activity A4.25.2 shows that, given a matrix with complex entries A, it's generally possible to find an infinite number of matrices B where AB has real entries only. Unfortunately, this means we can't talk about "the" conjugate of a matrix.

But let's consider a different approach. Remember that while we can't generally multiply a matrix A by itself, we can always multiply it by its transpose: AA^T and $A^T A$ always exists. This suggests the following definition:

Definition 4.17 (Conjugate Transpose)**.** *The conjugate transpose of* A*, designated* A^H*, is* $\overline{(A^T)}$.

The H in Definition 4.17 is in honor of Charles Hermite (1822-1901), who proved an important property of matrices with complex entries. We'll discuss Hermite's discovery later; in the meantime, we'll define:

Definition 4.18 (Hermitian). *A matrix is Hermitian if $A = A^H$.*

Example 4.11. *Let* $A = \begin{pmatrix} 2 & 3+2i & -i \\ i & 1-i & 5 \end{pmatrix}$. *Find* A^H.

Solution. *A^H is the conjugate of the transpose of A, so:*

$$\begin{aligned} A^H &= \overline{A^T} \\ &= \overline{\begin{pmatrix} 2 & i \\ 3+2i & 1-i \\ -i & 5 \end{pmatrix}} \\ &= \begin{pmatrix} 2 & -i \\ 3-2i & 1+i \\ i & 5 \end{pmatrix} \end{aligned}$$

Activity 4.26: Hermitian Matrices

A4.26.1 If possible, find a matrix satisfying the given properties. If not possible, explain why not.

a) A is a 2×2 Hermitian matrix, where all entries are real.
b) B is a 2×2 Hermitian matrix, with at least one complex entry.
c) C is a 2×3 Hermitian matrix, where all entries are real.
d) D is a 2×2 Hermitian matrix, with all entries complex.

A4.26.2 Prove/disprove: If the entries of A are real, then A is Hermitian.

A4.26.3 Suppose A is Hermitian.

a) What must be true about the dimensions of A? Explain why.
b) What must be true about the entries along the main diagonal of A? Explain why.
c) What can you say about the number of *complex* entries of A? Explain why.

A4.26.4 For each of the following, find A^H and AA^H. Also identify which matrices are Hermitian.

a) $A = \begin{pmatrix} 2 & 3 & 7 \\ 1 & 4 & 3 \end{pmatrix}$

b) $A = \begin{pmatrix} 2 & 3i & 1+i \\ i & -1 & 2+i \end{pmatrix}$

c) $A = \begin{pmatrix} 3+i & 5-i \\ 5+i & 2+7i \end{pmatrix}$

d) $A = \begin{pmatrix} 7 & 1-2i \\ 1+2i & 2 \end{pmatrix}$

A4.26.5 Prove/disprove: If A is Hermitian, then the entries of A are real.

A4.26.6 Consider the product AA^H.

a) Prove/disprove: AA^H is a real matrix.

b) Prove/disprove: If A is Hermitian, AA^H is a real matrix.

c) Prove/disprove: The diagonal entries of AA^H are real.

5

Vector Spaces

We'll start this section with a riddle: How many mathematicians does it take to screw in a light bulb? Just one: they give it to three philosophers, thereby reducing it to the previous riddle.

This riddle illustrates two things. First, it explains why I'm writing a math book and not doing stand-up comedy. And second, one of the constant goals of mathematics is to try and identify similarities between different types of problems, allowing one problem to be solved in the same way as another, seemingly unrelated problem.

5.1 Vector Spaces

We've already done a little bit of this, when we introduced the concept of a field: *any* set of objects where we can define two operations that meet certain properties is a field. Unfortunately, the only familiar fields are the set of real numbers and the set of complex numbers, using ordinary addition and multiplication.

Since then, we've been introduced to two new objects: vectors and matrices. But while the components of a vector are elements of a field, there seemed to be no easy way to define multiplication of two vectors (Activity 1.7), which means that the vectors don't form a field. And while multiplication is defined for matrices, we found that matrix multiplication was not in general commutative (Activity A3.19.3), which violates one of the requirements for being a field. Thus, neither vectors nor matrices form a field.

If they're not fields, what are they? We might begin by noting that there *is* a type of multiplication that is defined for both vectors and matrices: scalar multiplication. So let's investigate the properties of scalar multiplication.

Activity 5.1: Only So Many Symbols

In the following, we will not use notation like $\mathbf{a}$ or $\vec{a}$ to indicate whether a variable is a vector or a scalar, so unless otherwise indicated, a could be a vector or a scalar. However assume that if a is a vector it is a vector in $\mathbb{R}^n$;

while if a is a scalar, it is a real number. Also assume that the expressions use the standard notation for addition, scalar multiplication, and the dot product.

A5.1.1 Suppose $ab = b$, where b is a vector.

a) Is a a vector or a scalar? How do you know?

b) What must a be in order for $ab = b$?

A5.1.2 Suppose $a + b = a$, where a is a vector.

a) Is b a vector or a scalar?

b) What must b be in order for $a + b = a$?

A5.1.3 Each of the following expressions is a vector. Determine which of a, b, c are scalars, and which are vectors.

a) abc

b) $(a + b)c$

c) $a(b + c)$

In Activity 5.1, we saw that many of the properties of a field can be written as statements in vector arithmetic. However, if $(ab)c$ is a statement about arithmetic in a field, then a, b, c are all elements of the field. On the other hand, if $(ab)c$ is a statement about vector arithmetic, then c is a vector and a, b must be scalars.

If we take these changes into account, we can define a new type of object that includes both vectors and matrices, with operations of addition and scalar multiplication:

Definition 5.1 (Vector Space). *A **vector space** $\mathbf{V}$ over a field $\mathbb{F}$ is a set of objects (called vectors) and a field $\mathbb{F}$, where two operations are defined:*

- *For any two vectors $\mathbf{u}$ and $\mathbf{v}$ in $\mathbf{V}$, the **sum** $\mathbf{u} + \mathbf{v}$ is defined,*
- *For any scalar c in $\mathbb{F}$, the **scalar multiple** $c\mathbf{u}$ is defined.*

In addition, the following properties are required. For any vectors $\mathbf{u}$,$\mathbf{v}$, $\mathbf{w}$ in $\mathbf{V}$, and any scalars a, b, c:

- *Closure under addition: $\mathbf{u} + \mathbf{v}$ is in $\mathbf{V}$.*
- *Closure under scalar multiplication: $c\mathbf{u}$ is in $\mathbf{V}$.*
- *Commutativity of addition: $\mathbf{u} + \mathbf{v} = \mathbf{v} + \mathbf{u}$.*
- *Associativity of addition: $(\mathbf{u} + \mathbf{v}) + \mathbf{w} = \mathbf{u} + (\mathbf{v} + \mathbf{w})$.*
- *Multiplication by 1: For the multiplicative identity 1 in field $\mathbb{F}$, $1\mathbf{u} = \mathbf{u}$.*

- *Zero vector: There exists a* $\mathbf{0}$ *where* $\mathbf{u} + \mathbf{0} = \mathbf{u}$.
- *Additive inverse: For any vector* $\mathbf{u}$*, there is an additive inverse* $-\mathbf{u}$ *where* $\mathbf{u} + (-\mathbf{u}) = \mathbf{0}$.
- *Associativity of scalar multiplication: For all vectors* $\mathbf{u}$ *and all scalars* a, b, $a(b\mathbf{u}) = (ab)\mathbf{u}$.
- *Distributivity of scalar multiplication over vector addition:* $a(\mathbf{u} + \mathbf{v}) = a\mathbf{u} + a\mathbf{v}$.
- *Distributivity of scalar addition over scalar multiplication:* $(a+b)\mathbf{u} = a\mathbf{u} + b\mathbf{u}$.

Note that we designate our "vectors" using boldface type, since they might not be geometric vectors; instead, they could be anything on which addition and scalar multiplication is defined. Also note that we do not require the vector components come from the same field as the scalars.

This is a rather long laundry list of properties, so the requirements for being a vector space are very high. At the same time, anything that *is* a vector space has quite a bit of structure, which means we can do a great deal with them.

Example 5.1. *Consider the set* $\mathbb{Q}$*, the set of rational numbers. Does* $\mathbb{Q}$ *form a vector space over* $\mathbb{R}$*, the set of real numbers, using the ordinary rules of addition and multiplication?*

To decide, we go through our checklist. With some experience, you learn which items to check first.

Solution. *Note that our scalars come from* $\mathbb{R}$*, the set of real numbers, but our vectors are limited to* $\mathbb{Q}$*, the rational numbers.*

A vector space must be closed under scalar multiplication. But if $c = \sqrt{2}$*, then* cq*, where* q *is any rational number, is not a rational number. So* $\mathbb{Q}$ *does not form a vector space over* $\mathbb{R}$.

It's important to understand that no *set* forms a vector space: both the definitions of vector addition and scalar multiplication, and the underlying field, play a role in determining whether a set of vectors *over* a field form a vector space.

Example 5.2. *Does* $\mathbb{Q}$ *form a vector space over* $\mathbb{Q}$*, using the ordinary rules of addition and multiplication?*

Solution. *Let* p, q, r *be elements of* $\mathbb{Q}$*, and let* a, b *be other elements of* $\mathbb{Q}$*. By the standard properties of arithmetic, we know:*

- $p + q$ *is a rational number, so* $\mathbb{Q}$ *is closed under addition.*
- ap *is a rational number, so* $\mathbb{Q}$ *is closed under scalar multiplication.*

- *$p+q=q+p$, so $\mathbb{Q}$ exhibits commutativity of addition.*
- *$(p+q)+r=p+(q+r)$, so $\mathbb{Q}$ exhibits associativity of addition.*
- *1 is the multiplicative identity in $\mathbb{Q}$, and $1p=p$, so $\mathbb{Q}$ satisfies multiplication by 1.*
- *0 is a rational number, and $p+0=p$, so $\mathbb{Q}$ includes a zero vector.*
- *If p is rational, so is $-p$, and $p+(-p)=0$, so $\mathbb{Q}$ includes the additive inverses.*
- *$a(bp)=(ab)p$, so scalar multiplication is associative.*
- *$a(p+q)=ap+aq$, so we have distributivity of scalar multiplication over vector addition,*
- *$(a+b)p=ap+bp$, so we have distributivity of scalar addition over scalar multiplication.*

Since all requirements are met, $\mathbb{Q}$ forms a vector space over $\mathbb{Q}$.

Example 5.3. *Let $\mathbf{V}$ consist of all solutions $\langle x,y\rangle$ be solutions to the equation $5x+3y=15$. Does $\mathbf{V}$ form a vector space over $\mathbb{R}$, using the ordinary rule for vector addition and scalar multiplication?*

Solution. *Suppose two vectors in $\mathbf{V}$ are $\langle x_1,y_1\rangle$ and $\langle x_2,y_2\rangle$. We check closure: their sum is*

$$\langle x_1,y_1\rangle+\langle x_2,y_2\rangle=\langle x_1+x_2,y_1+y_2\rangle$$

To determine if this is in $\mathbf{V}$, we need to determine if it is a solution to $5x+3y=15$. In particular, we need to check if

$$5(x_1+x_2)+3(y_1+y_2)\stackrel{?}{=}15$$

We can always write one side of an equals, so we write:

$$5(x_1+x_2)+3(y_1+y_2)=$$

Since no better option presents itself, we can expand:

$$5(x_1+x_2)+3(y_1+y_2)=5x_1+5x_2+3y_1+3y_2$$

Since $\langle x_1,y_1\rangle$ and $\langle x_2,y_2\rangle$ are vectors in $\mathbf{V}$, we know that $5x_1+3y_1=15$ and $5x_2+3y_2=15$, which gives us:

$$\begin{aligned}&=5x_1+3y_1+5x_2+3y_2\\&=15+15\\&=30\end{aligned}$$

It follows that $5(x_1 + x_2) + 3(y_1 + y_2) \neq 15$, *which means that the sum of two vectors in* $\mathbf{V}$ *is not in* $\mathbf{V}$. *Thus* $\mathbf{V}$ *is not closed under vector addition, so it can't be a vector space.*

Sometimes it's convenient to focus on "part" of a vector space. We define:

Definition 5.2 (Subspace). *Let* $\mathbf{V}$ *a vector space. A subset* $\mathbf{V}'$ *of* $\mathbf{V}$ *is a vector space if* $\mathbf{V}'$ *meets all the requirements of being a vector space for the same operations defined for* $\mathbf{V}$.

Activity 5.2: Vector Spaces and Subspaces

A5.2.1 Determine whether the following are vector spaces.

a) The set of real numbers over the integers, using ordinary addition and multiplication.

b) The set of polynomials over the real numbers, using polynomial addition and multiplication by a real number.

c) The set of 2×2 matrices over the integers, using matrix addition and scalar multiplication.

A5.2.2 In computer graphics, the color of a pixel in an image is assigned a vector with three components, corresponding to the intensity of red, green, and blue light. The intensities range from 0 to 255; this is sometimes called a RGB vector. Using ordinary vector addition and scalar multiplication, do the RGB vectors form a vector space over the real numbers?

A5.2.3 Let $\mathbf{V}$ be a vector space over a field $\mathbb{F}$, and let $\mathbf{U}$ be some set of vectors in $\mathbf{V}$. Suppose we want to determine whether $\mathbf{U}$ is a subspace.

a) Because the vectors in $\mathbf{U}$ are vectors in $\mathbf{V}$, certain properties are "inherited." Which of the requirements for being a vector space are automatically satisfied by the vectors in $\mathbf{U}$ *because* these vectors are also in $\mathbf{V}$?

b) Explain why the only properties for $\mathbf{U}$ that need to be verified are closure and the presence of the additive inverse (for an vector $\mathbf{u}$ in $\mathbf{U}$, the inverse $-\mathbf{u}$ is also in $\mathbf{U}$).

A5.2.4 Consider the vectors in $\mathbb{R}^3$.

a) Show that these vectors form a vector space over the real numbers, using vector addition and scalar multiplication.

b) Consider the vectors of the form $\langle 2, 5, z \rangle$. Do these vectors form a subspace? Why/why not?

c) Consider the vectors of the form $\langle x, y, 0\rangle$. Do these vectors form a subspace? Why/why not?

d) Consider the vectors of the form $\langle x, y, 3x - 2y\rangle$. Do these vectors form a subspace? Why/why not?

A5.2.5 Let $\overrightarrow{v} = \langle v_1, v_2, v_3\rangle$ be vectors in $\mathbb{R}^3$, where the equation

$$v_1 x + v_2 y + v_3 z = 0$$

has solution $x = 3$, $y = 7$ (and z has some other value). Show that these vectors form a vector space under vector addition and scalar multiplication.

A5.2.6 Consider the vectors in $\mathbb{R}^n$, using ordinary vector addition and scalar multiplication.

a) Let $\overrightarrow{v} = \langle v_1, v_2, \ldots, v_n\rangle$. Explain why, in general, $\overrightarrow{v}$ *by itself* will not form a subspace of $\mathbb{R}^n$.

b) What other vectors must be included with $\overrightarrow{v}$ to form a subspace of $\mathbb{R}^n$? Defend your choice.

A5.2.7 Consider the vectors in $\mathbb{R}^n$, using standard vector addition and scalar multiplication. Does the zero vector $\overrightarrow{0}$ by *itself* form a subspace of $\mathbb{R}^n$?

It may not be immediately apparent why vector spaces are important. To understand their significance, we'll have to develop some mathematics.

First, suppose T is a linear transformation whose domain is some vector space $\mathbf{X}$. Is the codomain a vector space? As it turns out ... there's no way we can know! The problem is that we don't know enough about the codomain to verify whether it satisfies or fails the required properties for a vector space: all we know about the codomain is that it *contains* the range.

However, the range is another story: Because the range consists of all possible output vectors, we know that *every* vector in the range had a preimage in the domain. Thus we might see if the range is a vector space.

Activity 5.3: Vector Spaces and the Range

Suppose T is a linear transformation whose domain is some vector space $\mathbf{X}$ over some field $\mathbb{F}$. Let $\mathbf{Y}$ be the *range* (not codomain) of T, and suppose $\mathbf{y}_1$, $\mathbf{y}_2$, $\mathbf{y}_3$ are vectors in the range; let $\mathbf{0}$ be the appropriate zero vector.

In the following it is important to keep in mind that while we assume some form of vector addition $\mathbf{u} + \mathbf{v}$ and scalar multiplication $c\mathbf{v}$ exists for vectors in $\mathbf{Y}$, you may **NOT** assume anything about how these operations are defined.

A5.3.1 Explain why there must be vectors $\mathbf{x}_1, \mathbf{x}_2, \mathbf{x}_3$ where $T(\mathbf{x}_1) = \mathbf{y}_1$, $T(\mathbf{x}_2) = \mathbf{y}_2$, $T(\mathbf{x}_3) = \mathbf{y}_3$.

A5.3.2 The first requirement for being a vector space is that we have two operations: addition and scalar multiplication; and that the set of vectors is closed under these operations.

a) Prove: The range is closed under addition. Suggestion: You can always write down one side of an equality, so begin by writing down $\mathbf{y}_1 + \mathbf{y}_2 = \ldots$. What can you say it's equal to?

b) Prove: The range is closed under scalar multiplication.

A5.3.3 In order for the range to be a vector space, we must also have commutativity and associativity of vector addition.

a) Prove: If $\mathbf{y}_1$, $\mathbf{y}_2$ are vectors in the range, then $\mathbf{y}_1 + \mathbf{y}_2 = \mathbf{y}_2 + \mathbf{y}_1$.

b) Prove: If $\mathbf{y}_1$, $\mathbf{y}_2$, $\mathbf{y}_3$ are vectors in the range, then $(\mathbf{y}_1 + \mathbf{y}_2) + \mathbf{y}_3 = \mathbf{y}_1 + (\mathbf{y}_2 + \mathbf{y}_3)$.

A5.3.4 Let $1_{\mathbf{X}}$ be the multiplicative identity in the domain $\mathbf{X}$. Prove: There is a multiplicative identity in the range $\mathbf{Y}$.

A5.3.5 Let $\mathbf{0}_{\mathbf{X}}$ be the zero vector in the *domain* $\mathbf{X}$. Prove: $T(\mathbf{0}_{\mathbf{X}}) = \mathbf{0}_{\mathbf{Y}}$, where $\mathbf{0}_{\mathbf{Y}}$ is the zero vector in the *range* $\mathbf{Y}$.

A5.3.6 Suppose $T(\mathbf{x}) = \mathbf{y}$.

a) Find $T(-\mathbf{x})$. **NOTE**: You *cannot* use the property $T(a\mathbf{x}) = aT(\mathbf{x})$, since the $-$ in this case is the additive inverse and *not* a scalar multiplier. Suggestion: Find $T(\mathbf{x} + (-\mathbf{x}))$.

b) Why does this prove that if $\mathbf{y}$ is in the range, its additive inverse $-\mathbf{y}$ will also be in the range?

A5.3.7 A vector space also requires associativity of scalar multiplication. Let $\mathbf{y}$ be a vector in the range, and a, b scalars in $\mathbb{F}$. Prove: $a(b\mathbf{y}) = (ab)\mathbf{y}$. Use the following steps.

a) Prove: $a(b\mathbf{y}) = T(a(b\mathbf{x}))$, where $\mathbf{y} = T(\mathbf{x})$.

b) Explain why $T(a(b\mathbf{x})) = T((ab)\mathbf{x})$.

c) Why does this prove that $a(b\mathbf{y}) = (ab)\mathbf{y}$?

A5.3.8 A vector space must also have distributivity. Suppose a, b are elements of $\mathbb{F}$, and $\mathbf{y}_1$, $\mathbf{y}_2$ are in the range of T.

a) Prove $(a + b)\mathbf{y}_1 = a\mathbf{y}_1 + b\mathbf{y}_1$.

b) Prove $a(\mathbf{y}_1 + \mathbf{y}_2) = a\mathbf{y}_1 + a\mathbf{y}_2$.

A5.3.9 Suppose the codomain of T is a vector space. Show that the range is a subspace.

Activity 5.3 leads to:

Theorem 5.1. *Suppose T is a linear transformations whose domain is a vector space. The range of T is also a vector space; and if the codomain is a vector space, the range is a subspace.*

5.2 Kernels and Null Spaces

Let's consider a special element of the range: the zero vector. Every vector space and subspace is required to have a zero vector, so we might begin by wondering which (if any) of the vectors in the domain transform into the zero vector. We define:

Definition 5.3 (Kernel and Nullspace). *The kernel of a transformation T, written Ker T, consists of all elements in the domain whose image in the codomain is the zero vector.*

If the transformation can be represented by a matrix T, the kernel is also called the nullspace of T, and written Null T.

Because not all transformations are linear, kernel is a slightly more general concept. However, since we're focusing on linear transformations, we should use the term nullspace. But to reinforce the idea that they are the same thing, we will make no attempt to be consistent.

Activity 5.4: Null Space

A5.4.1 Consider the transformation T given by

$$x' = 3x - 2y + 4z$$
$$y' = x - 4y + 8z$$

a) Identify the domain and codomain of the transformation.

b) Write the matrix corresponding to the transformation.

c) Write and solve the system of equations is required to solve $T\vec{x} = \vec{0}$.

d) Express the solution in vector form.

A5.4.2 Consider the transformation R given by

$$x' = 2x + 3y$$
$$y' = x + y$$
$$z' = x - 2y$$

a) Identify the domain and codomain of the transformation.

b) Write the matrix corresponding to the transformation.

c) Write and solve the system of equations is required to solve $T\vec{x} = \vec{0}$.

d) Express the solution in vector form.

A5.4.3 Suppose A is a linear transformation. Prove/disprove: There is always at least one vector in the domain whose image is the zero vector.

In Activity 5.4, you showed that given any linear transformation T, there must be at least one vector in the domain that maps to the zero vector in the codomain. In fact, we can specify which vector:

Theorem 5.2 (Image of Zero)**.** *Let T be a linear transformation. Then $T(\mathbf{0}) = \mathbf{0}$, for the appropriate zero vectors.*

There might be other vectors that map to $\mathbf{0}$. To find them, we invoke our strategy of starting with a system of linear equations.

Example 5.4. *Let $T = \begin{pmatrix} 2 & 5 & 1 \\ 3 & 1 & 7 \end{pmatrix}$ be a transformation that acts on vectors with real components. Find the domain and codomain of T, then find all vectors that map to the zero vector in the codomain.*

Solution. *Since T is a 2×3 matrix, this means that the domain is $\mathbb{R}^3$ and the codomain is $\mathbb{R}^2$.*

We want to find vectors $\langle x, y, z\rangle$ where $T\langle x, y, z\rangle = \langle 0, 0\rangle$, so we set up our system of equations

$$\begin{aligned} 2x + 5y + z &= 0 \\ 3x + y + 7z &= 0 \end{aligned}$$

Since this is a homogeneous system, we can work with the coefficient matrix alone. Row reducing

$$\begin{pmatrix} 2 & 5 & 1 \\ 3 & 1 & 7 \end{pmatrix} \xrightarrow{2R_2 \to R_2} \begin{pmatrix} 2 & 5 & 1 \\ 6 & 2 & 14 \end{pmatrix} \xrightarrow{-3R_1 + R_2 \to R_2} \begin{pmatrix} 2 & 5 & 1 \\ 0 & -13 & 11 \end{pmatrix}$$

Parameterizing our solutions gives

$$x = -68t \qquad y = 22t \qquad z = 26t$$

or in vector form, $t\langle -68, 22, 26\rangle$.

Establishing the existence of something is important, because once we know something exists, we can ask about its properties.

Activity 5.5: Properties of the Nullspace

A5.5.1 Let T be a linear transformation, and suppose $\vec{u}$, $\vec{v}$ are in the null space. Let a, b be scalars. Find $T(a\vec{u} + b\vec{v})$.

A5.5.2 Find the nullspace for the given transformation.

a) M_y, the matrix representing the reflection of a point across the y-axis.

b) I, the identity transformation for vectors in $\mathbb{R}^5$.

c) S, the "shadow" transformation that takes a point (x, y, z) in $\mathbb{R}^3$ and maps it to its shadow (x, y) in $\mathbb{R}^2$.

d) T, the transformation represented by $\begin{pmatrix} 2 & 3 & 5 \\ 1 & 4 & 0 \end{pmatrix}$

e) V, the transformation represented by $\begin{pmatrix} 1 & 5 \\ -3 & 8 \\ -2 & -5 \end{pmatrix}$

A5.5.3 Let M be a $n \times m$ matrix representing a linear transformation, where $n < m$. Prove or disprove: There is a nonzero vector in the nullspace.

A5.5.4 Let P be a $n \times m$ matrix with rank $k < m$. Prove or disprove: There is a nonzero vector in the nullspace.

A5.5.5 Let Q be a $n \times m$ matrix with rank m. Prove or disprove: There is a nonzero vector in the nullspace.

In a linear transformation that acts on a vector space $\mathbb{F}^n$, the domain is easy to describe: it is the set of all vectors in $\mathbb{F}^n$. Similarly, the codomain would be easy to describe. But the range itself might be "smaller" than the codomain. We'd like to be able to describe the range in some meaningful fashion. To that end, we'll need to introduce a few more ideas.

5.3 Span

Suppose $\mathbf{u}$, $\mathbf{v}$ are two vectors in a vector space. Because a vector space must be closed under scalar multiplication and vector addition, then every linear combination $a\mathbf{u} + b\mathbf{v}$ must be in the vector space as well. We define:

Definition 5.4 (Span). *Let $\mathcal{V}$ be a set of some of the vectors in a vector space $\mathbf{V}$ over $\mathbb{F}$. The* ***span of*** $\mathcal{V}$*, written Span $\mathcal{V}$, is the set of all linear combinations of vectors in $\mathcal{V}$ with scalars from $\mathbb{F}$.*

Activity 5.6: The Ballad of East and West

Map coordinates are usually given in terms of the four cardinal compass directions: north, south, east, and west. Assume that locations are to be placed on the following map, showing the origin and the cardinal directions:

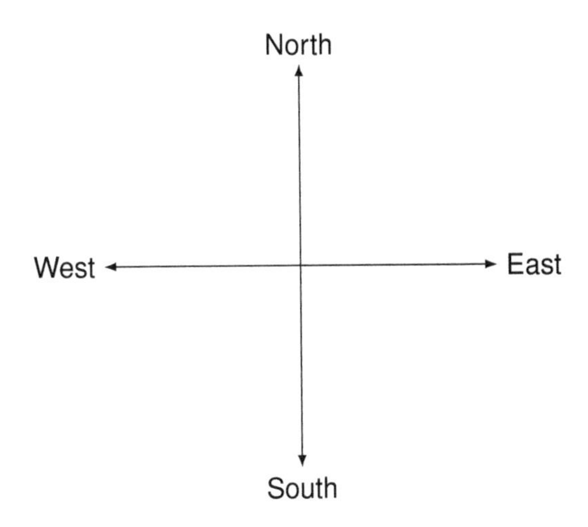

A5.6.1 Locate the following points on the map:

a) East 3, South 5.
b) West 3, South 5.
c) North 3, East 5.
d) West 3, North 5.

A5.6.2 Let $\overrightarrow{N} = \langle 0, 1 \rangle$, $\overrightarrow{S} = \langle 0, -1 \rangle$, $\overrightarrow{E} = \langle 1, 0 \rangle$, $\overrightarrow{W} = \langle -1, 0 \rangle$, and let $\mathcal{R} = \left\{ \overrightarrow{N}, \overrightarrow{S}, \overrightarrow{E}, \overrightarrow{W} \right\}$. Describe each of the points above as a linear combination of the vectors in $\mathcal{R}$.

Because every vector in Span $\mathcal{V}$ is a linear combination of the vectors in $\mathcal{V}$, it follows that we can describe any vector in the span by referencing the scalars that form the linear combination. Since these scalars are from a field, and the order in which we list them makes a difference, we could describe them as a vector. But then we'd have two different types of vectors floating around: the vectors in Span $\mathcal{V}$; and the vectors consisting of the coefficients of the linear combination. Instead, we'll define:

Definition (Coordinates, Provisional). *Suppose* $\mathcal{V} = \{\mathbf{v}_1, \mathbf{v}_2, \ldots, \mathbf{v}_n\}$, *and*

$$\mathbf{x} = x_1\mathbf{v}_1 + x_2\mathbf{v}_2 + \ldots + x_n\mathbf{v}_n$$

We say that $(x_1, x_2, \ldots, x_n)$ *are the* ***coordinates of*** $\mathbf{x}$ ***with respect to*** $\mathcal{V}$

We note two things:

- The coordinates are designated using *parentheses*, to distinguish them from the vectors,
- The definition is provisional, since we'll want to modify it to make it more useful.

Activity 5.7: Coordinates

A5.7.1 Let

$$\vec{v}_1 = \langle 1, 5, 3 \rangle \qquad \vec{v}_2 = \langle 2, -4, 1 \rangle$$

a) Suppose $\vec{x}$ has coordinates (x_1, x_2) with respect to $\mathcal{V} = \{\vec{v}_1, \vec{v}_2\}$. Find $\vec{x}$.

b) How many *components* does $\vec{x}$ have?

A5.7.2 Let

$$\vec{v}_1 = \langle 6, 3 \rangle \qquad \vec{v}_2 = \langle 1, 2 \rangle$$
$$\vec{v}_3 = \langle -2, 5 \rangle \qquad \vec{v}_4 = \langle 1, 1 \rangle$$

a) Suppose $\vec{x}$ has coordinates $(3, 1, -2, 5)$ with respect to $\mathcal{V} = \{\vec{v}_1, \vec{v}_2, \vec{v}_3, \vec{v}_4\}$. Find $\vec{x}$.

b) How many *components* does $\vec{x}$ have?

A5.7.3 Suppose $\mathcal{V}$ is a set of k vectors.

a) Prove/disprove: Any vector in Span $\mathcal{V}$ has k *components.*

b) Prove/disprove: Any vector in Span $\mathcal{V}$ has k *coordinates.*

If you have the coordinates of a vector in Span $\mathcal{V}$, it's easy to determine the actual vector. But what if we have a vector and want to find the coordinates?

Activity 5.8: Column Space

A5.8.1 Suppose $\vec{v}_1 = \langle 2, 5 \rangle$, $\vec{v}_2 = \langle -1, 6 \rangle$, and let $\mathcal{V} = \{\vec{v}_1, \vec{v}_2\}$.

a) Consider $\vec{b} = \langle 5, 12 \rangle$. If it is in Span $\mathcal{V}$, find the coordinates with respect to $\mathcal{V}$; if not, explain why no such coordinates can exist.

b) What vectors $\vec{b} = \langle b_1, b_1 \rangle$ are in Span $\mathcal{V}$? Explain how you know.

A5.8.2 Suppose $\vec{v}_1 = \langle 1, 5, 3 \rangle$ and $\vec{v}_2 = \langle -3, -1, 4 \rangle$, and let $\mathcal{V} = \{\vec{v}_1, \vec{v}_2\}$.

a) Is $\langle 1, 4, -9 \rangle$ in Span $\mathcal{V}$? How do you know?

b) What vectors $\langle b_1, b_2, b_3 \rangle$ are in Span $\mathcal{V}$?

A5.8.3 Suppose $\vec{v}_1 = \langle 1, 5 \rangle$, $\vec{v}_2 = \langle -1, -4 \rangle$, $\vec{v}_3 = \langle 2, 8 \rangle$. Let $\mathcal{V} = \{\vec{v}_1, \vec{v}_2, \vec{v}_3\}$. Find all vectors $\vec{b}$ in Span $\mathcal{V}$.

A5.8.4 Suppose $\mathcal{V} = \{\vec{v}_1, \vec{v}_2, \ldots, \vec{v}_n\}$. How would you go about finding all vectors $\vec{b}$ in Span $\mathcal{V}$?

Activity 5.8 shows the problem of whether a vector is in the Span of $\mathcal{V}$ is exactly the same as the problem of deciding whether a vector is in the range of a linear transformation, where the columns of the transformation matrix are the vectors in $\mathcal{V}$. Since every vector in Span $\mathcal{V}$ is a linear combination of the vectors in $\mathcal{V}$, this leads to the following definition:

Definition 5.5 (Column Space). *Let T be a matrix. The* ***column space of*** *T is Span $\mathcal{V}$, where the vectors of $\mathcal{V}$ are the columns of T.*

Suppose $\mathbf{v}$ is a vector in Span $\mathcal{V}$. Finding its coordinates will be a matter of solving a system of linear equations. However, there are some potentially disturbing features about coordinates.

Activity 5.9: Coordinates

A5.9.1 Let $\vec{v}_1 = \langle 2, 1 \rangle$, $\vec{v}_2 = \langle 1, 4 \rangle$, and let $\mathcal{V} = \{\vec{v}_1, \vec{v}_2\}$.

a) If possible, find the coordinates of $\langle 5, 7 \rangle$ with respect to $\mathcal{V}$; if not, explain why not.

b) Can you find another set of coordinates for $\langle 5, 7 \rangle$ with respect to $\mathcal{V}$? Why/why not?

A5.9.2 Let $\vec{v}_1 = \langle 5, 3 \rangle$, $\vec{v}_2 = \langle 1, -2 \rangle$, $\vec{v}_3 \langle -5, 1 \rangle$, and $\vec{v}_4 = \langle 1, 5 \rangle$, and let $\mathcal{V} = \{\vec{v}_1, \vec{v}_2, \vec{v}_3, \vec{v}_4\}$.

a) If possible, find a set of coordinates for $\langle 2, 7 \rangle$ with respect to $\mathcal{V} = \{\vec{v}_1, \vec{v}_2, \vec{v}_3, \vec{v}_4\}$; if not possible, explain why not.

b) Can you find another set of coordinates for $\langle 2, 7 \rangle$? Why/why not?

Activity 5.9 reveals a problematic feature of our (provisional) definition of coordinates: it's possible for a vector to have more than one set of coordinates.

To see why this happened, let's take a closer look at the span of a set of vectors.

Activity 5.10: Spanning Set

In the following, assume all vectors are in some vector space.

A5.10.1 Let $\vec{v}_1 = \langle 5, 3\rangle$, $\vec{v}_2 = \langle 1, 4\rangle$, $\vec{v}_3 = \langle -2, 5\rangle$, and let $\mathcal{V} = \{\vec{v}_1, \vec{v}_2, \vec{v}_3\}$.

a) Find x_1, x_2, x_3 that solves $\langle 3, 7\rangle = x_1\vec{v}_1 + x_2\vec{v}_2 + x_3\vec{v}_3$.

b) If possible, find a different solution.

c) Since $\langle 3, 7\rangle = \langle 3, 7\rangle$, use the two different solutions to express $\langle 0, 0\rangle$ as a linear combination of $\vec{v}_1$, $\vec{v}_2$, $\vec{v}_3$.

A5.10.2 Let $\mathcal{V} = \{\mathbf{v}_1, \mathbf{v}_2, \ldots, \mathbf{v}_n\}$, and suppose $\mathbf{x}$ has two different sets of coordinates, $(a_1, a_2, \ldots, a_n)$ and $(b_1, b_2, \ldots, b_n)$.

a) Express $\mathbf{x}$ as a linear combination of the vectors in $\mathcal{V}$ using the coordinates $(a_1, a_2, \ldots, a_n)$.

b) Express $\mathbf{x}$ as a linear combination of the vectors in $\mathcal{V}$ using the coordinates $(b_1, b_2, \ldots, b_n)$.

c) Using the fact that both expressions equal $\mathbf{x}$ to find an expression for $\mathbf{0}$ as a linear combination of $\mathcal{V}$. (We'll call this a **linear combination equal to $\mathbf{0}$**)

d) How do you know that some the coefficients in the linear combination equal to $\mathbf{0}$ are not equal to the scalar 0?

A5.10.3 Let $\mathcal{V} = \{\mathbf{v}_1, \mathbf{v}_2, \ldots, \mathbf{v}_n\}$, and suppose

$$\mathbf{0} = a_1\mathbf{v}_1 + a_2\mathbf{v}_2 + \ldots + a_n\mathbf{v}_n$$

where not all of the a_is are equal to 0.

a) Suppose $\mathbf{x}$ has coordinates $(x_1, x_2, \ldots, x_n)$. Find a second set of coordinates for $\mathbf{x}$. Suggestion: What is the significance of the zero vector?

b) Why does this show that any vector $\mathbf{x}$ will have an *infinite* number of sets of coordinates?

Remember that we said a linear combination of vectors was *trivial* if all the scalar multipliers were zero; it is *nontrivial* otherwise. This allows us to state the two main results of Activity 5.10 as follows. Suppose we have some set of vectors $\mathcal{V}$.

- If $\mathbf{x}$ in Span $\mathcal{V}$ has more than one set of coordinates, then there is a nontrivial linear combination equal to $\mathbf{0}$,
- If there is a nontrivial linear combination equal to $\mathbf{0}$, then $\mathbf{x}$ in Span $\mathcal{V}$ has more than one set of coordinates.

Carefully compare the two sentences. Notice that the first has the construction "If A, then B," while the second has the construction "If B, then A." When we swap the order in an "if ... then ... " statement, we produce what is known as the **converse**, and we can combine both into a single statement:

- If $\mathbf{x}$ in Span $\mathcal{V}$ has more than one set of coordinates, then there is a nontrivial linear combination equal to $\mathbf{0}$, and conversely.

Ending the sentence with "and conversely" means that the converse statement is also true. However, while ending a sentence this way may be grammatically correct, it leaves the listener waiting to hear what the converse is, so another way to join a statement and its converse is to use the phrasing "A if and only if B." This gives us the preferred statement of our result:

Theorem 5.3. *Let $\mathcal{V}$ be a set of vectors. A vector $\mathbf{x}$ in Span $\mathcal{V}$ has more than one set of coordinates if and only if there is a nontrivial linear combination of the vectors in $\mathcal{V}$ equal to $\mathbf{0}$.*

5.4 Linear Independence and Dependence

Because it's inconvenient to have multiple sets of coordinates for any given vector, it seems desirable to require the *non*existence of a nontrivial linear combination of vectors equal to $\mathbf{0}$. This leads to:

Definition 5.6 (Independence). *A set of vectors $\mathcal{V}$ is **independent** if there is no nontrivial linear combination equal to $\mathbf{0}$. Otherwise, the vectors are dependent.*

If $\mathcal{V}$ is an independent set of vectors, then any vector $\vec{x}$ in Span $\mathcal{V}$ will have a unique set of coordinates. This allows us to define coordinates usefully:

Definition 5.7 (Coordinates). *Let $\mathcal{V}$ be a set of independent vectors, and $\mathbf{x}$ a vector in Span $\mathcal{V}$. The **coordinates of $\mathbf{x}$ with respect to $\mathcal{V}$** are the coefficients of the unique linear combination equal to $\mathbf{x}$.*

Suppose $\mathcal{V} = \{\mathbf{v}_1, \mathbf{v}_2, \ldots, \mathbf{v}_n\}$, and the coordinates of $\mathbf{x}$ are $(x_1, x_2, \ldots, x_n)$. By definition,

$$\mathbf{x} = x_1\mathbf{v}_1 + x_2\mathbf{v}_2 + \ldots + x_n\mathbf{v}_n$$

We can express this relationship in matrix form as

$$\mathbf{x} = A\begin{pmatrix} x_1 \\ x_2 \\ \vdots \\ x_n \end{pmatrix}$$

where A is a matrix whose columns are the basis vectors $\mathbf{v}_1$, $\mathbf{v}_2$, ..., $\mathbf{v}_n$.

While the notion of independence is a powerful one, it's a little abstract. Moreover, you might wonder why the term *independence* is used. We'll explore that next.

Activity 5.11: Dependence

A5.11.1 Let $\mathcal{V} = \{\mathbf{v}_1, \mathbf{v}_2, \ldots, \mathbf{v}_n\}$.

a) Suppose there is a nontrivial linear combination of the vectors equal to $\mathbf{0}$. Explain why this means you can express at least one of the vectors in $\mathcal{V}$ as a linear combination of the others.

b) Suppose you can express one of the vectors as a linear combination of the others. Explain why this means you can find a nontrivial linear combination of the vectors that is equal to $\mathbf{0}$.

A5.11.2 Let $\vec{v}_1 = \langle -1, 1\rangle$, $\vec{v}_2 = \langle 3, 5\rangle$, $\vec{v}_3 = \langle -2, 10\rangle$

a) Find a nontrivial linear combination of $\vec{v}_1$, $\vec{v}_2$, $\vec{v}_3$ that is equal to $\vec{0}$.

b) If possible, express $\vec{v}_1$ in terms of the other vectors.

c) If possible, express $\vec{v}_2$ in terms of the other vectors.

d) If possible, express $\vec{v}_3$ in terms of the other vectors.

The words "independence" and "dependence" suggest a different relationship among a set of vectors. Suppose we can write one vector in $\mathcal{V}$ as a linear combination of the other vectors in $\mathcal{V}$. It makes sense to say that the vector is dependent on the rest. "There are only so many words," so we'll distinguish (for now) between the two by indicating whether we're talking about a *set* of vectors being dependent, in the sense of Definition 5.6, or an individual *vector* being dependent, in the sense that we can write it in terms of the remaining vectors of our set.

In Activity A5.11.1, you showed:

- If a set of vectors is dependent, then at least one vector in the set can be written as a linear combination of the others,

- If at least one vector in a set can be written as a linear combination of the others, then the set of vectors is dependent.

Previously we wrote paired conditionals using "if and only if," so we might say that a set of vectors is dependent if and only if at least one of the vectors can be written as a linear combination of the others. But since we've already defined dependence of a set of vectors, this means we can give a more natural equivalent definition:

Definition 5.8 (Independence, Equivalent Definition). *A set of vectors is dependent if a vector in the set can be written as a linear combination of the others; it is independent otherwise.*

Activity A5.11.2 points to a complicating factor: we might be able to express *any* of the vectors in terms of the remaining vectors. Consequently, it doesn't make sense to talk about a *vector* being independent or dependent: independence or dependence is a property of the *set* of vectors.

Suppose we have a set of dependent vectors. Since we can write some of the vectors in terms of the others, it follows that we don't need all the vectors. But which ones are redundant?

Activity 5.12: Steps Towards Independence

In the following, let $\mathbf{V}$ be the span of some set of vectors $\mathcal{V}$.

A5.12.1 Suppose $\mathcal{V}$ is a *de*pendent set of vectors, where some vector $\mathbf{v}$ can be written in terms of the remaining vectors. Show that $\mathbf{V}$ is the span of $\mathcal{V}'$, where $\mathcal{V}'$ consists of all vectors in $\mathcal{V}$ *except* $\mathbf{v}$.

A5.12.2 Suppose $\mathcal{V}'$ is a *de*pendent set of vectors, where some vector $\mathbf{u}$ can be written in terms of the remaining vectors. Show that $\mathbf{V}$ is the span of $\mathcal{V}''$, where $\mathcal{V}''$ consists of all vectors in $\mathcal{V}$ *except* $\mathbf{u}$.

A5.12.3 Suppose we continue this process: If our set is dependent, we'll find a vector that can be written in terms of the others, and eliminate it to produce a new set. Explain why we must eventually arrive at a set of vectors that is independent.

A5.12.4 Let $\mathbf{x} = \langle 1, -1, 3, 5 \rangle$, $\mathbf{y} = \langle 2, -2, 6, 10 \rangle$, $\mathbf{z} = \langle 3, 7, -3, -5 \rangle$, $\mathbf{w} = \langle 5, 5, 3, 5 \rangle$, and let $\mathbf{V} = \text{Span } \{\mathbf{x}, \mathbf{y}, \mathbf{z}, \mathbf{w}\}$. Which of the vectors are needed to span $\mathbf{V}$?

Activity 5.12 leads to the following idea. Suppose we have a dependent set of vectors $\mathcal{V}$ that spans some vector space $\mathbf{V}$. Since $\mathcal{V}$ is a dependent set, then at least one of its vectors can be expressed in terms of the others. Thus we can eliminate this vector from $\mathcal{V}$ to produce a smaller set of vectors $\mathcal{V}'$, which still span $\mathbf{V}$. "Lather, rinse, repeat," so if $\mathcal{V}'$ is dependent, we can eliminate another vector, and so on. Eventually, we will arrive at a set of independent vectors that spans $\mathbf{V}$. Hence we define:

Definition 5.9 (Basis). *Let* $\mathbf{V}$ *be some vector space. A* ***basis for*** $\mathbf{V}$ *is a set of independent vectors* $\mathcal{V}$ *whose span is* $\mathbf{V}$.

In some sense, the basis is a "minimal" set of vectors that span the vector space.

As we saw, we can find a basis for the vector space spanned by any given set of vectors. But having to determine which vectors can be expressed as linear combinations of the others, and then checking to see if the remaining vectors are independent, is very tedious. Let's try to find a more efficient approach.

Activity 5.13: Gaining Independence

A5.13.1 In the following, suppose $\mathcal{V} = \{\mathbf{v}_1, \mathbf{v}_2, \ldots, \mathbf{v}_n\}$ is a dependent set of nonzero vectors.

a) Explain how you know there are values x_1, x_2, ..., x_n, not all equal to 0, for which

$$x_1\mathbf{v}_1 + x_2\mathbf{v}_2 + \ldots + x_n\mathbf{v}_n = \mathbf{0}$$

b) Suppose $x_i \neq 0$. Explain why you can solve for $\mathbf{v}_i$ in terms of the other vectors.

A5.13.2 Let $\mathcal{V} = \{\vec{v}_1, \vec{v}_2, \vec{v}_3, \vec{v}_4, \vec{v}_5\}$, where the nontrivial solutions to

$$x_1\vec{v}_1 + x_2\vec{v}_2 + x_3\vec{v}_3 + x_4\vec{v}_4 + x_5\vec{v}_5 = \vec{0}$$

can be found by row reducing a coefficient matrix to

$$\begin{pmatrix} 2 & 3 & -4 & 1 & 5 \\ 0 & 0 & 3 & -5 & 1 \end{pmatrix}$$

a) Find the free and basic variables, and parameterize the solutions to the vector equation.

b) Explain why you can always solve for the vectors corresponding to the free variables.

c) If possible, write a vector equation with exactly one of the vectors corresponding to the free variables. If not possible, explain why not. (The equation can have vectors corresponding to the basic variables.)

d) What does this say about the relationship between the vectors corresponding to the free variables and the vectors corresponding to the basic variables?

e) Find a basis for Span $\mathcal{V}$.

A5.13.3 Let $\vec{x} = \langle 2, 5\rangle$, $\vec{y} = \langle 1, 4\rangle$, $\vec{z} = \langle 1, 1\rangle$, and $\vec{w} = \langle 4, 8\rangle$.

a) Solve

$$x_1\vec{x} + x_2\vec{y} + x_3\vec{z} + x_4\vec{w} = \vec{0}$$

b) Is $\{\vec{x}, \vec{y}, \vec{z}, \vec{w}\}$ an independent set of vectors? Why/why not?

c) Find a basis for the vector space spanned by $\{\vec{x}, \vec{y}, \vec{z}, \vec{w}\}$

Activity 5.13 gives us a strategy for identifying a basis. Suppose $\mathcal{V} = \{\mathbf{v}_1, \mathbf{v}_2, \ldots, \mathbf{v}_n\}$. To find a basis for Span $\mathcal{V}$, look for a solution to

$$x_1\mathbf{v}_1 + x_2\mathbf{v}_2 + \ldots + x_n\mathbf{v}_n = \mathbf{0}$$

The vectors corresponding to the free variables are redundant, since they can always be expressed in terms of the vectors corresponding to the basic variables.

Example 5.5. *Let* $\mathbf{V}$ *be Span* $\{\langle 1,4,5\rangle, \langle -3,5,8\rangle, \langle 4,-1,-3\rangle, \langle 7,-6,-11\rangle\}$. *Find a basis for* $\mathbf{V}$.

Solution. *We look for solutions to*

$$x_1\langle 1,4,5\rangle + x_2\langle -3,5,8\rangle + x_3\langle 4,-1,-3\rangle + x_4\langle 7,-6,-11\rangle = \langle 0,0,0\rangle$$

This corresponds to the system

$$\begin{aligned} x_1 - 3x_2 + 4x_3 + 7x_4 &= 0 \\ 4x_1 + 5x_2 - 1x_3 - 6x_4 &= 0 \\ 5x_1 + 8x_2 - 3x_3 - 11x_4 &= 0 \end{aligned}$$

Since this is a homogeneous system, we can row reduce the coefficient matrix;

$$\begin{pmatrix} 1 & -3 & 4 & 7 \\ 4 & 5 & -1 & -6 \\ 5 & 8 & -3 & -11 \end{pmatrix} \to \begin{pmatrix} 1 & 0 & 1 & 1 \\ 0 & 1 & -1 & -2 \\ 0 & 0 & 0 & 0 \end{pmatrix}$$

So x_3, x_4 *are the free variables. If we let* $x_4 = 1$, $x_3 = 0$, *we find* $x_2 = 2$, $x_1 = -1$, *which gives us*

$$-\langle 1,4,5\rangle + 2\langle -3,5,8\rangle + \langle 7,-6,-11\rangle = \langle 0,0,0\rangle$$

so $\langle 7,-6,-11\rangle = \langle 1,4,5\rangle - 2\langle -3,5,8\rangle$. *If we let* $x_4 = 0$, $x_3 = 1$, *we find* $x_2 = 1$, $x_1 = -1$, *which gives*

$$-\langle 1,4,5\rangle + \langle -3,5,8\rangle + \langle 4,-1,-3\rangle = \langle 0,0,0\rangle$$

so $\langle 4,-1,-3\rangle = \langle 1,4,5\rangle - \langle -3,5,8\rangle$. *Consequently these two vectors are redundant, and our basis consists of the two vectors* $\langle 1,4,5\rangle$ *and* $\langle -3,5,8\rangle$.

There is one disconcerting feature of this approach.

Activity 5.14: Dimension

Let $\mathcal{V} = \{\langle 1,4,5\rangle, \langle -3,5,8\rangle, \langle 4,-1,-3\rangle, \langle 2,8,10\rangle\}$ span vector space $\mathbf{V}$.

A5.14.1 Consider the equation

$$y_1\langle 1,4,5\rangle + y_2\langle 4,-1,-3\rangle + y_3\langle -3,5,8\rangle + y_4\langle 7,-6,-11\rangle = \langle 0,0,0\rangle$$

a) Solve.

b) What does this suggest as a basis for the vector space $\mathbf{V}$?

A5.14.2 Consider the equation

$$z_1\langle 7,-6,-11\rangle + z_2\langle 4,-1,-3\rangle + z_3\langle 1,4,5\rangle + z_4\langle -3,5,8\rangle = \langle 0,0,0\rangle$$

a) Solve.

b) What does this suggest as a basis for the vector space $\mathbf{V}$?

A5.14.3 A useful strategy in mathematical research is to "look for commonalities." While the preceding shows that $\mathbf{V}$ could have several different bases, what common feature do all the bases share?

As Activity 5.14 shows, changing the order we list the vectors will change the basis we obtain. But all the basis we obtained had the same number of basis vectors. This suggests the importance of:

Definition 5.10 (Dimension). *The dimension of a vector space is the number of vectors in a basis.*

However, we should be cautious: In one case, we found that the number of basis vectors was the same, no matter how we listed the vectors. But that's no guarantee that *any* set of vectors will reduce down to the same number. Should we be concerned that the two different bases have *different* numbers of vectors?

Activity 5.15: A Basis Exchange

In the following, suppose $\mathcal{W} = \{\mathbf{w}_1, \mathbf{w}_2, \ldots, \mathbf{w}_n\}$ is a set of n vectors whose span is $\mathbf{V}$, and that $\mathcal{V} = \{\mathbf{v}_1, \mathbf{v}_2, \ldots, \mathbf{v}_k\}$ is an independent set of k vectors in $\mathbf{V}$.

A5.15.1 Consider the equation

$$\mathbf{v}_1 = x_1\mathbf{w}_1 + x_2\mathbf{w}_2 + \ldots x_n\mathbf{w}_n$$

a) Explain why this equation must have a solution $x_1, x_2, \ldots, x_n$, and at least one of the values must be nonzero.

b) A common phrase in mathematical proofs is "without loss of generality." Explain why, "without loss of generality," we can assume the nonzero value is x_1. Suggestion: $\mathbf{w}_1$ is the first vector in $\mathcal{W}$ because we wrote it first. Did we have to write it first?

c) Explain why this means that $\mathbf{w}_1$ is in the span of $\{\mathbf{v}_1, \mathbf{w}_2, \ldots, \mathbf{w}_n\}$.

d) Explain why this means $\{\mathbf{v}_1, \mathbf{w}_2, \ldots, \mathbf{w}_n\}$ also spans $\mathbf{V}$. (We say that we have exchanged $\mathbf{v}_1$ for $\mathbf{w}_1$.)

A5.15.2 Consider the equation

$$\mathbf{v}_2 = x_1\mathbf{v}_1 + x_2\mathbf{w}_2 + \ldots + x_n\mathbf{w}_n$$

a) Explain why this equation must have a solution $x_1, x_2, \ldots, x_n$, and at least one of the values must be nonzero.

b) Explain why x_1 can't be the *only* nonzero value. (In other words, why can't x_2, x_3, ..., x_n all be zero?)

c) Explain why, without loss of generality, we may assume x_2 is nonzero.

d) Explain why $\{\mathbf{v}_1, \mathbf{v}_2, \mathbf{w}_3, \ldots, \mathbf{w}_n\}$ also spans $\mathbf{V}$.

A5.15.3 "Lather, rinse, repeat." Explain why $\{\mathbf{v}_1, \mathbf{v}_2, \mathbf{v}_3, \mathbf{w}_4, \ldots, \mathbf{w}_n\}$ also spans $\mathbf{V}$.

A5.15.4 If $\mathcal{V}$ and $\mathcal{W}$ have different numbers of vectors, one of them will have more.

a) Why do you know that

$$\mathbf{v}_{n+1} = x_1\mathbf{v}_1 + x_2\mathbf{v}_2 + \ldots + x_n\mathbf{v}_n$$

has no solution for any value n?

b) Explain why this requires $k \leq n$. Suggestion: If $k > n$, we can keep exchanging $\mathbf{v}_i$ for $\mathbf{w}_i$ until we run out of $\mathbf{w}_i$. What happens then?

A5.15.5 Suppose $\mathcal{A} = \{\mathbf{a}_1, \mathbf{a}_2, \ldots, \mathbf{a}_k\}$ and $\mathcal{B} = \{\mathbf{b}_1, \mathbf{b}_2, \ldots, \mathbf{b}_n\}$ are two independent sets of vectors that spans $\mathbf{V}$.

a) Since $\mathcal{A}$ spans $\mathbf{V}$ and $\mathcal{B}$ is independent, what do you know about the relationship between k and n?

b) Since $\mathcal{B}$ spans $\mathbf{V}$ and $\mathcal{A}$ is independent, what do you know about the relationship between k and n?

c) What does this say about k and n?

The first part of Activity 5.15 proves the following:

Lemma 1 (Steinitz Exchange Lemma). *Let $\mathcal{V} = \{\mathbf{v}_1, \mathbf{v}_2, \ldots, \mathbf{v}_k\}$ be an independent set of vectors in the vector space $\mathbf{W}$ spanned by $\mathcal{W} = \{\mathbf{w}_1, \mathbf{w}_2, \ldots \mathbf{w}_n\}$. Then:*

- *$k \leq n$,*

- $\{\mathbf{v}_1, \mathbf{v}_2, \ldots, \mathbf{v}_k, \mathbf{w}_{k+1}, \mathbf{w}_{k+2}, \ldots \mathbf{w}_n\}$ *also spans* $\mathbf{W}$ *(possibly after reordering the* $\mathbf{w}_i$*)*

The second part of Activity 5.15 gives an important result:

Theorem 5.4 (Dimension Invariance). *The dimension of a vector space does not depend on how we choose the basis vectors.*

As there's some work involved in finding which vectors of a set are independent, we might wonder if it's worth it. After all, if the *only* advantage of a basis is that the coordinates of a vector are unique, we might be able to live with nonuniqueness: a person could be addressed by several different names, depending on the context, and while it may be inconvenient to have to remember that Chris, Kit, and Red are the same person, the name we use when working with them is irrelevant. So why bother with finding an independent set of vectors?

Activity 5.16: Transformation Basis

Suppose T is a linear transformation with domain $\mathbf{V}$ and codomain $\mathbf{W}$, and let

$$\mathcal{V} = \{\mathbf{v}_1, \mathbf{v}_2, \ldots, \mathbf{v}_n\}$$

be a basis for $\mathbf{V}$, and let

$$\mathbf{x} = x_1\mathbf{v}_1 + x_2\mathbf{v}_2 + \ldots + x_n\mathbf{v}_n$$

A5.16.1 Find the dimension of $\mathbf{V}$.

A5.16.2 Suppose $\mathbf{w}_i = T(\mathbf{v}_i)$ for all i. Find $T(\mathbf{x})$ in terms of $\mathbf{w}_i$.

Activity 5.16 shows that if T is a linear transformation that takes vectors in $\mathbf{V}$ to vectors in $\mathbf{W}$, then T applied to *any* vector in $\mathbf{V}$ can be determined by its effect on the vectors in a basis for $\mathbf{V}$.

We can go a little farther.

Activity 5.17: Nothing Counts

Suppose T is a linear transformation with domain $\mathbf{V}$ and codomain $\mathbf{W}$. Let

$$\mathcal{V} = \{\mathbf{k}_1, \mathbf{k}_2, \ldots, \mathbf{k}_n, \mathbf{v}_1, \mathbf{v}_2, \ldots, \mathbf{v}_m\}$$

be a basis for $\mathbf{V}$, where the $\mathbf{k}_i$s form a basis for the nullspace of T.

A5.17.1 Find $T(x_1\mathbf{k}_1 + x_2\mathbf{k}_2 + \ldots + x_n\mathbf{k}_n)$.

A5.17.2 Suppose $T(\mathbf{v}_i) = \mathbf{w}_i$.

a) Show that the $\mathbf{w}_i$s are independent.

b) Show that the $\mathbf{w}_i$s form a basis for the range of T. Suggestion: Once you've shown they're independent, you need to show that every vector in the range can be expressed as a linear combination of the $\mathbf{w}_i$s.

A5.17.3 Find the following.

a) The dimension of the domain of $\mathbf{V}$.

b) The dimension of the range of $\mathbf{V}$.

c) The dimension of the nullspace of T.

d) The rank of T. Suggestion: How would you find the nullspace of T, and what would its dimension be?

The result of Activity 5.17 can be stated as:

Theorem 5.5 (Rank-Nullity). *Let T be a matrix corresponding to a linear transformation acting on vectors in a n-dimensional vector space. If T has rank r and nullspace with dimension k, then*

$$n = r + k$$

5.5 Change of Basis

In general, given any vector space $\mathbf{V}$, we can find *some* basis. Moreover, our choice of basis vectors does not fundamentally alter any feature of the vector space. However, not all bases are the same: some may be better than others.

Activity 5.18: Good Basis, Bad Basis

A5.18.1 Suppose $\vec{v}_1 = \langle 1, 7 \rangle$, $\vec{v}_2 = \langle 5, 36 \rangle$. Let $\mathcal{V} = \{\vec{v}_1, \vec{v}_2\}$.

a) Show that $\mathcal{V}$ is a basis for vectors in $\mathbb{R}^2$.

b) Find the coordinates of $\langle 23, 165 \rangle$, $\langle 22, 165 \rangle$, and $\langle 23, 166 \rangle$ with respect to $\mathcal{V}$. Suggestion: Use a "double wide" augmented coefficient matrix.

c) In what sense might $\mathcal{V}$ be regarded as a *bad* basis for $\mathbb{R}^2$? Defend your conclusion based on the coordinates you found for the vectors $\langle 23, 165 \rangle$, $\langle 22, 165 \rangle$, and $\langle 23, 166 \rangle$. Suggestion: Notice the given vectors are very nearly equal. How do their coordinates compare?

As Activity 5.18 demonstrates, a basis might give very similar vectors radically different coordinates with respect to the basis. This can be bad, if we're using the coordinates as a way to compare the vectors; however, it turns out we can use this as a way to maintain the confidentiality of private information; we'll take a look at that in Activity 5.35.

For now, the problem we're interested in is this: Suppose we have a basis $\mathcal{V}$ for some set of vector space $\mathbf{V}$. Can we find a different basis $\mathcal{W}$ for the same set of vectors that is, in some sense, a better basis?

The exchange lemma suggests a way to proceed, while illustrating an important part of mathematical process. First, the exchange lemma tells us that if both $\mathcal{V}$ and $\mathcal{W}$ are bases for the same vector space, then we can replace the vectors in one with the vectors in the other and still have a basis. The result is rather useless, because if we replace the vectors in $\mathcal{W}$ with the vectors in $\mathcal{V}$, we get $\mathcal{V}$.

However, if we look at the *proof* of the exchange lemma, we saw that it relied on replacing the vectors one-by-one. In other words, the value of the exchange lemma is *not* the lemma itself, but the process by which we arrived at the result. Thus:

Strategy. *It's the journey, not the destination.*

This suggests a way we can find a new basis for a vector space: we can choose vectors that a linear combinations of the existing vectors and, provided that we obtain a linearly independent set, we'll create a new basis.

Activity 5.19: Change of Basis

A5.19.1 Suppose $\mathcal{V} = \{\mathbf{v}_1, \mathbf{v}_2\}$ and $\mathcal{W} = \{\mathbf{w}_1, \mathbf{w}_2\}$ are two different bases for $\mathbf{V}$.

a) Suppose (x_1, x_2) are the coordinates of $\mathbf{x}$ with respect to $\mathcal{W}$. Express $\mathbf{x}$ in terms of the $\mathbf{w}_i$s.

b) Suppose (y_1, y_2) are the coordinates of $\mathbf{x}$ with respect to $\mathcal{V}$. Express $\mathbf{x}$ in terms of the $\mathbf{v}_i$s.

c) Let T be a transformation that takes the coordinates (x_1, x_2) as input and produces the coordinates (y_1, y_2) as output. Show that T is a linear transformation.

A5.19.2 Suppose

$$\mathbf{w}_1 = a\mathbf{v}_1 + b\mathbf{v}_2$$
$$\mathbf{w}_2 = c\mathbf{v}_1 + d\mathbf{v}_2$$

a) Let (x_1, x_2) be the coordinates of $\mathbf{x}$ with respect to $\mathcal{W}$. Find the coordinates (y_1, y_2) of $\mathbf{x}$ with respect to $\mathcal{V}$.

b) Find the transformation matrix T that takes the coordinates (x_1, x_2) and produces the coordinates (y_1, y_2). In particular, find a matrix T where

$$T\begin{pmatrix} x_1 \\ x_2 \end{pmatrix} = \begin{pmatrix} y_1 \\ y_2 \end{pmatrix}$$

A5.19.3 Suppose $\mathcal{V} = \{\mathbf{v}_1, \mathbf{v}_2, \ldots, \mathbf{v}_n\}$ and $\mathcal{W} = \{\mathbf{w}_1, \mathbf{w}_2, \ldots, \mathbf{w}_n\}$, where

$$\begin{aligned}
\mathbf{w}_1 &= a_{11}\mathbf{v}_1 + a_{21}\mathbf{v}_2 + \ldots + a_{n1}\mathbf{v}_n \\
\mathbf{w}_2 &= a_{12}\mathbf{v}_1 + a_{22}\mathbf{v}_2 + \ldots + a_{n2}\mathbf{v}_n \\
&\vdots \\
\mathbf{w}_n &= a_{1n}\mathbf{v}_1 + a_{2n}\mathbf{v}_2 + \ldots + a_{nn}\mathbf{v}_n
\end{aligned}$$

Find the transformation matrix T that takes the coordinates $(x_1, x_2, \ldots, x_n)$ and produces the coordinates $(y_1, y_2, \ldots, y_n)$. In other words, find matrix T so

$$T\begin{pmatrix} x_1 \\ x_2 \\ \vdots \\ x_n \end{pmatrix} = \begin{pmatrix} y_1 \\ y_2 \\ \vdots \\ y_n \end{pmatrix}$$

Example 5.6. *Suppose $\mathcal{V} = \{\mathbf{v}_1, \mathbf{v}_2\}$ is a basis for a vector space $\mathbf{V}$. Let*

$$\begin{aligned}
\mathbf{w}_1 &= 3\mathbf{v}_1 + 2\mathbf{v}_2 \\
\mathbf{w}_2 &= 5\mathbf{v}_1 + 3\mathbf{v}_2
\end{aligned}$$

If possible, find T that transforms the coordinate (x_1, x_2) of a vector with respect to $\mathcal{V}$ into the coordinates (y_1, y_2) with respect to $\mathcal{W} = \{\mathbf{w}_1, \mathbf{w}_2\}$.

Solution. *If the coordinates with respect to $\mathcal{V}$ are x_1, x_2, and the coordinates with respect to $\mathcal{W}$ are y_1, y_2, then we want*

$$\begin{aligned}
x_1\mathbf{v}_1 + x_2\mathbf{v}_2 &= y_1\mathbf{w}_1 + y_2\mathbf{w}_2 \\
&= y_1(3\mathbf{v}_1 + 2\mathbf{v}_2) + y_2(5\mathbf{v}_1 + 3\mathbf{v}_2) \\
&= (3y_1 + 5y_2)\mathbf{v}_1 + (2y_1 + 3y_2)\mathbf{v}_2
\end{aligned}$$

To make the two sides to be equal, we make the coefficients of $\mathbf{v}_1$, $\mathbf{v}_2$ equal:

$$\begin{aligned}
x_1 &= 3y_1 + 5y_2 \\
x_2 &= 2y_1 + 3y_2
\end{aligned}$$

which we can solve by row reducing

$$\left(\begin{array}{cc|c} 3 & 5 & x_1 \\ 2 & 3 & x_2 \end{array}\right) \rightarrow \left(\begin{array}{cc|c} 1 & 0 & -3x_1 + 5x_2 \\ 0 & 1 & 2x_1 - 3x_2 \end{array}\right)$$

So $y_1 = -3x_1 + 5x_2$, $y_2 = 2x_1 - 3x_2$.

One of the places where a change of basis is useful is computer graphics.

Activity 5.20: Rotations in $\mathbb{R}^3$

To describe rotations in $\mathbb{R}^3$, we specify an **axis of rotation**, typically given as a vector $\overrightarrow{u}$, and an angle of rotation θ.

A5.20.1 Suppose we use $\langle 1, 0\rangle$, $\langle 1, 1\rangle$ as our basis for vectors in $\mathbb{R}^2$.

a) Find (x', y'), the coordinates of (x, y) with respect to this basis.

b) In $\mathbb{R}^2$ using standard coordinate, the image of (x, y) when it is rotated around the origin through an angle of $90°$ is $(-y, x)$. Explain *geometrically* why $(-y', x')$ is **NOT** the image of (x', y') when the point is rotated around the origin through an angle of $90°$. Suggestion: Remember the coordinate are the coefficients of the linear combinations.

A5.20.2 Suppose we use $\langle 5, 0\rangle$, $\langle 0, 3\rangle$ as our basis for vectors $\mathbb{R}^2$.

a) Find (x', y'), the coordinates of (x, y) with respect to this basis.

b) In $\mathbb{R}^2$ using standard coordinate, the image of (x, y) when it is rotated around the origin through an angle of $90°$ is $(-y, x)$. Explain *geometrically* why $(-y', x')$ is **NOT** the image of (x', y') when the point is rotated around the origin through an angle of $90°$.

A5.20.3 The standard basis for $\mathbb{R}^2$ is $\langle 1, 0\rangle$, $\langle 0, 1\rangle$.

a) What features do the standard basis vectors have that are *absent* in the alternate basis vectors of Activity A5.20.1 and Activity A5.20.2?

b) What does this suggest about the choice of basis vectors, if you want to perform a rotation?

A5.20.4 Suppose we want to rotate points by an angle of $90°$ counterclockwise around an axis of rotation specified by $\overrightarrow{u} = \langle 3, -4, 5\rangle$.

a) Explain why this rotation can be described by the rotation of P around a point on a plane perpendicular to $\overrightarrow{u}$.

b) Find any unit vector $\overrightarrow{x} = \langle x_1, x_2, x_3\rangle$ perpendicular to $\overrightarrow{u}$.

c) Find any unit vector $\overrightarrow{y} = \langle y_1, y_2, y_3\rangle$ perpendicular to $\overrightarrow{u}$ and $\overrightarrow{x}$.

d) Let $P = (x, y, z)$, using the standard coordinate system. Find the coordinates of the point with respect to $\{\overrightarrow{u}, \overrightarrow{x}, \overrightarrow{y}\}$.

e) Find the coordinates of (x', y', z'), corresponding to the point P rotated around the axis specified by $\overrightarrow{u}$.

5.6 Orthogonal Bases

As we saw in Activity A5.20.3, having the basis vectors be mutually perpendicular and normal helps to avoid certain "distortions" of the coordinates under a transformation. We also saw that in Activity 5.18, the "wrong" basis can cause two vectors that are in fact very close together can have drastically different coordinates. How can we avoid such distortions?

To begin with, let's focus on the simplest case: the magnitude of a vector. Recall that if $\mathbf{v}$ is a vector in $\mathbb{R}^n$, the magnitude $||\mathbf{x}|| = \sqrt{\mathbf{v} \cdot \mathbf{v}}$. But what if we had the coordinates of the vector with respect to some basis, and not the actual components?

Activity 5.21: Distance Formulas

In the following, suppose $\mathcal{V} = \{\vec{v}_1, \vec{v}_2\}$ is a basis for a set of vectors in $\mathbb{R}^n$.

A5.21.1 Suppose $\vec{v}$ is a vector in Span $\mathcal{V}$. Let (x, y) be the *coordinates* of $\vec{v}$ with respect to $\mathcal{V}$.

a) Find an expression for $\vec{v}$ in terms of x, y and the basis vectors.

b) Find an expression for $|\vec{v}|$ in terms of x, y and the basis vectors.

A5.21.2 Suppose $\mathcal{W} = \{\vec{w}_1, \vec{w}_2, \vec{w}_3\}$ is a basis for a vector space $\mathbf{W}$, and let $\vec{y}$ be a vector in $\mathbf{W}$. Find an expression for $|\vec{y}|$ in terms of its coordinates and the basis vectors.

A5.21.3 Suppose $\mathbf{Z}$ has basis $\mathcal{Z}$ with n basis vectors. Find an expression for the magnitude of a vector in $\mathbf{Z}$ in terms of its coordinates and the basis vectors.

In Activity 5.21, you found that if the basis vectors are $\mathbf{v}_1$, $\mathbf{v}_2$, and the coordinates of a vector $\mathbf{v}$ are (x, y), then

$$||\mathbf{v}|| = \sqrt{x^2 + y^2 + 2xy\mathbf{v}_1 \cdot \mathbf{v}_2}$$

Ideally, the magnitude of the vector should only depend on its coordinates x and y. However, our formula for the magnitude also seems to depend on the dot product $\mathbf{v}_1 \cdot \mathbf{v}_2$. If we want the magnitude to depend *only* on the coordinates x, y, then we need $\mathbf{v}_1 \cdot \mathbf{v}_2 = 0$.

Since two vectors whose dot product is 0 are said to be orthogonal, this leads to:

Definition 5.11. *A basis $\mathcal{V}$ is* ***orthogonal*** *if $\mathbf{v}_i \cdot \mathbf{v}_j = 0$ for all vectors in $\mathcal{V}$. The basis is* ***orthonormal*** *if the vectors are also unit vectors.*

Now let's consider the problem of finding an orthogonal basis for a vector space.

Activity 5.22: Orthogonal Bases

A5.22.1 Let $\mathcal{V} = \{\vec{v}_1, \vec{v}_2, \ldots, \vec{v}_n\}$ be a basis for vector space $\mathbf{V}$.

a) Let $\vec{w}_1 = \vec{v}_1$, and $\vec{w}_2 = x_1\vec{w}_1 + x_2\vec{v}_2$. Show that as long as $x_2 \neq 0$, we can replace $\vec{v}_1$ and $\vec{v}_2$ with $\vec{w}_1$ and $\vec{w}_2$ in $\mathcal{V}$ and still have a set of independent vectors that span $\mathbf{V}$.

b) What equation(s) must x_1, x_2 satisfy in order for $\vec{w}_1$ and $\vec{w}_2$ to be orthogonal?

c) Suppose $\vec{w}_1$, $\vec{w}_2$ are defined as above, and $\vec{w}_3 = y_1\vec{w}_1 + y_2\vec{w}_2 + y_3\vec{v}_3$. Show that as long as $y_3 \neq 0$, we can replace $\vec{v}_1$, $\vec{v}_2$, $\vec{v}_3$ in $\mathcal{V}$ with $\vec{w}_1$, $\vec{w}_2$, $\vec{w}_3$ and still have a set of independent vectors that span $\mathbf{V}$.

d) What equation(s) must y_1, y_2, y_3 satisfy in order for $\vec{w}_1$ to be orthogonal to both $\vec{w}_2$ and $\vec{w}_1$?

A5.22.2 Let $\vec{v}_1 = \langle 2, 5, -3, 4\rangle$, $\vec{v}_2 = \langle 1, 1, 4, -8\rangle$, and $\vec{v}_3 = \langle 0, 0, 1, 5\rangle$.

a) Let $\vec{w}_1 = \vec{v}_1$. Find x_1, x_2 where $\vec{w}_2 = x_1\vec{w}_1 + x_2\vec{v}_2$ is orthogonal to $\vec{w}_1$.

b) Find x_1, x_2, x_3 where $\vec{w}_3 = x_1\vec{w}_1 + x_2\vec{w}_2 + x_3\vec{v}_3$ is orthogonal to both $\vec{w}_1$ and $\vec{w}_2$.

c) Explain why $\vec{w}_1$, $\vec{w}_2$, $\vec{w}_3$ form an orthogonal basis for the vector space spanned by $\vec{v}_1$, $\vec{v}_2$, $\vec{v}_3$.

Activity 5.22 leads to what is known as the **Gram-Schmidt** algorithm to produce an orthogonal basis from any basis for a vector space.

Example 5.7. *Find an orthogonal basis for the vector space spanned by*

$$\mathbf{v}_1 = \langle 5, 3, 1, 4\rangle, \mathbf{v}_2 = \langle 1, 1, 6, 3\rangle, \mathbf{v}_3 = \langle 1, 4, 7, -8\rangle$$

Solution. *The first basis vector will be the first vector in our basis:* $\mathbf{w}_1 = \langle 5, 3, 1, 4\rangle$.

To find a second basis vector, we let

$$\mathbf{w}_2 = x_1\mathbf{w}_1 + x_2\mathbf{v}_2$$

The dot product with $\mathbf{w}_1$ *gives us:*

$$\mathbf{w}_1 \cdot \mathbf{w}_2 = x_1\mathbf{w}_1 \cdot \mathbf{w}_1 + x_2\mathbf{w}_1 \cdot \mathbf{v}_2$$

Since the basis vectors $\mathbf{w}_1$, $\mathbf{w}_2$ *are orthogonal,* $\mathbf{w}_1 \cdot \mathbf{w}_2 = 0$*. We can compute the other dot products directly:*

$$0 = x_1(51) + x_2(26)$$

Since there's nothing gain by working with fractions, we'll find integer solution $x_2 = 51$, $x_1 = -26$, *and so:*

$$\begin{aligned} \mathbf{w}_2 &= -26\mathbf{w}_1 + 51\mathbf{v}_2 \\ &= \langle -79, -27, 280, 49 \rangle \end{aligned}$$

To find a third basis vector, we let

$$\mathbf{w}_3 = x_1\mathbf{w}_1 + x_2\mathbf{w}_2 + x_3\mathbf{v}_3$$

The dot product with $\mathbf{w}_1$ *gives us:*

$$\mathbf{w}_1 \cdot \mathbf{w}_3 = x_1\mathbf{w}_1 \cdot \mathbf{w}_1 + x_2\mathbf{w}_1 \cdot \mathbf{w}_2 + x_3\mathbf{w}_1 \cdot \mathbf{v}_3$$

where our orthogonality assumption tells us $\mathbf{w}_1 \cdot \mathbf{w}_3 = 0$ *and* $\mathbf{w}_1 \cdot \mathbf{w}_2 = 0$*, and we compute the remaining dot product directly:*

$$0 = 51x_1 - 8x_3$$

The dot product of our equation with $\mathbf{w}_2$ *gives us:*

$$\mathbf{w}_2 \cdot \mathbf{w}_3 = x_1\mathbf{w}_2 \cdot \mathbf{w}_1 + x_2\mathbf{w}_2 \cdot \mathbf{w}_2 + x_3\mathbf{w}_2 \cdot \mathbf{v}_3$$

Again our orthogonality assumption gives us$\mathbf{w}_2 \cdot \mathbf{w}_3 = 0$, $\mathbf{w}_2 \cdot \mathbf{w}_1 = 0$*, and we can compute the remaining dot products directly:*

$$0 = 87771x_2 + 1381x_3$$

This gives us the system of equations

$$\begin{aligned} 51x_1 - 8x_3 &= 0 \\ 87771x_2 + 1381x_3 &= 0 \end{aligned}$$

which is already in row echelon form with free variable x_3*. Again, there's little to be gained by working with fractions, so we let* $x_3 = -87771 \times 51 = 4476321$*, giving us* $x_2 = -1381 \times 51 = -70431$*, and* $x_1 = 702168$*, and so*

$$\begin{aligned} \mathbf{w}_3 &= 702168\mathbf{w}_1 - 70431\mathbf{w}_2 - 4476321\mathbf{v}_3 \\ &= \langle 13551210, 21913425, 12315735, -36453015 \rangle \end{aligned}$$

Note that the price for not dealing with fractions is dealing with large integers.

If we wanted to, we could produce an orthonormal basis by normalizing each of the orthogonal vectors.

5.7 Normed Vector Spaces

For vectors in $\mathbb{R}^n$, we were able to define the magnitude of a vector by drawing an analogy with Euclidean geometry: If $\mathbf{v} = \langle v_1, v_2, \ldots, v_n \rangle$, then $||\mathbf{v}|| = \sqrt{v_1^2 + v_2^2 + \ldots + v_n^2}$. "Math ever generalizes," so perhaps we can define magnitudes in other types of vectors spaces. We define:

Definition 5.12. *Given a vector space* $\mathbf{V}$*, a* ***norm*** *is a function that assigns each vector* $\mathbf{u}$ *to a real number* $||\mathbf{u}||$*, where:*

- $||\mathbf{u}|| = 0$ *if and only if* $\mathbf{u} = \mathbf{0}$*, the zero vector in* $\mathbf{V}$*,*
- $||\mathbf{u}|| > 0$ *for all other vectors,*
- $||c\mathbf{u}|| = |c|||\mathbf{u}||$ *for all scalars c,*
- $||\mathbf{u} + \mathbf{v}|| \leq ||\mathbf{u}|| + ||\mathbf{u}||$

A vector space with a norm is a **normed vector space**.

Example 5.8. *Show the Euclidean norm is a norm.*

Solution. *Suppose* $\mathbf{u} = \langle u_1, u_2, \ldots, u_n \rangle$*. Then*

$$||\mathbf{u}|| = \sqrt{u_1^2 + u_2^2 + \ldots + u_n^2}$$

Since the sum of the squares of real numbers is 0 *only if all the numbers are* 0*, then* $||\mathbf{u}|| = 0$ *if and only if* $\mathbf{u} = \mathbf{0}$*. If any of these are nonzero, the sum is positive, and so* $||\mathbf{u}|| > 0$.

We have:

$$\begin{aligned} ||c\mathbf{u}|| &= ||\langle cu_1, cu_2, \ldots, cu_n \rangle|| \\ &= \sqrt{c^2u_1^2 + c^2u_2^2 + \ldots + c^2u_n^2} \\ &= |c|\sqrt{u_1^2 + u_2^2 + \ldots + u_n^2} \\ &= |c|||\mathbf{u}|| \end{aligned}$$

The requirement that $||\mathbf{u} + \mathbf{v}|| \leq ||\mathbf{u}|| + ||\mathbf{u}||$ *follows from the triangle inequality.*

Thus the Euclidean norm meets all the requirements for a norm.

What other norms might exist? Consider two feature vectors, say the grade distributions for two classes or the document vectors for two classes (see Activity 1.2). If we wish to compare them, one potential problem is that the class sizes or the documents might be very different in size; consequently, the components of one feature vector might be much larger than the other. While we could use the Euclidean norm, it's tempting to consider another approach.

Activity 5.23: Another Norm

A5.23.1 Suppose $\vec{v} = \langle 6, 8, 17, 3, 1\rangle$ is a feature vector describing the grade distribution for a class, with the first component giving the number of As; the second component giving the number of Bs; and so on. Let N be the sum of the vector components.

a) Find N.
b) What is the significance of N?
c) Interpret the components of $\frac{1}{N}\vec{v}$.

A5.23.2 Suppose $\vec{v} = \langle 157, 55, 25, 18, 37\rangle$ and $\vec{u} = \langle 12, 6, 3, 9, 8\rangle$ are two document vectors. Let N_1 be the sum of the components of $\vec{v}$ and N_2 be the sum of the components of $\vec{u}$.

a) Find N_1 and N_2.
b) What information does N_1 and N_2 reveal about $\vec{v}$ and $\vec{u}$?
c) Find $\frac{1}{N_1}\vec{v}$ and $\frac{1}{N_2}\vec{u}$.
d) Suppose you the fifth keyword is "bear." Which of the two documents might be more relevant if you are looking for information about bears? Defend your conclusion.

A5.23.3 In Activity 1.11, we described a directed distance in terms of a horizontal and a vertical motion. Suppose you could *only* move horizontally or vertically.

a) Find the vector corresponding to the directed distance from the point $(3, 5)$ to the point $(5, 8)$.
b) Under the assumption you could only move horizontally or vertically, what would be an appropriate magnitude for this vector?
c) Find the vector corresponding to the directed distance from the point $(4, 7)$ to the point $(1, 9)$.
d) Under the assumption you could only move horizontally or vertically, what would be an appropriate magnitude for this vector?
e) What does this suggest about a definition for the magnitude of the vector $\langle v_1, v_2\rangle$, under the assumption that only horizontal or vertical motion is allowed?

Activity 5.23 suggests that the sum of the vector components may be of interest. While feature vectors don't generally have negative components, other vectors might, so we define:

Definition 5.13 (1-Norm)**.** *Let* $\mathbf{v} = \langle v_1, v_2, \ldots, v_n\rangle$ *be a vector in* $\mathbb{R}^n$*. The* 1*-norm of a vector is the value*

$$||\mathbf{v}||_1 = |v_1| + |v_2| + \ldots + |v_n|$$

In Activity A5.23.1, you found that the 1-norm of the vector gave the total number of grades. Thus when each component of the vector was divided by the 1-norm, the resulting value was the frequency of a particular grade. This leads to the following definition:

Definition 5.14 (Frequency Vector). *Let* $\mathbf{v}$ *be a vector. The* ***frequency vector*** *is* $\frac{1}{||\mathbf{v}||_1}\mathbf{v}$.

If our vector represents the grade distribution, the frequency vector corresponds to the percentage of students assigned each grade.

In Activity A5.23.3, you found you could interpret the 1-norm as the distance you'd to travel between two points if you could only move parallel to the coordinate axes. This is the situation we would face if we were trying to go from one point to another in a city, and our travel was limited to the city streets; for this reason, the 1-norm is sometimes called the **Manhattan metric** or the **taxicab distance**.

Activity 5.24: The Secret Life of Norms

Over the past twenty years, a number of matchmaking services have arisen, promising to match a client to an ideal mate. While the exact algorithms used by these companies are proprietary secrets, publicly available information, such as advertising claims like "29 dimensions of compatibility" or submitted patent applications, suggest most rely on the following approach:

- The client's characteristics are converted into a vector $\mathbf{c}_i$. The vector components could include biometric data, such as age and weight; and scale data, such as a client's feelings about pets or children or how frequently they smoke (these values would be recorded as numbers on a scale, for example 0 to 10).
- The client describes an "ideal mate" by describing desirable or undesirable characteristics; these are also converted into a vector $\mathbf{m}$.

A search is then made through all the clients, to find a client $\mathbf{c}_j$ that is closest to $\mathbf{m}$.

A5.24.1 Explain why cosine similarity might *not* be a suitable method to match a client to an ideal mate. In other words, why two vectors $\mathbf{c}_i$, $\mathbf{c}_j$ that have a high cosine similarity might not, in fact, be "close" vectors.

A5.24.2 Explain how you could use the 1-norm to determine how "close" two vectors are.

A5.24.3 Since different aspects of one's personality may be deemed more or less important by the client, a further modification is a **weighted 1-norm**, where a sequence of weights w_i are assigned; the weighted 1-norm would then be

$$||\mathbf{v}||_{1'} = |w_1v_1| + |w_2v_2| + \ldots + |w_nv_n|$$

Show that the weighted 1-norm is a norm.

What if we have vectors in $\mathbb{C}^n$?

Activity 5.25: Complex Norms

Remember that for real numbers, $|a|$ corresponds to the distance of a from 0 on the numberline. If z is a number in the complex plane, define $|z|$ as the distance of z from the origin (see Activity 9.5).

A5.25.1 Remember the complex numbers *include* the real numbers. Show that defining $|z|$ as the distance of z from the origin is consistent with the usual definition of $|a|$, where a is a real number.

A5.25.2 Find the following.

a) $|3i|$
b) $|3 + 4i|$
c) $|12 - 5i|$

A5.25.3 Prove the following. Assume z_1, z_2 are complex numbers

a) $|z_1z_2| = |z_1||z_2|$.
b) $|z_1| = |\overline{z_1}|$.

A5.25.4 Suppose $z = a + bi$.

a) Give a formula for $|z|^2$ in terms of a, b.
b) Give a formula for $|z|^2$ in terms of z.

A5.25.5 Let $\overrightarrow{v} = \langle v_1, v_2, \ldots, v_n \rangle$.

a) Explain why, if the v_is are real numbers, $|v_i|^2 = v_i^2$.
b) What does this suggest about a definition of $||\overrightarrow{v}||$, if the v_is are complex numbers?

For the 1-norm of a vector with complex components, we can still use the notation $|v_1|$, as long as we use the extended definition of the absolute value of a component $|z| = \sqrt{z\overline{z}}$. For the Euclidean norm, we need to redefine it:

Definition 5.15. *Let* $\mathbf{v} = \langle v_1, v_2, \ldots, v_n \rangle$. *Then*

$$||\mathbf{v}|| = \sqrt{|v_1|^2 + |v_2|^2 + \ldots + |v_n|^2}$$

"Math ever generalizes." This redefinition gives us two ways to generalize the idea of a norm. First, our definitions for the 1-norm and the Euclidean norm suggest the following generalization:

Definition 5.16 (k-norm). *Let* $\mathbf{v} = \langle v_1, v_2, \ldots, v_n \rangle$ *be a vector in* $\mathbb{F}^n$. *The* k*-norm is*

$$||\mathbf{v}||_k = \sqrt[k]{|v_1|^k + |v_2|^k + \ldots + |v_n|^k}$$

The other generalization possible is that if we view the v_is as *components* of the vector, then we can extend the concept of norms to *any* vector space.

Activity 5.26: Even More Norms

A5.26.1 Let $\vec{v} = \langle -3, 5, 1 \rangle$.

a) Find $||\vec{v}||_1$.
b) Find $||\vec{v}||_2$.
c) Find $||\vec{v}||_3$.
d) Find $||\vec{v}||_5$.
e) Find $||\vec{v}||_{10}$.
f) Find $||\vec{v}||_{20}$.

A5.26.2 Let A be a $n \times m$ matrix. How might you define $||A||_1$ and $||A||_2$?

Activity 5.26 leads to the following:

Definition 5.17. *Let* $||\mathbf{u}||_p$ *be defined. The* ***infinity norm*** *is*

$$||\mathbf{u}||_\infty = \lim_{p \to \infty} ||\mathbf{u}||_p$$

We also have:

Definition 5.18 (Frobenius Norm). *Let* A *be a matrix. The* ***Frobenius norm*** *is*

$$||A||_F = \sqrt{\sum_{i=1}^{n} \sum_{j=1}^{m} |a_{ij}|^2}$$

(the square root of the sum of the absolute squares of the entries)

It's important to remember that $||\mathbf{v}||$ could refer to any norm, so it's important specify *which* norm we're using in a given context.

5.8 Inner Product Spaces

Can we generalize the concept of the dot product? We'll define:

Definition 5.19 (Inner Product). *Let* $\mathbf{u}$, $\mathbf{v}$, $\mathbf{w}$ *be vectors in vector space* $\mathbf{V}$. *An* ***inner product*** *is a function that assigns each pair* $\mathbf{u}$, $\mathbf{v}$ *of vectors to a scalar, where the inner product* $\langle \mathbf{u}, \mathbf{v} \rangle$ *satisfies the following properties:*

- $\langle \mathbf{u}, \mathbf{u} \rangle = 0$ *if and only if* $\mathbf{u} = \mathbf{0}$, *the zero vector in* $\mathbf{V}$.
- $\langle \mathbf{u}, \mathbf{u} \rangle > 0$ *for all other* $\mathbf{u}$ *in* $\mathbf{V}$.
- $\langle \mathbf{u}, \mathbf{v} \rangle = \overline{\langle \mathbf{v}, \mathbf{u} \rangle}$.
- $\langle \mathbf{u}, \mathbf{v} + \mathbf{w} \rangle = \langle \mathbf{u}, \mathbf{v} \rangle + \langle \mathbf{u}, \mathbf{w} \rangle$.
- $\langle \mathbf{u}, c\mathbf{v} \rangle = c\langle \mathbf{u}, \mathbf{v} \rangle$ *for any scalar* c.

If vector space $\mathbf{V}$ *has an inner product, we call it an* ***inner product space***

Example 5.9. *Show that the dot product of vectors in* $\mathbb{R}^n$ *is an inner product.*

Solution. *Let* $\mathbf{u} = \langle u_1, u_2, \ldots, u_n \rangle$, $\mathbf{v} = \langle v_1, v_2, \ldots, v_n \rangle$, *and* $\mathbf{w} = \langle w_1, w_2, \ldots, w_n \rangle$.

Note that $\mathbf{u} \cdot \mathbf{u} = u_1^2 + u_2^2 + \ldots + u_n^2$. *The only time a sum of squares can be* 0 *is when all of the terms are* 0, *so* $\mathbf{u} \cdot \mathbf{u} = 0$ *if and only if* $\mathbf{u} = \mathbf{0}$, *meeting the first requirement.*

Moreover, if $u_i \neq 0$ *for any* i, $\mathbf{u} \cdot \mathbf{u} > 0$, *meeting the second requirement.*

For the third requirement, suppose $\mathbf{u} \cdot \mathbf{v} = c$. *Since* $\mathbf{u}$, $\mathbf{v}$ *are vectors in* $\mathbb{R}^n$, *then* c *must be a real number. Consequently* $\overline{c} = c$. *Reversing the order of the dot product gives us:*

$$\begin{aligned} \mathbf{v} \cdot \mathbf{u} &= c \\ &= \overline{c} \end{aligned}$$

so we have $\mathbf{u} \cdot \mathbf{v} = \overline{\mathbf{v} \cdot \mathbf{u}}$ *as required.*

For the fourth requirement, we want to show that

$$\mathbf{u} \cdot (\mathbf{v} + \mathbf{w}) = \mathbf{u} \cdot \mathbf{v} + \mathbf{u} \cdot \mathbf{w}$$

This follows from the properties of the dot product (Theorem 1.7).

For the fifth requirement, we have:

$$\mathbf{u} \cdot c\mathbf{v} = c\mathbf{v} \cdot \mathbf{u}$$

by the commutativity of the dot product, then:

$$= c(\mathbf{v} \cdot \mathbf{u})$$

by the scalar associativity property of the dot product. Commutativity gives us:

$$= c(\mathbf{u} \cdot \mathbf{v})$$

and so the dot product satisfies the requirement that $\langle \mathbf{u}, c\mathbf{v} \rangle = c\langle \mathbf{u}, \mathbf{v} \rangle$.

Before we try to find inner products beyond the dot product of vectors with real components, let's consider the properties of inner products.

Activity 5.27: Properties of the Inner Product

In the following, suppose $\langle \mathbf{u}, \mathbf{v} \rangle$ is an inner product.

A5.27.1 Suppose $\langle \mathbf{u}, \mathbf{0} \rangle = k$.

a) Find $\langle \mathbf{u}, c\mathbf{0} \rangle$.
b) Why does this show $\langle \mathbf{u}, \mathbf{0} \rangle = 0$?
c) Find $\langle \mathbf{0}, \mathbf{u} \rangle$.

A5.27.2 Suppose $\langle \mathbf{u}, \mathbf{v} \rangle = a$, $\langle \mathbf{u}, \mathbf{w} \rangle = b$.

a) Express $\langle \mathbf{u}, \mathbf{v} + \mathbf{w} \rangle$ in terms of a and b.
b) Let $\langle \mathbf{v} + \mathbf{w}, \mathbf{u} \rangle = k$. Find a relationship between k, a, and b. (Remember that we *don't* require additivity in the first component!)
c) What does your result say about the relationship between $\langle \mathbf{v}+\mathbf{w}, \mathbf{u} \rangle$ and the sum $\langle \mathbf{v}, \mathbf{u} \rangle + \langle \mathbf{w}, \mathbf{u} \rangle$?

A5.27.3 Suppose $\langle \mathbf{u}, \mathbf{v} \rangle = a$. Find $\langle c\mathbf{u}, \mathbf{v} \rangle$ in terms of a.

Activity 5.27 leads to the following:

Theorem 5.6. *Let* $\langle \mathbf{u}, \mathbf{v} \rangle$ *be an inner product. Then:*

- $\langle \mathbf{v}, \mathbf{0} \rangle = 0$,
- $\langle \mathbf{0}, \mathbf{v} \rangle = 0$,
- $\langle \mathbf{v} + \mathbf{w}, \mathbf{u} \rangle = \langle \mathbf{v}, \mathbf{u} \rangle + \langle \mathbf{w}, \mathbf{u} \rangle$,
- $\langle c\mathbf{u}, \mathbf{v} \rangle = \overline{c}\langle \mathbf{u}, \mathbf{v} \rangle$.

The first important generalization of the inner product will be to vectors in $\mathbb{C}^n$.

Activity 5.28: Inner Products

In the following, suppose $\vec{u} = \langle u_1, u_2, \ldots, u_n \rangle$, $\vec{v} = \langle v_1, v_2, \ldots, v_n \rangle$, where $\vec{u}$, $\vec{v}$ are vectors in $\mathbb{C}^n$.

A5.28.1 If $\vec{u}$, $\vec{v}$ were vectors in $\mathbb{R}^n$, we defined the dot product as the sum of the componentwise products:

$$\vec{u} \cdot \vec{v} = u_1 v_1 + u_2 v_2 + \ldots + u_n v_n$$

In the following, we'll show that if our components are complex numbers, this definition fails *three* of the requirements for an inner product.

a) The first requirement is that the inner product $\langle \mathbf{u}, \mathbf{u} \rangle = 0$ if and only if $\mathbf{u} = \mathbf{0}$. Find a nonzero vector $\mathbf{u}$ in $\mathbb{C}^2$ for which the sum of the componentwise products of the vector with itself is 0.

b) The second requirement is that the inner product $\langle \mathbf{u}, \mathbf{u} \rangle > 0$ for all other vectors. Find a nonzero vector $\mathbf{u}$ in $\mathbb{C}^2$ for which the sum of the componentwise products of the vector with itself is less than 0.

c) The third requirement is that the inner product of $\langle \mathbf{u}, \mathbf{v} \rangle = \overline{\langle \mathbf{v}, \mathbf{u} \rangle}$. Find vectors $\mathbf{u}$, $\mathbf{v}$ where the sum of the componentwise products of the two vectors fails this requirement.

A5.28.2 Suppose instead we define the dot product as

$$\vec{u} \cdot \vec{v} = u_1 \overline{v}_1 + u_2 \overline{v}_2 + \ldots + u_n \overline{v}_n$$

a) Explain why, if $\vec{u}$, $\vec{v}$ are vectors in $\mathbb{R}^n$, this new definition of the dot product is the same as our original definition.

b) Prove that this definition of the dot product satisfies the requirements of an inner product for vectors in $\mathbb{C}^n$.

Activity 5.28 means that we should revise our definition of the dot product:

Definition 5.20 (Dot Product, Revised). *Let* $\mathbf{u} = \langle u_1, u_2, \ldots, u_n \rangle$, $\mathbf{v} = \langle v_1, v_2, \ldots, v_n \rangle$ *be vectors in* $\mathbb{F}^n$, *where* $\mathbb{F}$ *is either the field of real numbers or the field of complex numbers. The dot product is defined as*

$$\vec{u} \cdot \vec{v} = u_1 \overline{v}_1 + u_2 \overline{v}_2 + \ldots + u_n \overline{v}_n$$

This definition incorporates our original definition of the dot product and expands it to vectors with complex components in a way that makes it an inner product.

On the one hand, our ability to expand the dot product to handle vectors with complex components is good. But this required altering the definition of the dot product, which means that all of our previously established theorems about the dot product might be compromised.

Activity 5.29: Complexities of the Dot Product

In the following, assume the vectors have complex components.

A5.29.1 Prove/disprove: For any real *or* complex value c, $(c\mathbf{u}) \cdot \mathbf{v} = c(\mathbf{u} \cdot \mathbf{v})$.

A5.29.2 Prove/disprove: $\mathbf{u} \cdot (\mathbf{v} + \mathbf{w})$.

A5.29.3 Prove/disprove: $\mathbf{u} \cdot \mathbf{v} = \mathbf{v} \cdot \mathbf{u}$.

Activity 5.29 shows that while distributivity and associativity of scalar multiplication hold for the generalized dot product, commutativity does *not*.

What if our vector components aren't real or complex numbers?

Activity 5.30: More Inner Products

A5.30.1 Suppose f, g are continuous functions of x with domain $\mathbb{R}$ and codomain $\mathbb{R}$. Let $\langle f, g\rangle = \int\limits_0^1 f(x)g(x)\,dx$. Prove: $\langle f, g\rangle$ is an inner product.

A5.30.2 Suppose A, B are $m \times n$ matrices in $\mathbb{F}^n$, where $\mathbb{F}$ is either $\mathbb{R}$ or $\mathbb{C}$.

a) Can we define the inner product $\langle A, B\rangle = AB$, the product of the two matrices? Why/why not?

b) Can we define the inner product $\langle A, B\rangle = \sum\limits_{i=1}^{m}\sum\limits_{j=1}^{n} a_{ij}\bar{b}_{ij}$ (the sum of the componentwise products of the matrix entries)? Why/why not?

From Theorem 1.5, we know we can express the magnitude of a vector in $\mathbb{R}^n$ using the dot product as: $||\mathbf{v}|| = \sqrt{\mathbf{v} \cdot \mathbf{v}}$. This suggests:

Definition 5.21. *Given an inner product space* $\mathbf{V}$ *with inner product* $\langle \mathbf{u}, \mathbf{v}\rangle$*, the* ***induced norm*** *is*

$$||\mathbf{u}|| = \sqrt{\langle \mathbf{u}, \mathbf{u}\rangle}$$

Activity 5.31: Induced Norms

A5.31.1 Let A, B be two $m \times n$ matrices. Show that the Frobenius norm is the norm induced by the inner product $\langle A, B\rangle = \sum\limits_{i=1}^{m}\sum\limits_{j=1}^{n} a_{ij}\bar{b}_{ij}$.

A5.31.2 Let $\langle \mathbf{u}, \mathbf{v}\rangle$ be an inner product, and $||\mathbf{u}|| = \sqrt{\langle \mathbf{u}, \mathbf{u}\rangle}$, the induced norm. Define $f(x) = ||\mathbf{u} + x\mathbf{v}||^2$, where x is a real number. Prove $\langle \mathbf{u}, \mathbf{v}\rangle^2 \leq ||\mathbf{u}||^2||\mathbf{v}||^2$.

A5.31.3 Let $\langle \mathbf{u}, \mathbf{v} \rangle$ be an inner product, and $||\mathbf{u}|| = \sqrt{\langle \mathbf{u}, \mathbf{u} \rangle}$, the induced norm.

a) Prove: $||\mathbf{u}|| = 0$ if and only if $\mathbf{u} = \mathbf{0}$.

b) Prove: $||\mathbf{u}|| > 0$ for all other vectors $\mathbf{u}$.

c) Prove: $||c\mathbf{u}|| = |c|\,||\mathbf{u}||$ for all scalars c.

d) Prove: $||\mathbf{u} + \mathbf{v}|| \leq ||\mathbf{u}|| + ||\mathbf{v}||$.

Activity A5.31.2 generalizes the Cauchy-Bunyakovsky-Schwarz inequality for inner products in general:

Theorem 5.7 (Cauchy-Bunyakovsky-Schwarz). *Let $\langle \mathbf{u}, \mathbf{v} \rangle$ be an inner product with induced norm $||\mathbf{u}|| = \langle \mathbf{u}, \mathbf{v} \rangle$. Then for any vectors $\mathbf{u}, \mathbf{v}$,*

$$\langle \mathbf{u}, \mathbf{v} \rangle^2 \leq ||\mathbf{u}||^2 ||\mathbf{v}||^2$$

We can then prove (Activity A5.31.3):

Theorem 5.8. *The induced norm is a norm.*

Because the Cauchy-Bunyakovsky-Schwarz theorem holds for inner products, and the induced norm is a norm, we can define the angle between two vectors:

Definition 5.22. *Let $\mathbf{V}$ be an inner product space. Given two vectors $\mathbf{u}$, $\mathbf{v}$, the angle θ between them is the real value such that*

$$\cos\theta = \frac{\langle \mathbf{u}, \mathbf{v} \rangle}{||\mathbf{u}||\ ||\mathbf{v}||}$$

While it's obvious what we mean by the angle between two vectors in $\mathbb{R}^n$, it's not as clear what we mean when our vectors are in some more abstract vector space. However, we can focus on one important possibility. When two vectors in $\mathbb{R}^n$ had dot product 0, we said the vectors were orthogonal. Even if we don't have a geometric meaning of orthogonality in more abstract vector spaces, we can still speak of orthogonal vectors.

Activity 5.32: Orthogonal Functions

As long as we can determine whether two vectors are orthogonal, we can find an orthogonal basis for a vector space.

A5.32.1 Consider the set of functions $\mathbf{F}$ of the form $f(x) = a_0 + a_1 + a_2x^2$, where the a_is are real numbers.

a) Show that the $\mathbf{F}$ is an inner product space under ordinary polynomial addition and scalar multiplication by a real number, where $\langle f,g\rangle = \int_0^1 f(x)g(x)\,dx$.
b) Would $\mathbf{F}$ still be an inner product space if we defined $\langle f,g\rangle = \int_{-1}^1 f(x)g(x)\,dx$?
c) Find a basis for $\mathbf{F}$.

A5.32.2 Consider the vector space $\mathbf{F}$ from Activity A5.32.1, with inner product defined as $\langle f,g\rangle = \int_0^1 f(x)g(x)\,dx$. We wish to produce an orthogonal basis $\{\mathbf{v}_1, \mathbf{v}_2, \mathbf{v}_3\}$.

a) Show that $\mathcal{F} = \{1, x, x^2\}$ is a basis for $\mathbf{F}$.
b) Find an orthogonal basis for $\mathbf{F}$.

A5.32.3 Consider the vector space $\mathbf{F}$ from Activity A5.32.1, but this time define the inner product defined as $\langle f,g\rangle = \int_{-1}^1 f(x)g(x)\,dx$.

a) Consider the basis $\{\mathbf{v}_1, \mathbf{v}_2, \mathbf{v}_3\}$ that you found in Activity A5.32.2. Prove/disprove: This is still a basis for $\mathbf{F}$.
b) Find an orthogonal basis for $\mathbf{F}$ using the new definition of inner product.

5.9 Applications

As before, we might want to compare two feature vectors to see which are most similar. For example, we might want to compare the document vector for a search query "What is a feature vector?" to the document vectors for each of a billion internet pages, and put the pages that are most similar to the query at the top of our search results.

Theorem 1.11 states that the greater the dot product of normalized vectors, the more similar the vectors. However, normalizing a vector requires dividing by the square root of the sum of the components. Might we be able to use the frequency vectors instead of the normalized vectors?

Activity 5.33: Dot Products and Frequency Vectors

In the following, let $\vec{u} = \langle 6, 4\rangle$ and $\vec{v} = \langle 8, 2\rangle$.

A5.33.1 Find $\vec{\hat{u}}$, $\vec{\hat{v}}$, the corresponding frequency vectors.

A5.33.2 Which vector, $\vec{u}$ or $\vec{v}$, should be "most similar" to $\vec{u}$?

A5.33.3 Find $\vec{u} \cdot \vec{u}$ and $\vec{u} \cdot \vec{v}$.

A5.33.4 Assess the claim: "The greater the dot product of the *frequency* vectors, the more similar the two vectors."

Activity 5.33 shows that we *can't* simply use the dot product of frequency vectors to measure similarity.

An important use of a basis change involves printing color images.

Activity 5.34: Color Images

A digital image can be viewed as a collection of colored points called pixels (probably a contraction of "picture element"). Each pixel has a color specified by a vector $\langle r, g, b \rangle$, which identify the intensity of red, green, and blue in the pixel, where 0 is "zero intensity" or off, with higher values corresponding to greater intensity; this is usually referred to as RGB color. Some examples:

- A pixel with color $\langle 0, 0, 0 \rangle$ has 0 intensity in all colors: it appears black.
- A pixel with color $\langle 255, 255, 255 \rangle$ has equal and maximal intensity in all colors: it appear white.

In practice, the actual values of the components are limited to the range from 0 to 255, so strictly speaking the color is not a true vector. However, it's convenient to treat it as one, so in the following, disregard the limitation to this range.

A5.34.1 If all three intensities are equal, the pixel will appear as some shade of gray (with black at one extreme and white at the other).

a) Explain why the shades of gray correspond to a one-dimensional vector space. What is the basis vector?

b) A "grayscale image" is an image whose colors correspond to different shades of gray. Explain a grayscale picture can be represented by a matrix.

A5.34.2 Suppose the intensities can be any value, including negative and fractional values.

a) Explain why color is a three-dimensional vector space.

b) What is the simplest possible basis?

c) Express the color $\langle 15, 150, 87 \rangle$ as a linear combination of the basis vectors.

A5.34.3 In order to produce more realistic colors, some manufacturers sell products that display images in RGBY color, where the color of a pixel is produced by varying intensities of red, green, blue, and yellow. The simplest basis for RGBY color space is $r = \langle 1, 0, 0, 0 \rangle$, $g = \langle 0, 1, 0, 0 \rangle$, $b = \langle 0, 0, 1, 0 \rangle$, and $y = \langle 0, 0, 0, 1 \rangle$; and the simplest basis for RGB color space is $r = \langle 1, 0, 0 \rangle$, $g = \langle 0, 1, 0 \rangle$, and $b = \langle 0, 0, 1 \rangle$. In order to produce a RGBY image from a RGB image, some transformation T must be applied to a RGB color vector to produce a RGBY color vector.

a) What should the values of $T\langle 1, 0, 0 \rangle$, $T\langle 0, 1, 0 \rangle$, and $T\langle 0, 0, 1 \rangle$ be? Defend your conclusion.

b) In color theory, yellow is a combination of red and green in equal amounts. Find $\overrightarrow{x}$ so that $T\overrightarrow{x} = \langle 0, 0, 0, 1 \rangle$.

c) Can T be a linear transformation? Why/why not?

d) Is it true that a picture displayed on a RGBY devices will be "more colorful" than the same picture displayed on a RGB devices? Why/why not? (In particular: RGB images can product $256 \times 256 \times 256 = 16,777,216$ distinct colors. Can RGBY devices produce more colors from the same image?)

A5.34.4 The color of a pixel in a *printed* images must be specified using a different vector $\langle c, m, y \rangle$, where c, m, and y represent the density of cyan, magenta, yellow, and black inks; this is known as CMYK color. In color theory, cyan is a mixture of equal parts of green and blue; magenta is a mixture of equal parts of red and blue; and yellow is a mixture of equal parts of red and green.

a) Find a linear transformation that takes CMY coordinates (x, y, z) into RGB coordinates (x', y', z'). Then find the RGB color of a pixel with CMY coordinates $(75, 85, 45)$.

b) Find a linear transformation that takes RGB coordinates (x', y', z') into CMY coordinates (x, y, z). Then find the CMY color of a pixel with RGB coordinates $(140, 80, 114)$.

A5.34.5 Suppose (x, y, z) are the coordinates of a pixel's RGB color, and (x', y', z') are the coordinates of the pixel's CMY color.

a) Find the CMY coordinates (x', y', z') for a "pure red" $(x, 0, 0)$.

b) What difficulties arise when trying to produce a "pure red" from CMY color?

c) Do these difficulties also arise when trying to produce a "pure green" or "pure blue"?

A5.34.6 To help resolve this issue, a fourth color (black) is used to produce CMYK color. To compute the CMYK values, the following process is used. First, we compute

$$\begin{aligned} x &= 255 - r \\ y &= 255 - g \\ z &= 255 - b \end{aligned}$$

Then we compute:

$$\begin{aligned} k &= \text{Minimum value of } x, y, z \\ c &= x - k \\ m &= y - k \\ y &= z - k \end{aligned}$$

Suppose C is a transformation that acts on (x, y, z) to produce (c, m, y, k). Prove/disprove: C is a linear transformation.

Ordinarily, we consider all linear combinations of the basis vectors; this gives us the span, and allows us to treat our vectors as living in some vector space. But what if we require our scalar multipliers to be integers? We define:

Definition 5.23 (Lattice). *Let $\mathcal{V}$ be a set of vectors. The linear combinations of vectors in $\mathcal{V}$ whose coefficients are integers forms the lattice over $\mathcal{V}$.*

We usually say that $\mathcal{V}$ generates the lattice.

Activity 5.35: Lattices

A5.35.1 Let $\mathcal{V}$ be a set of vectors over $\mathbb{F}^n$, and let $\Lambda_{\mathcal{V}}$ be a lattice over $\mathcal{V}$.

a) Prove/disprove: $\Lambda_{\mathcal{V}}$ is closed under vector addition.
b) Prove/disprove: $\Lambda_{\mathcal{V}}$ is closed under scalar multiplication (where our scalars are integers).
c) Prove/disprove: $\Lambda_{\mathcal{V}}$ is a vector space.

A5.35.2 Let $\vec{v}_1 = \langle 3, 5\rangle$ and $\vec{v}_2 = \langle 5, 8\rangle$. Prove/disprove: If y_1, y_2 are integers, there are integers x_1, x_2 where

$$\langle y_1, y_2\rangle = x_1\vec{v}_1 + x_2\vec{v}_2$$

A5.35.3 Let $\vec{v}_1 = \langle 3, 5\rangle$ and $\vec{v}_2 = \langle 8, 20\rangle$. Prove/disprove: If y_1, y_2 are integers, there are integers x_1, x_2 where

$$\langle y_1, y_2\rangle = x_1\vec{v}_1 + x_2\vec{v}_2$$

A5.35.4 Let $\vec{v}_1 = \langle a_{11}, a_{21}\rangle$, $\vec{v}_2 = \langle a_{12}, a_{22}\rangle$. If y_1, y_2 are integers, what must be required of the a_{ij}s in order for there to be integer solutions x_1, x_2 where

$$\langle y_1, y_2\rangle = x_1\vec{v}_1 + x_2\vec{v}_2$$

Lattices are most useful when the lattice vectors do *not* include all vectors with integer components. Of particular importance is having two sets, $\mathcal{V}$ and $\mathcal{W}$, which span the same lattice.

Activity 5.36: More Lattices

A5.36.1 Suppose $\mathcal{V} = \{\vec{v}_1, \vec{v}_2\}$ and $\mathcal{W} = \{\vec{w}_1, \vec{w}_2\}$ are independent sets of vectors, with

$$\begin{aligned}\vec{w}_1 &= a_{11}\vec{v}_1 + a_{12}\vec{v}_2\\ \vec{w}_2 &= a_{21}\vec{v}_1 + a_{22}\vec{v}_2\end{aligned}$$

where the a_{ij}s are integers.

a) Let $\vec{x}$ have integer coordinates (x_1, x_2) with respect to $\mathcal{W}$. Does $\vec{x}$ has integer coordinates with respect to $\mathcal{V}$? Why/why not?
b) Let $\vec{y}$ have integer coordinates (y_1, y_2) with respect to $\mathcal{V}$. Does $\vec{y}$ has integer coordinates with respect to $\mathcal{W}$? Why/why not?
c) What must be required for $\vec{z}$ to have integer coordinates with respect to both $\mathcal{V}$ *and* $\mathcal{W}$?

A useful feature of lattices is that with a "bad" basis, vectors that are very nearly equal will have radically different coordinates. This leads to an entirely new method of concealing information.

Activity 5.37: Lattice Cryptography

To implement a **lattice cryptosystem**, we use a "good" basis $\mathcal{V}$, and a "bad basis" $\mathcal{W}$, both spanning the same lattice $\mathcal{L}$. The bad basis is known to everybody; the good basis is known only to the intended recipient.

In the following, let the "good basis" be $\mathcal{V}$, where $\vec{v}_1 = \langle 1, 11\rangle$, $\vec{v}_2 = \langle 12, 1\rangle$; and the "bad basis" be $\mathcal{W}$, where $\vec{w}_1 = \langle 85, 18\rangle$, $\vec{w}_2 = \langle 437, 91\rangle$.

A5.37.1 Show that if $\vec{x}$ has integer coordinates with respect to $\mathcal{V}$, it will also have integer coordinates with respect to $\mathcal{W}$.

A5.37.2 Show that if $\vec{x}$ has integer coordinates with respect to $\mathcal{W}$, it will also have integer coordinates with respect to $\mathcal{V}$.

A5.37.3 Let (x, y) be a set of coordinates with respect to $\mathcal{W}$. We'll call the vector $\vec{p}$ corresponding to those coordinates the *plaintext vector.*

a) Find the plaintext vector corresponding to the coordinates $(16, 74)$ with respect to $\mathcal{W}$.
b) Change one of the components of the plaintext vector *slightly* (add or subtract 1): we'll call this the encrypted vector. (This is the vector that will be communicated.)
c) Find the coordinates of the encrypted vector with respect to $\mathcal{W}$.
d) Since the coordinates of any vector in the lattice must be integers, what does this suggest for the coordinates of the plaintext vector with respect to $\mathcal{W}$?
e) Find the coordinates of the encrypted vector with respect to $\mathcal{V}$.
f) Since the coordinates of any vector in the lattice must be integers, what does this suggest for the coordinates of the plaintext vector with respect to $\mathcal{V}$?
g) What does this suggest about the plaintext vector itself?
h) Remember that only the intended recipient knows the good basis $\mathcal{V}$. Explain why the fact that *only* the intended recipient knows the basis $\mathcal{V}$ means that *only* the intended recipient can correctly recover the coordinates of the plaintext vector with respect to $\mathcal{W}$.

Naturally, any method of encrypting information invites the creation of a method of breaking the encryption. In this case, an attack on lattice cryptography relies on constructing a quasiorthogonal basis for the lattice: a set of basis vectors that are "as orthogonal as possible."

Activity 5.38: Quasiorthogonal Basis

Let $\mathcal{V} = \{\mathbf{v}_1, \mathbf{v}_2, \ldots, \mathbf{v}_n\}$ be a "good" basis for a lattice $\mathcal{L}$ while $\mathcal{W} = \{\mathbf{w}_1, \mathbf{w}_2, \ldots, \mathbf{w}_n\}$ be a "bad" basis for the lattice. One of the distinguishing features between a "good" basis and a "bad" basis is that in a good basis, the vectors are nearly orthogonal; while in a bad basis, the vectors are nearly parallel. Thus $\langle 1, 91\rangle$, $\langle -92, 1\rangle$ is a good basis, since the vectors are nearly orthogonal; but $\langle 1598, -12071\rangle$, $\langle -757, 5713\rangle$ are not.

A5.38.1 Explain why you *can't* in general produce an orthogonal basis for a lattice.

A5.38.2 Let $\mathbf{v}_2' = \mathbf{v}_2 + x_1\mathbf{v}_1$, where x_1 is an integer.

a) Prove: If $\mathbf{x}$ can be expressed as a linear combination of $\mathbf{v}_1$, $\mathbf{v}_2$, ..., $\mathbf{v}_n$ with integer coefficients, then $\mathbf{x}$ can be expressed as a linear combination of $\mathbf{v}_1$, $\mathbf{v}_2'$, ..., $\mathbf{v}_n$ with integer coefficients.

b) Prove: If $\mathbf{x}$ can be expressed as a linear combination of $\mathbf{v}_1, \mathbf{v}'_2, \ldots, \mathbf{v}_n$ with integer coefficients, then $\mathbf{x}$ can be expressed as a linear combination of $\mathbf{v}_1, \mathbf{v}_2, \ldots, \mathbf{v}_n$ with integer coefficients.

c) Why does this mean we can replace $\mathbf{v}_2$ with $\mathbf{v}'_2$ in the basis for our lattice?

A5.38.3 Consider the lattice spanned by $\langle -2164, 13736 \rangle$ and $\langle 1025, -6501 \rangle$.

a) Let $\mathbf{v}_1 = \langle -21614, 13736 \rangle$. Find x_1 where

$$\mathbf{v}_2 = \langle 1025, -6501 \rangle + x_1 \mathbf{v}_1$$

is orthogonal to $\mathbf{v}_1$. **Note**: x_1 need not be an integer.

b) Remember if x_1 is an integer, we can replace $\langle 1025, -6501 \rangle$ with $\mathbf{v}_2$ in our basis. Does rounding your value of x_1 to the nearest integer provide you with a suitable replacement vector $\mathbf{v}_2$? Why/why not?

c) Suppose instead we let $\mathbf{v}_1 = \langle 1025, -6501 \rangle$ and try to find x_1 where

$$\mathbf{v}_2 = \langle -2164, 13736 \rangle + x_1 \mathbf{v}_1$$

is orthogonal to $\mathbf{v}_1$. If x_1 is rounded to the nearest integer, will $\mathbf{v}_2$ be a suitable replacement vector?

d) "Lather, rinse, repeat." If the new basis vectors aren't sufficiently orthogonal, repeat the process until you obtain a pair of nearly orthogonal basis vectors. (This is called a **quasiorthogonal basis**.)

5.10 Least Squares

An important problem connected with the ideas of span is the following. Suppose you observe a set of output values $y_1, y_2, \ldots, y_m$. Find a function f of n inputs $x_1, x_2, \ldots, x_n$ that predicts the outputs y.

As always, any problem in linear algebra starts with a system of linear equations. So to solve this problem, we have to identify a system of linear equations a solution would satisfy.

Activity 5.39: Predictions and Observations

In the following, it will be important to distinguish between output values *predicted* (i.e., computed) by a function of the input values; and output values that are *observed* along with an input value or values.

A5.39.1 In **linear regression**, we want to find m, b that give a linear function $y = mx + b$ that predicts an output value y from an input value x. Let a set of observed input/output pairs (x, y) be

$$(2,1), (5,11), (6,12), (7,14), (10,25)$$

a) In the data given, what are the input values?
b) What are the *observed* output values? (These are the values we hope to predict by our formula.)
c) Write down a system of equations that give the *predicted* (not observed!) output values y_1, y_2, y_3 from the *observed* input values x_1, x_2, x_3, where the predicted output values are computed using a linear function of the input values.
d) What do we want the predicted values to be?
e) What are the unknowns in the system?
f) Write the system of equations in the form $A\vec{x} = \vec{y}$, where $\vec{y}$ is the vector of the predicted output values. Clearly identify A, $\vec{x}$, and $\vec{y}$.
g) Let $\vec{b}$ be the vector of the *observed* output values. Is the system $A\vec{x} = \vec{b}$ solvable? Why/why not?

A5.39.2 In **quadratic regression**, we want to find coefficients a, b, c for a quadratic function $y = ax^2 + bx + c$ that predicts an output value y from input value x. Suppose our observed input/output sets are

$$(0,8), (1,7), (2,11), (3,18), (4,26)$$

a) Identify the input values and the observed output values.
b) Write down a system of equations that give the predicted output values y_i as a quadratic function of the from the actual input values x_i. What are the unknowns in the system of equations?
c) Write the system of equations in the form $A\vec{x} = \vec{y}$, where $\vec{y}$ is the vector of the predicted output values. Clearly identify A, $\vec{x}$, and $\vec{y}$.
d) Let $\vec{b}$ be the vector of the *observed* output values. Is the system $A\vec{x} = \vec{b}$ solvable? Why/why not?

A5.39.3 In **exponential regression**, we want to find coefficients a, b that give an exponential function $y = ae^{bx}$ that predicts an output y from an input x. Suppose our input/output sets are

$$(0,3), (1,5), (2,10), (3,21), (4,38)$$

a) Write down a system of equations that give the predicted output values y_i as an exponential function of the from the actual input values x_i. What are the unknowns in the system of equations?

b) Explain why you **CAN'T** write this system in matrix form $A\vec{x} = \vec{b}$.

c) Rewrite your system of equations so it is a system of *linear* equations. Suggestion: What is $\ln(ae^{bx})$?

d) Write the system of equations in the form $A\vec{x} = \vec{y}$, where $\vec{y}$ is the vector of the predicted output values. Clearly identify A, $\vec{x}$, and $\vec{y}$.

e) Let $\vec{b}$ be the vector of the *observed* output values. Is the system $A\vec{x} = \vec{b}$ solvable? Why/why not?

A5.39.4 In **power regression**, we want to find coefficients a, n that give a power function $y = ax^n$ that predicts an output y from an input x. Suppose our input/output sets are

$$(1, 3), (2, 25), (3, 85), (4, 200), (5, 360)$$

Write a system of equations in the form $A\vec{x} = \vec{y}$, where $\vec{y}$ is the vector of the predicted output values. Clearly identify A, $\vec{x}$, and $\vec{y}$.

A5.39.5 In **bilinear regression**, we want to find coefficients a, b of a linear function $z = ax + by$ that predicts an output z from inputs values x,. Suppose

$$(1, 1, 5), (2, 1, 7), (2, 2, 10), (2, 4, 15), (4, 4, 18)$$

is a set of input/output values (x, y, z).

a) Identify the input values and observed output values from the data.

b) Write down a system of equations that give the *predicted* output values from the observed input values.

c) Write the system of equations in the form $A\vec{x} = \vec{y}$, where $\vec{y}$ is the vector of the predicted output values. Clearly identify A, $\vec{x}$, and $\vec{b}$.

Activity 5.39 suggests that for a wide range of functions, we can find a system of linear equations that corresponds to the solution to the best fit problem. In particular, *all* the best-fit problems seem to be related to a system of equations we can write as $A\vec{x} = \vec{y}$, where A is a coefficient matrix, $\vec{x}$ is a vector corresponding to the parameters of our function, and $\vec{y}$ is a vector of *predicted* output values.

Example 5.10. *Suppose you want to find a linear function $y = mx + b$ that predicts the output values from the input values, when observed input/output pairs are*

$$(2, 7), (5, 24), (9, 27)$$

Express the function in the form $A\vec{x} = \vec{y}$, where $\vec{y}$ is the vector of predicted output values. What do we want $A\vec{x}$ to approximate?

Solution. *Our inputs are 2, 5, and 9, which means our outputs will be*

$$2m + b = y_1$$
$$5m + b = y_2$$
$$9m + b = y_3$$

This corresponds to the system $A\vec{x} = \vec{y}$, *where*

$$A = \begin{pmatrix} 2 & 1 \\ 5 & 1 \\ 9 & 1 \end{pmatrix}, \vec{x} = \begin{pmatrix} m \\ b \end{pmatrix}, \vec{y} = \begin{pmatrix} y_1 \\ y_2 \\ y_3 \end{pmatrix}$$

where y_1, y_2, y_3 *are the predicted vales.*

Based on our observed input/output pairs, we want the input value 2 *to yield output* 7*; our input* 5 *to yield output* 24*; and our input* 9 *to yield output* 27, *so we want* $A\vec{x} = \langle 7, 24, 27 \rangle$.

Of course, our predicted values $\vec{y}$ won't generally be our observed values $\vec{b}$. What we *want* to solve is the system $A\vec{x} = \vec{b}$. But in general, this system is *not* solvable. Moreover, the reason it's not solvable is because we have more observed values than parameters. The obvious solution is to ignore some of the data. However, this is a terrible solution for real-world problems.

Instead, we consider the following: Rather than trying to *solve* $A\vec{x} = \vec{b}$, we find $\vec{x}$ where $A\vec{x}$ is *nearly* $\vec{b}$. This raises an important question: What do we mean by "nearly"?

Activity 5.40: Squared Deviations

Suppose we want to find m, b where $y = mx + b$ predicts the outputs y from inputs x. Moreover, suppose we observe the following pairs of values of the input and output:

$$(2, 1), (5, 11), (6, 12), (7, 14), (10, 25)$$

A5.40.1 The *deviation* or the *error* is the difference between the predicted values for an input, and the observed values.

a) Find the deviations from the mean.

b) Is the sum of the deviations from the mean a good measure of the accuracy of an approximation? Why/why not?

c) In practice, it's common to use the sum of the **squared deviations** or the **squared errors** from the mean. Find the sum of the squared deviations from the mean.

d) Suppose we use $y = 3x - 5$ to predict the output values y from the input values x. Find the sum of the squared deviations from the predicted values.

e) Suppose we use $y = x + 6$ to predict the output values y from the input values x. Find the sum of the squared deviations from the predicted values.

f) Suppose we use $y = 2x$ to predict the output values y from the input values x. Find the sum of the squared deviations from the predicted values.

g) Which of the linear equations is the best predictor of the outputs from the inputs?

A5.40.2 (Requires multivariable calculus.) Find m, b that minimizes the sum of the squared deviations between the line $y = mx + b$ and the data values

$$(2, 1), (5, 11), (6, 12), (7, 14), (10, 25)$$

Activity 5.40 leads to the following: The solution to the best-fit problem will be the vector $\vec{x}$ that minimizes the sum of the squares of the components of $A\vec{x} - \vec{b}$, where $A\vec{x}$ corresponds to the *predicted* outputs, and $\vec{b}$ corresponds to the *observed* outputs.

While we can solve this problem using multivariable calculus, let's see if we can find an easier solution by considering the problem from the viewpoint of a vector space.

Activity 5.41: Close Approximations

In the following, let A be a $m \times n$ matrix that acts on vectors with real components. Let $\vec{x}$ be a vector in the domain of A, and $\vec{b}$ a vector in the codomain (*not* necessarily the range) of A. Assume $m > n$.

A5.41.1 Find the domain and codomain of A.

A5.41.2 Interpret $A\vec{x}$ as a vector in a subspace. Identify the basis for the subspace, and where the subspace "lives."

A5.41.3 Suppose $\vec{b}$ is not in the subspace. Interpret $A\vec{x} - \vec{b}$, and $||A\vec{x} - \vec{b}||^2$.

Activity 5.41 leads to the following ideas. First, we can view $A\vec{x}$ as a vector in the column space of A. If $\vec{b}$ is not in the column space, then $A\vec{x} - \vec{b}$ is the vector corresponding to the difference between the predicted values $A\vec{x}$, and the observed values $\vec{b}$. Moreover, $||A\vec{x} - \vec{b}||^2$ is the sum of the squares of these differences: it's exactly the quantity we wish to minimize.

How do we minimize this value?

Activity 5.42: Minimizing

A5.42.1 Let $P = (3, 5, 1)$, $Q = (1, 1, 2)$, and let O be the origin.

a) Let $\overrightarrow{y}$ be any vector in the plane OPQ. Express $\overrightarrow{y}$ as a linear combination of the vectors $\overrightarrow{OP}$ and $\overrightarrow{OQ}$.
b) Interpret the plane OPQ as a subspace of $\mathbb{R}^3$. What are a set of basis vectors for the plane OPQ?
c) Let $\overrightarrow{b}$ be any vector *not* in the plane OPQ. What is the significance of $\overrightarrow{y} - \overrightarrow{b}$?
d) Suppose you want $\overrightarrow{y} - \overrightarrow{b}$ to be as small as possible. What must be the relationship between $\overrightarrow{y} - \overrightarrow{b}$ and the vectors $\overrightarrow{OP}$, $\overrightarrow{OQ}$?

Activity 5.42 leads to an important idea: Suppose the column space of A defines a subspace of $\mathbb{R}^m$. If $\overrightarrow{b}$ is a vector *not* in the subspace, then the distance $||A\overrightarrow{x} - \overrightarrow{b}||$ between $\overrightarrow{b}$ and any vector $A\overrightarrow{x}$ in the subspace will be minimized when $A\overrightarrow{x} - \overrightarrow{b}$ is orthogonal to all vectors in the subspace. We illustrate the situation where $\mathbf{V}$ is a 1-dimensional subspace of $\mathbb{R}^2$:

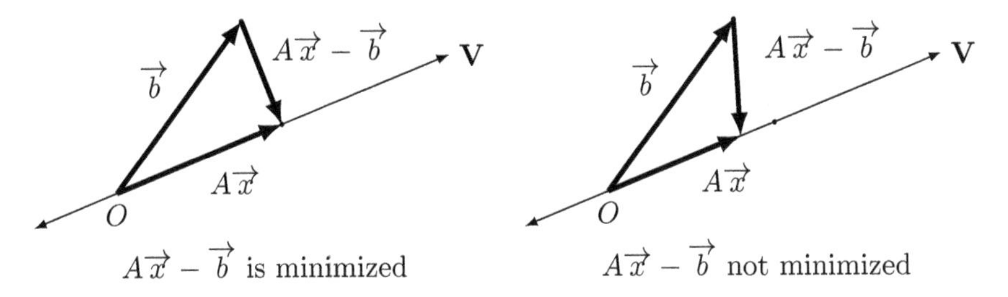

$A\overrightarrow{x} - \overrightarrow{b}$ is minimized $\qquad$ $A\overrightarrow{x} - \overrightarrow{b}$ not minimized

This geometric insight provides an approach to solving the least squares problem.

Activity 5.43: Least Squares

A5.43.1 Suppose we want $A\overrightarrow{x} - \overrightarrow{b}$ to be orthogonal to all vectors in the column space of A.

a) Explain why it is sufficient to require that $A\overrightarrow{x} - \overrightarrow{b}$ be orthogonal to the basis vectors.
b) Explain why $A^T(A\overrightarrow{x} - \overrightarrow{b})$ is a compact way of expressing the dot product of $A\overrightarrow{x} - \overrightarrow{b}$ with all the basis vectors of the column space of A.
c) Explain why the requirement that $A\overrightarrow{x} - \overrightarrow{b}$ be orthogonal to the basis vectors corresponds to the requirement that $A^TA\overrightarrow{x} = A^T\overrightarrow{b}$.

Activity 5.43 leads to the following: Suppose we wish to model a set of data using a function with parameters $\vec{x}$. Find matrix A where $A\vec{x}$ gives a predicted output vector $\vec{y}$. Then the parameter values $\vec{x}$ that provide the best fit to the data will be the solution to $A^T A\vec{x} = A^T\vec{b}$, where $\vec{b}$ is the vector of observed output values.

Example 5.11. *Find a linear function $y = mx + b$ that best predicts the output values from the input values, when observed input/output pairs are*

$$(2,7),(5,24),(9,27)$$

Solution. *These are the same observed values from Example 5.10, so we know the corresponding matrices A and $\vec{x}$ are*

$$A = \begin{pmatrix} 2 & 1 \\ 5 & 1 \\ 9 & 1 \end{pmatrix}, \vec{x} = \begin{pmatrix} m \\ b \end{pmatrix}, \vec{y} = \begin{pmatrix} y_1 \\ y_2 \\ y_3 \end{pmatrix}$$

Meanwhile, $\vec{b}$ is the column vector consisting of the observed outputs (what we meant y_1, y_2, y_3 to equal): $\vec{b} = \begin{pmatrix} 7 \\ 24 \\ 25 \end{pmatrix}$.

We want to find $\vec{x}$ where $A^T A\vec{x} = A^T\vec{b}$ so we have:

$$A^T A\vec{x} = A^T\vec{b}$$

$$\begin{pmatrix} 2 & 5 & 9 \\ 1 & 1 & 1 \end{pmatrix}\begin{pmatrix} 2 & 1 \\ 5 & 1 \\ 9 & 1 \end{pmatrix}\begin{pmatrix} m \\ b \end{pmatrix} = \begin{pmatrix} 2 & 5 & 9 \\ 1 & 1 & 1 \end{pmatrix}\begin{pmatrix} 7 \\ 24 \\ 27 \end{pmatrix}$$

$$\begin{pmatrix} 110 & 16 \\ 16 & 3 \end{pmatrix}\begin{pmatrix} m \\ b \end{pmatrix} = \begin{pmatrix} 377 \\ 58 \end{pmatrix}$$

Row reducing, we obtain solution $m \approx 2.743$, $b \approx 4.703$, giving us best-fit line $y = 2.743x + 4.703$.

This approach generalizes to all functions we can describe as a linear function of parameters.

Activity 5.44: Best Fit Curves

A5.44.1 In **linear regression**, we want to find m, b that give a linear function $y = mx + b$ that predicts an output value y from an input value x. Let a set of observed input/output pairs (x, y) be

$$(2,1),(5,11),(6,12),(7,14),(10,25)$$

a) Write down a system of equations in m, b that predicts the output values y from the input values x. Express this system in matrix form $A\vec{x} = \vec{y}$, where $\vec{y}$ are the predicted output values.

b) Rewrite this in the form $A^T A\vec{x} = A^T\vec{b}$ and solve for $\vec{x}$. What is the corresponding equation of the line?

A5.44.2 Suppose we want to find a linear function $mx + b$ that approximates the square root function $\sqrt{x}$.

a) Find a set of three input/output pairs for the square root function. (In other words, find three pairs (x, y), where $y = \sqrt{x}$). The inputs should be in the interval $0 \leq x \leq 10$, and your output values should be exact numbers and not decimal approximations.

b) Write down a system of equations in m, b that predict an output y from the input x. Express this system in matrix form $A\vec{x} = \vec{y}$, where $\vec{y}$ are the predicted output values.

c) Rewrite this in the form $A^T A\vec{x} = A^T\vec{b}$ and solve for $\vec{x}$. What is the corresponding equation of the line?

d) Use your linear function to approximate $\sqrt{5}$. Compare your result to the actual value of $\sqrt{5}$.

A5.44.3 In **quadratic regression**, we want to find coefficients a, b, c for a quadratic function $y = ax^2 + bx + c$ that predicts an output value y from input value x. Suppose our observed input/output sets are

$$(0, 8), (1, 7), (2, 11), (3, 18), (4, 26)$$

a) Write down a system of equations in a, b, c where $ax^2 + bx + c = y$. Also write this system in matrix form $A\vec{x} = \vec{b}$.

b) Rewrite this in the form $A^T A\vec{x} = A^T\vec{b}$ and solve for $\vec{x}$. What is the corresponding function $ax^2 + bx + c$?

A5.44.4 Suppose we want to find a quadratic function that models $\sin x$.

a) Find a set of four input/output values for $\sin x$. Your input values should be in the interval $0 \leq x \leq \pi$, and your output values should be exact expressions.

b) Write down a system of equations in a, b, c where $ax^2 + bx + c = y$, the predicted output. Also write this system in matrix form $A\vec{x} = \vec{b}$, where $\vec{b}$ is the vector of observedoutputs.

c) Rewrite this in the form $A^T A\vec{x} = A^T\vec{b}$ and solve for $\vec{x}$. What is the corresponding function $ax^2 + bx + c$?

d) Use your function to approximate $\sin x$ over the interval $0 \leq x \leq \pi$ (use a graph). How good is the approximation?

A5.44.5 In **bilinear regression**, we want to find coefficients a, b of a linear function $z = ax + by$ that predicts an output z from inputs values $x,$. Suppose

$$(1,1,5),(2,1,7),(2,2,10),(2,4,15),(4,4,18)$$

is a set of input/output values (x, y, z).

a) Write down a system of equations in a, b where $ax + by = z$. Also write this system in matrix form $A\overrightarrow{x} = \overrightarrow{b}$.

b) Rewrite this in the form $A^T A\overrightarrow{x} = A^T\overrightarrow{b}$ and solve for $\overrightarrow{x}$. What is the corresponding function $z = ax + by$?

A5.44.6 In **exponential regression**, we want to find coefficients a, b that give an exponential function $y = ae^{bx}$ that predicts an output y from an input x. Suppose our input/output sets are

$$(0,3),(1,5),(2,10),(3,21),(4,38)$$

a) Write down a system of equations in a, b where $ae^{bx} = y$.

b) Explain why you **CAN'T** write this system in matrix form $A\overrightarrow{x} = \overrightarrow{b}$.

c) Rewrite your system of equations so it is a system of *linear* equations. Suggestion: What is $\ln(ae^{bx})$?

d) Solve for a, b.

A5.44.7 In **power regression**, we want to find coefficients a, n that give a power function $y = ax^n$ that predicts an output y from an input x. Suppose our input/output sets are

$$(1,3),(2,25),(3,85),(4,200),(5,360)$$

Find a, b.

One use of the best-fit curve is to provide recommendations of similar products to consumers.

Activity 5.45: "You Might Also Like ..."

In recent years, automatic recommendation systems have become a ubiquitous part of our lives. These systems will recommend products we might like, or suggest other products that go along with the ones we've purchased.

One way to implement a recommendation system is to use a set of *proxy observers*. These observers assign ranks to all objects in a set. For example, they might be a group of movie critics, ratings movies on a scale from 0 to 10.

To make a prediction, a consumer's ratings are recorded, and compared to those of the proxy observers. From this data, a composite observer is created that will predict the consumer's ratings. For example, a consumer's ranking might be predicted as the sum of 70% of the ranking given by one proxy observer, 50% of the raking given by a second, and 10% of the ranking given by a third (the percentages do not need to add to 100%, since they are percentages of different quantities).

In the following, suppose there are three proxy observers, who assigned the following ranks to the listed products; the consumer's ranking of the products is included.

	Product A	Product B	Product C	Product D
Observer 1	5	3	7	2
Observer 2	1	7	3	9
Observer 3	7	4	4	8
Consumer	4	5	5	7

A5.45.1 Let x, y, z be a set of weights assigned to each of the observers.

a) Set up a system of equations corresponding to the prediction of the consumer's rating, based on the observer's ratings.

b) Find the weights that minimize the sum of the squared deviations.

c) Suppose a product has received ratings of 6, 8, and 3 from the observers. What rating would be predicted for the user?

A5.45.2 Suppose the consumer eventually gave the product a rating of 7. Incorporate the new proxy and consumer ratings and find new weights that minimize the squared deviations.

A5.45.3 How might you change the approach to reflect the possibility a user's preferences would change over time?

6

Determinants

This chapter deals with a concept in linear algebra known as the *determinant of a matrix*. As we proceed, it's important to remember:

> ***You should almost never use the determinant to solve a problem.***

In particular, almost every problem in linear algebra that can be solved *with* the determinant can be solved faster, more easily, and more transparently *without* the determinant.

So why study the determinant? There are several reasons. First, determinants have embedded themselves in most standard introductions to linear algebra, so in every course after linear algebra, it's assumed you know what it is and how to find it. Thus, every introductory linear algebra course discusses the determinant.

Second, while the determinant is not needed to solve *most* problems in linear algebra, it is extremely useful for *some* problems. For example, finding the inverse of a matrix with integer entries is a problem we can solve without the determinant. But finding a matrix with integer entries whose inverse is *also* a matrix with integer entries is far more challenging; the determinant will give us insight into how to produce such a matrix (see Activity 6.40).

Third, the determinant is useful for theoretical arguments. In fact, we can go further: a proper study of the determinant is a good way to hone your ability to prove statements. To that end, we'll start with an investigation of some of the properties of 2×2 matrices.

6.1 Linear Equations

We introduced matrices in the context of solving systems of linear equations. Let's take a closer look at these systems. An important question is: When does a solution exist?

Activity 6.1: Solving Systems of Equations

A6.1.1 Consider the system of equations

$$\begin{aligned} ax + by &= f \\ cx + dy &= g \end{aligned}$$

where x, y are the variables.

a) Why may we assume "without loss of generality" that $a \neq 0$? (In particular: Suppose $a = 0$. Show that there must be an equivalent system

$$\begin{aligned} a'x + b'y &= f' \\ c'x + d'y &= g' \end{aligned}$$

where $a' \neq 0$.)

b) Express this system as an augmented coefficient matrix, then solve. Express your solutions in terms of a, b, c, d, f, g.

c) What relationship between a, b, c, d, f, g must hold in order for a unique solution to exist?

A6.1.2 In the following, you are given the coefficient matrix of a system of equations. Determine whether the corresponding system of homogeneous equations has a unique solution.

a) $\begin{pmatrix} 1 & 5 \\ 2 & 7 \end{pmatrix}$

b) $\begin{pmatrix} 3 & 8 \\ -17 & 5 \end{pmatrix}$

c) $\begin{pmatrix} 2 & 5 \\ 4 & 10 \end{pmatrix}$

d) $\begin{pmatrix} 0 & 3 \\ 1 & 4 \end{pmatrix}$

In Activity 6.1, you found that if a system of equations had augmented coefficient matrix

$$\left(\begin{array}{cc|c} a & b & f \\ c & d & g \end{array}\right)$$

then the solutions to the system could be expressed as

$$x_1 = \frac{fd - bg}{ad - bc} \qquad x_2 = \frac{ag - fc}{ad - bc}$$

Notice that the expression $ad - bc$ appears as a divisor of both expressions.

Because it's a divisor, there are two possibilities:

- If $ad - bc = 0$, then the expressions for x_1 and x_2 are undefined, so there can be no solution to the inhomogeneous system,
- If $ad - bc \neq 0$, then the expressions for x_1 and x_2 are completely defined, so there is a unique solution.

We might say that the expression $ad - bc$ determines whether the system of equations has a unique solution, or no solution.

6.2 Transformations

We also used matrices to represent geometric transformations. Geometric transformations can be divided into two broad categories. First, there are transformations, like reflections and rotations, that preserve lengths and angles. These transformations are known as **isometries**. Since they preserve lengths and angles, they also preserve areas. Other types of transformations will alter areas.

Activity 6.2: Transformation of Areas

A6.2.1 Let $ABCD$ be the unit square with opposite vertices $(0, 0)$ and $(1, 1)$.

a) Find the other vertices.
b) Find the area of the square $ABCD$.

A6.2.2 If we treat the coordinates of the points A, B, C, D as vectors with the same components, we can apply a linear transformation T to each point to get A', B', C', D'. Find the area of the figure produced by applying the transformation $T = \begin{pmatrix} 2 & 1 \\ 3 & 8 \end{pmatrix}$ to the unit square $ABCD$. Suggestion: $A'B'C'D'$ can fit into a rectangle with opposite vertices A' and C'. Find the area of the rectangle, then subtract the areas of the parts of the rectangle that aren't in $A'B'C'D'$.

A6.2.3 Let $T = \begin{pmatrix} a & b \\ c & d \end{pmatrix}$, and assume $a, b, c, d > 0$.

a) Apply the transformation T to the vertices A, B, C, D to get the vertices A', B', C', D'.
b) Find the area of the quadrilateral $A'B'C'D'$ as a formula using a, b, c, d.

In Activity 6.2, you found that $T = \begin{pmatrix} a & b \\ c & d \end{pmatrix}$ will transform a unit square into a quadrilateral with an area $ad - bc$. The formula should cause some concern: Because it's the difference between two products, it's possible you might get a negative value.

Activity 6.3: Orientation

A polygon $PQR \cdots Z$ is said to be **positively oriented** if, as you travel through the vertices P, Q, R, ..., Z, the polygon is always on your left; it is **negatively oriented** if, as you travel through the vertices P, Q, R, ..., Z the polygon is on your right.

A transformation is **orientation preserving** if the transformed figure has the same orientation as the original; and **orientation reversing** if the transformed figure has the opposite orientation as the original.

In the following, assume $ABCD$ is a positively oriented unit square with $A = (0, 0)$ and $C = (1, 1)$.

A6.3.1 Suppose $ABCD$ is the positively oriented unit square with $A = (0, 0)$ and $C = (1, 1)$.

a) Find the coordinates of B and D.

b) Find the area.

c) What is the orientation of $BCDA$?

d) What is the orientation of $ADCB$?

A6.3.2 In Activity 6.2, you found that $T = \begin{pmatrix} a & b \\ c & d \end{pmatrix}$ would transform a square into a figure with area $ad - bc$.

a) Let $T = \begin{pmatrix} 1 & 3 \\ 2 & 5 \end{pmatrix}$. Let T be applied to the vertices A, B, C, D to produce the transformed vertices A', B', C', D'. Find the coordinates of the vertices.

b) Is T an orientation preserving transformation, or an orientation reversing transformation?

c) Find the actual area of the figure $A'B'C'D'$.

d) How does the actual area compare to the area computed by the formula $ad - bc$?

A6.3.3 Suppose a, b, c, d are all positive. If $T = \begin{pmatrix} a & b \\ c & d \end{pmatrix}$ is applied to the positively oriented unit square $PQRS$, it can produce either of two parallelograms, depending on the relative magnitudes of a, b, c, d:

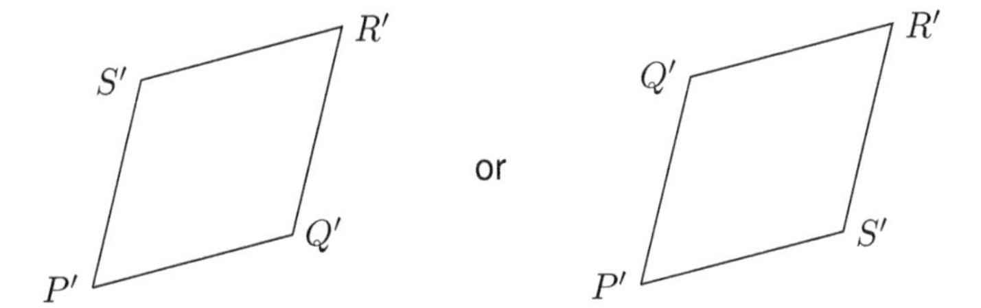

a) Which transformation is orientation preserving, and which is orientation reversing?
b) Consider the orientation preserving transformation. Based on the *picture*, which is greater: the slope of $\overleftrightarrow{P'Q'}$ or the slope of $\overleftrightarrow{P'S'}$?
c) Consider the orientation reversing transformation. Based on the *picture*, which is greater: the slope of $\overleftrightarrow{P'Q'}$ or the slope of $\overleftrightarrow{P'S'}$?
d) How can you use the values of a, b, c, d to decide whether T is orientation preserving or orientation reversing?

A6.3.4 If possible, find a transformation $T = \begin{pmatrix} a & b \\ c & d \end{pmatrix}$ with the following properties, where a, b, c, d are nonzero.

a) An orientation preserving transformation that transforms a unit square into a figure with an area of 3.
b) An orientation reversing transformation that transforms a unit square into a figure with an area of 1.
c) A transformation that transforms a unit square into a figure with an area of 0.

Activities 6.3 leads to the following result: Suppose a, b, c, d are positive. If $T = \begin{pmatrix} a & b \\ c & d \end{pmatrix}$, the "difference of the cross products" $ad - bc$ will be positive if the transformation is orientation preserving, and negative if it is orientation reversing. In either case, the area of a unit square under the transformation will be $|ad - bc|$, and the value $ad - bc$ will determine whether or not the transformation is orientation preserving.

What if some or all of a, b, c, d are zero or negative?

Activity 6.4: More Orientation

A6.4.1 Let R be a rotation.

a) Is R is orientation preserving? Defend your conclusion.

b) Suppose T is orientation preserving. Are RT and TR orientation preserving? Defend your conclusion.

c) Suppose T is orientation reversing. Are RT and TR orientation reversing? Defend your conclusion.

A6.4.2 Let $T = \begin{pmatrix} a & b \\ c & d \end{pmatrix}$, where a, b, c, d could be positive, negative, or zero.

a) Explain why T can be expressed as RT', where R is a matrix representing a rotation and $T' = \begin{pmatrix} a' & b' \\ c' & d' \end{pmatrix}$ with $a', c' > 0$.

b) Show that if T' is orientation preserving, $a'd' - b'c' > 0$, and if T' is orientation reversing, $a'd' - b'c' < 0$.

c) Explain geometrically why $a'd' - b'c' = ad - bc$.

Thus:

Proposition 2. *Let* $T = \begin{pmatrix} a & b \\ c & d \end{pmatrix}$. *Then T is orientation preserving or orientation reversing, depending on whether $ad - bc$ is positive or negative.*

6.3 Inverses

More generally, we can view a matrix T as a transformation, and consider the inverse transformation T^{-1}. Again, let's take a closer look at that process.

Activity 6.5: The Inverse of a Matrix

Let $A = \begin{pmatrix} a & b \\ c & d \end{pmatrix}$, and remember you can find A^{-1} by row reducing the double wide matrix

$$\left(\begin{array}{cc|cc} a & b & 1 & 0 \\ c & d & 0 & 1 \end{array}\right) \rightarrow \left(\begin{array}{cc|cc} 1 & 0 & x_1 & x_3 \\ 0 & 1 & x_2 & x_4 \end{array}\right)$$

where $A^{-1} = \begin{pmatrix} x_1 & x_3 \\ x_2 & x_4 \end{pmatrix}$

A6.5.1 Suppose $a \neq 0$.

a) Find A^{-1}.

b) What might prevent you from finding A^{-1}? (In particular: Among the *computations* that must be performed to find A^{-1}, is there any situation that would result in an undefined answer?)

A6.5.2 Suppose $a = 0$.

a) Explain why, if A^{-1} exists, $c \neq 0$.
b) Find A^{-1} under the assumption $a = 0$, $c \neq 0$. Suggestion: Do **NOT** replace a with 0, but leave it as a.
c) What might prevent you from finding A^{-1}? (In particular: Among the *computations* that must be performed to find A^{-1}, is there any situation that would result in an undefined answer?)

Once again, we see that if $A = \begin{pmatrix} a & b \\ c & d \end{pmatrix}$, then $ad - bc$ determines whether the inverse exists.

6.4 The Determinant

There's an old saying: Once is an accident, twice is a coincidence, but three times is a conspiracy. In mathematics, we can interpret this as follows: If you keep running into a certain expression in different contexts, there's something about that expression that's important. For the 2×2 matrix $\begin{pmatrix} a & b \\ c & d \end{pmatrix}$, we found the difference of the cross products $ad - bc$ is an important quantity that determines many features about the matrix, either viewed as a coefficient matrix; or as a geometric transformation; or as a more general linear transformation.

Since this difference of the cross products seems to determine many features of the matrix viewed as a linear transformations, mathematicians (who are terrible at coming up with new names) call it the *determinant*, and we define:

Definition 6.1 (Determinant of a 2×2 Matrix). *The* ***determinant*** *of the 2×2 matrix $\begin{pmatrix} a & b \\ c & d \end{pmatrix}$ is $ad - bc$.*

"Math ever generalizes," so the natural question to ask at this point is whether we can extend the concept of the determinant to matrices of different sizes. We'll use the notation Determ A to refer to the determinant of some matrix A.

Activity 6.6: Determinants for Nonsquare Matrices?

In the following, assume A is matrix with more columns than rows; and B is a matrix with more rows than columns.

A6.6.1 One use of the determinant is to decide whether a system of equations has a unique solution.

a) Suppose A represents a system of equations. Based on your experiences, will the corresponding system of equations have a unique solution?

b) Suppose B represents a system of equations. Based on your experiences, will the corresponding system of equations have a unique solution?

c) "Information is an answer you didn't already know." Would a "determinant" of A or B provide information?

A6.6.2 Another use of the determinant is to decide whether a matrix has an inverse.

a) Do A or B have an inverse? Why/why not?

b) Would a "determinant" of A or B provide information?

A6.6.3 A third use of the determinant is to determine the content of a unit n-cube under the transformation. (A unit n-cube is a n-dimensional geometric figure with n orthogonal sides, each of which have length 1: a unit 1-cube is a line segment; a unit 2-cube is a square, a unit 3-cube is a cube, and so on. Meanwhile the content of a n-dimensional object corresponds to the number of n-cubes contained by it: this would be length of a 1-dimensional object, the area of a 2-dimensional object, the volume of a 3-dimensional object, and so on)

a) Do A or B transform a unit n-cube into a n-dimensional object? Why/why not?

b) Would a "determinant" of A or B be meaningful in this case?

Activity 6.6 suggests that even if we could define a "determinant" for a nonsquare matrix, that determinant would be useless: it would tell us nothing we didn't already know about the matrix. Thus, we'll focus on the determinants of square matrices.

One way to generalize the determinant is to decide what properties we *want* the determinant to have, and then define it so that it has all the properties we want. To do that, it helps to have a good, concrete view of what we want the determinant to do.

First, we have a formula for the determinant of a 2×2 matrix, so let's explore the *algebraic* properties of the determinant. The natural question to ask (in linear algebra, at least) is whether the determinant is a linear transformation of the entries of the matrix.

Activity 6.7: Algebraic Properties of the Determinant

In the following, assume A is a 2×2 matrix, and Determ A is computed by the formula in Definition 6.1.

A6.7.1 We can view Determ A as a transformation Determ A : $\mathbb{F}^n \to \mathbb{F}^m$.

a) What are the domain and codomain of Determ A?

b) Prove/disprove: Determ A is a linear transformation.

A6.7.2 Rather than considering Determ A to act on vectors in $\mathbb{F}^n$, we could consider it to act on the column vectors $\mathbf{u}$, $\mathbf{v}$, and write Determ $\begin{pmatrix}\mathbf{u} & \mathbf{v}\end{pmatrix} = D(\mathbf{u}, \mathbf{v})$. We say that D is a function of the column vectors of A.

a) Prove/disprove: $D(\mathbf{u}, c\mathbf{v}) = cD(\mathbf{u}, \mathbf{v})$, where c is some scalar.

b) Prove/disprove: $D(\mathbf{u}, \mathbf{v} + \mathbf{w}) = D(\mathbf{u}, \mathbf{v}) + D(\mathbf{u}, \mathbf{w})$.

c) Prove/disprove: Determ A is a linear transformation on the second column of A.

A6.7.3 Prove/disprove: If we view Determ A as a function of the row vectors of A, then Determ A is a linear transformation on the second row of A.

Activity 6.7 reveals two features of Determ A. First, it is *not* a linear transformation on the entries of A. However:

Proposition 3. *Determ A is a linear transformation on the last column of A, and on the last row of A.*

What other algebraic properties does Determ A possess?

Activity 6.8: More Algebraic Properties of the Determinant

A6.8.1 Suppose I is the 2×2 identity matrix. Find Determ I.

A6.8.2 Let $A = \begin{pmatrix} a & b \\ c & d \end{pmatrix}$. Find the following.

a) Determ B, the matrix where the two rows of A have been switched.

b) Determ C, the matrix where the two columns of A have been switched.

c) Suppose D is the matrix A where two rows *or* two columns have been switched. What is the relationship between Determ A and Determ D?

Activity 6.8 suggests two more properties of Determ A:

Proposition 4 (Identity). *If I is the 2×2 identity matrix, Determ $I = 1$.*

Moreover:

Proposition 5 (Row or Column Interchanges). *Let A be a 2×2 matrix, and let B be a matrix formed by switching two rows or two columns of A. Then Determ $A = -$Determ B.*

Putting our results together:

Theorem 6.1. *Let $A = \begin{pmatrix} a & b \\ c & d \end{pmatrix}$, and define Determ $A = ad - bc$. Determ A has the following properties:*

- *Determ A is a linear transformation on the last column of A, and on the last row of A.*
- *Determ $I = 1$, where I is the 2×2 identity matrix,*
- *Switching two rows or two columns changes the sign of Determ A.*

Before proceeding, we might want to take a look at these properties from a geometric point of view. Remember that if A is a 2×2 matrix, the determinant of A told us the area and orientation of a unit square transformed by A.

Activity 6.9: Geometry and the Determinant

For the following, let $T = \begin{pmatrix} a & b \\ c & d \end{pmatrix}$ be an orientation preserving transformation; we may assume a, c are positive (see Activity 6.4). We wish to see the effect of an elementary row operation on the value of the determinant of T.

In the following, do **NOT** compute a determinant using any formula. Instead, base your conclusions on geometric considerations.

A6.9.1 One of the elementary row operations is switching two rows.

a) *Without* computing the determinant, show that B, the matrix obtained by switching the *columns* of T, is orientation reversing.

b) *Without* computing the determinant, show that A, the matrix obtained by switching the rows of T, is orientation reversing.

c) Remember the determinant provides information about the area and orientation of a unit square under a transformation. What is the relationship between the determinants of T, A, and B?

A6.9.2 Let $T = \begin{pmatrix} a & b \\ c & d \end{pmatrix}$, $V = \begin{pmatrix} a & e \\ c & f \end{pmatrix}$, and $W = \begin{pmatrix} a & b+e \\ c & d+f \end{pmatrix}$. Assume that T, V preserve the orientation of the unit square.

a) **Without** computing the determinant, show that W preserves the orientation of the unit square.

b) **Without** computing the determinant, express the relationship between the determinants of T, V, and W. Suggestion: Remember $|T|$ will be the area of the figure produced by transforming the unit square by T.

A6.9.3 Let $T = \begin{pmatrix} a & b \\ c & d \end{pmatrix}$, $V = \begin{pmatrix} a & b \\ e & f \end{pmatrix}$, and $W = \begin{pmatrix} a & b \\ c+e & d+f \end{pmatrix}$. Assume that T, V preserve the orientation of the unit square.

a) **Without** computing the determinant, show that W preserves the orientation of the unit square.

b) **Without** computing the determinant, express the relationship between the determinants of T, V, and W.

A6.9.4 Suppose I is the 2×2 identity matrix. What should $|I|$ be? Base your answer on the *geometric* meaning of the determinant.

Activity 6.9 shows that the *algebraic* properties of the determinant, found from the formula, are consistent with the *geometric* properties of the determinant. This suggests that when we try to generalize the determinant to larger matrices, we can use the algebraic properties as our guide.

There's one problem we have to resolve. In Activity 6.8, you showed that switching the rows or columns of a 2×2 matrix would change the sign of the determinant. However, since a 2×2 matrix only has two rows and two columns, switching *any* two rows or columns is the same as switching *adjacent* rows or columns. For larger matrices, we can switch two nonadjacent rows. Thus we need to decide whether the sign changing property of the determinant is based on switching *any* rows or columns, or *adjacent* rows or columns.

Activity 6.10: Switching Rows and Columns

A6.10.1 Suppose $f(M)$ is a function of the entries of a matrix, with the following property: If N is formed from M by switching *any* two rows or columns, $f(M) = -f(N)$.

a) Let A be some matrix with at least three rows. Consider the row operations:

$$A \xrightarrow{R_1 \leftrightarrow R_3} B$$

What is the relationship between $f(A)$ and $f(B)$?

b) Consider the row operations:

$$A \xrightarrow{R_3 \leftrightarrow R_2} C$$

What is the relationship between $f(A)$ and $f(C)$?

c) Show that

$$C \xrightarrow{R_1 \leftrightarrow R_2} B$$

d) What is the relationship between $f(C)$ and $f(B)$?

e) If possible, find $f(A)$, $f(B)$, $f(C)$. If not possible, explain why.

A6.10.2 Suppose $g(M)$ is a function of the entries of a matrix, with the following property: If N is formed from M by switching *adjacent* rows or columns, $g(M) = -g(N)$.

a) Using the matrices A, B, C above, what are the relationships between $g(A)$, $g(B)$, and $g(C)$?

b) If possible, find $g(A)$, $g(B)$, $g(C)$. If not possible, explain why.

Activity 6.10 leads to an important conclusion: If the sign of the determinant changes for *any* row or column switching, then the determinant for a matrix with 3 or more rows and columns must be 0. But remember Shannon's definition of information: If the question "What is the determinant of a matrix?" always going to be 0, then the determinant provides us with no information.

Thus if we want the determinant to be useful, we can only require that the sign changes when *adjacent* rows or columns are switched.

Put together, this suggests the following generalization of the determinant:

Definition 6.2 (Determinant)**.** *The determinant of a square matrix M, written $|M|$, is a linear transformation of its last column and its last row, with the following properties:*

- *The determinant of the identity matrix is* 1*,*
- *Switching two adjacent rows or columns will change the sign of the determinant,*

The linearity property of the determinant takes a little getting used to, since it does *not* involve the addition or scalar multiplication of matrices.

Example 6.1. *Use linearity to express as a sum of determinants:* $\begin{vmatrix} 1 & 3 & 5 \\ 2 & -3 & 9 \\ 1 & 4 & 3 \end{vmatrix}$

Solution. *The last column can be written as a sum of three column vectors:*

$$\begin{pmatrix} 5 \\ 9 \\ 3 \end{pmatrix} = \begin{pmatrix} 5 \\ 0 \\ 0 \end{pmatrix} + \begin{pmatrix} 0 \\ 9 \\ 0 \end{pmatrix} + \begin{pmatrix} 0 \\ 0 \\ 3 \end{pmatrix}$$

Replacing the last column of the original matrix with these column vectors, but keeping the other columns the same, gives us

$$\begin{vmatrix} 1 & 3 & 5 \\ 2 & -3 & 9 \\ 1 & 4 & 3 \end{vmatrix} = \begin{vmatrix} 1 & 3 & 5 \\ 2 & -3 & 0 \\ 1 & 4 & 0 \end{vmatrix} + \begin{vmatrix} 1 & 3 & 0 \\ 2 & -3 & 9 \\ 1 & 4 & 0 \end{vmatrix} + \begin{vmatrix} 1 & 3 & 0 \\ 2 & -3 & 0 \\ 1 & 4 & 3 \end{vmatrix}$$

We could also write the last row as

$$\langle 1, 4, 3 \rangle = \langle 1, 0, 0 \rangle + \langle 0, 4, 0 \rangle + \langle 0, 0, 3 \rangle$$

Replacing the last row of the original with these row vectors, but keeping the others the same, gives us:

$$\begin{vmatrix} 1 & 3 & 5 \\ 2 & -3 & 9 \\ 1 & 4 & 3 \end{vmatrix} = \begin{vmatrix} 1 & 3 & 5 \\ 2 & -3 & 9 \\ 1 & 0 & 0 \end{vmatrix} + \begin{vmatrix} 1 & 3 & 5 \\ 2 & -3 & 9 \\ 0 & 4 & 0 \end{vmatrix} + \begin{vmatrix} 1 & 3 & 5 \\ 2 & -3 & 9 \\ 0 & 0 & 3 \end{vmatrix}$$

We say we have **expanded along the last column** or **expanded along the last row.**

Activity 6.11: Multilinearity of the Determinant

In the following, assume that W is a square matrix, and that A, B are matrices of the appropriate size, and $\mathbf{u}$, $\mathbf{v}$, $\mathbf{w}$ are column vectors of the appropriate size.

A6.11.1 Suppose W is a square matrix that can be written in block form as $W = \left(\begin{array}{c|c|c} A & \mathbf{u}+\mathbf{v} & B \end{array} \right)$ where A, B are appropriately sized matrices, and $\mathbf{u}$, $\mathbf{v}$ are appropriately sized column vectors.

a) Using **ONLY** the properties in the definition of the determinant, show that

$$|W| = \begin{vmatrix} A & \mathbf{u} & B \end{vmatrix} + \begin{vmatrix} A & \mathbf{v} & C \end{vmatrix}$$

b) Why does this mean that the determinant is a linear transformation of the entries in *any* column?

A6.11.2 Using **ONLY** the properties in the definition of the determinant, show that the determinant is a linear transformation of the entries in any row.

A6.11.3 Suppose $W = \begin{pmatrix} A & \mathbf{u} & B \end{pmatrix}$. Using **ONLY** the properties in the definition of the determinant, show that $\begin{vmatrix} A & c\mathbf{v} & B \end{vmatrix} = c|W|$.

Activity 6.11 gives us two key properties of the determinant:

Proposition 6 (Multilinearity). *The determinant is a linear transformation of the entries in its rows and columns.*

We say that the determinant is a **multilinear transformation** on the rows and columns of a matrix. This means we can expand along any row or column we choose.

Example 6.2. *Expand* $\begin{vmatrix} a & b & c \\ d & e & f \\ g & h & j \end{vmatrix}$ *along the second row.*

Solution. *We can express the second row of our matrix as the sum*

$$\langle d, e, f \rangle = \langle d, 0, 0 \rangle + \langle 0, e, 0 \rangle + \langle 0, 0, f \rangle$$

Using these as the second rows in our expansion gives us

$$\begin{vmatrix} a & b & c \\ d & e & f \\ g & h & j \end{vmatrix} = \begin{vmatrix} a & b & c \\ d & 0 & 0 \\ g & h & j \end{vmatrix} + \begin{vmatrix} a & b & c \\ 0 & e & 0 \\ g & h & j \end{vmatrix} + \begin{vmatrix} a & b & c \\ 0 & 0 & f \\ g & h & j \end{vmatrix}$$

Activity A6.11.3 proves:

Proposition 7 (Constant Multiple). *Let W' be a matrix formed by multiplying any column of W by a scalar c. Then $|W'| = c|W|$.*

Since this is based on the column linearity of the determinant, and the determinant also has row linearity, we also have:

Proposition 8 (Constant Multiple). *Let W' be a matrix formed by multiplying any row of W by a scalar c. Then $|W'| = c|W|$.*

Notice that the proofs of these properties were based entirely on the definition of the determinant, regardless of how it might be computed.

6.5 A Formula for the Determinant

Unfortunately, while our definition of the determinant describes its *properties*, it falls short of giving us a way to calculate the determinant. To develop a formula for the determinant, let's begin with some matrices whose determinants we can calculate directly.

One way to simplify a matrix is to have a row or column of 0s, or two have two rows or two columns identical. What can we say about the determinants of such matrices?

Activity 6.12: Determinant Properties

For the following, use **ONLY** the properties of the determinant; do **NOT** use any formula for computing the determinant.

A6.12.1 Prove the following, using **ONLY** the properties of the determinant; do **NOT** compute any determinants.

a) $\begin{vmatrix} a & b \\ c & d \end{vmatrix} = \begin{vmatrix} a & b \\ 0 & 0 \end{vmatrix} + \begin{vmatrix} a & 0 \\ 0 & d \end{vmatrix} + \begin{vmatrix} 0 & b \\ c & 0 \end{vmatrix} + \begin{vmatrix} 0 & b \\ 0 & d \end{vmatrix}$

b) $\begin{vmatrix} a & b \\ c & d \end{vmatrix} = \begin{vmatrix} a & b \\ 0 & d \end{vmatrix} - \begin{vmatrix} c & d \\ 0 & b \end{vmatrix}$

A6.12.2 Consider the equation $|A \quad \mathbf{v}| = |A \quad \mathbf{0}| + |A \quad \mathbf{v}|$, where A is a $n \times (n-1)$ matrix, and $\mathbf{0}$ and $\mathbf{v}$ are $n \times 1$ column vectors.

a) Prove this equation is true, using only the properties of the determinant. Do **NOT** compute any determinants.
b) What does this say about $|A \quad \mathbf{0}|$?

A6.12.3 Consider the equation $\begin{vmatrix} A \\ c(\mathbf{0}) \end{vmatrix} = c \begin{vmatrix} A \\ \mathbf{0} \end{vmatrix}$ where A is a $(n-1) \times n$ matrix, $\mathbf{0}$ is $1 \times n$ a row vector, and c is a scalar, with $c \neq 1$.

a) Prove this equation is true.
b) What does this say about $\begin{vmatrix} A \\ \mathbf{0} \end{vmatrix}$?

A6.12.4 Suppose A is a $n \times n$ matrix.

a) Let A' be the matrix formed by switching the last two columns of A. What is the relationship between the determinants $|A|$ and$|A'|$?
b) Suppose the last two columns of A are the same. What is the relationship between the matrices A' and A?
c) Prove/disprove: If any two or columns of A are the same, $|A| = 0$.

Activities A6.12.2 and A6.12.3 prove

Proposition 9. *If a square matrix A has a row or column of* 0*s*, $|A| = 0$.

Similarly, Activity A6.12.4 proves

Proposition 10. *If two rows or two columns of a square matrix are equal,* $|A| = 0$.

The next step up in complexity is a matrix with entries only along the main diagonal.

Activity 6.13: The Determinant of a Diagonal Matrix

A6.13.1 Let A be a $n \times n$ diagonal matrix, $A = \begin{pmatrix} a_{11} & 0 & \dots & 0 \\ 0 & a_{22} & \dots & 0 \\ \vdots & \vdots & \ddots & \vdots \\ 0 & 0 & \dots & a_{nn} \end{pmatrix}$ where all entries off the main diagonal are zero.

a) Prove: $|A| = a_{11} \begin{pmatrix} 1 & 0 & \dots & 0 \\ 0 & a_{22} & \dots & 0 \\ \vdots & \vdots & \ddots & \vdots \\ 0 & 0 & \dots & a_{nn} \end{pmatrix}$

b) Prove: $|A| = a_{11}a_{22}\cdots a_{nn}I$, where I is the identity matrix of the appropriate size.

c) Find $|A|$.

Activity 6.13 proves:

Proposition 11. *The determinant of a diagonal matrix is the product of the entries on the diagonal.*

Example 6.3. *Find* $\begin{vmatrix} 1 & 0 & 0 \\ 0 & 5 & 0 \\ 0 & 0 & -2 \end{vmatrix}$

Solution. *Since this is a diagonal matrix,*

$$\begin{vmatrix} 1 & 0 & 0 \\ 0 & 5 & 0 \\ 0 & 0 & -2 \end{vmatrix} = 1 \cdot 5 \cdot (-2) = -10$$

The next step up in complexity is a triangular matrix.

Activity 6.14: Determinants of Triangular Matrices

In the following, use **ONLY** the results of Propositions9, 10, and 11, and the definition of the determinant.

A6.14.1 Let $T = \begin{pmatrix} a & b & c \\ 0 & e & f \\ 0 & 0 & i \end{pmatrix}$

a) Using only the properties and formulas allowed above, show $\begin{vmatrix} a & b & c \\ 0 & e & f \\ 0 & 0 & i \end{vmatrix} = \begin{vmatrix} a & 0 & c \\ 0 & e & f \\ 0 & 0 & i \end{vmatrix}$.

b) Using only the properties and formulas allowed above, show $\begin{vmatrix} a & 0 & c \\ 0 & e & f \\ 0 & 0 & i \end{vmatrix} = \begin{vmatrix} a & 0 & 0 \\ 0 & e & 0 \\ 0 & 0 & i \end{vmatrix}$

c) What does this suggest about the determinant of an upper triangular matrix?

A6.14.2 Consider matrices like $\begin{pmatrix} a & b & c \\ d & e & 0 \\ f & 0 & 0 \end{pmatrix}$ and $\begin{pmatrix} 0 & 0 & a \\ 0 & b & c \\ d & e & f \end{pmatrix}$. We'll call these reversed triangular matrices.

a) Find a formula for the determinant of a reversed triangular matrix.

b) Use your formula to find $\begin{vmatrix} 3 & 1 \\ 5 & 0 \end{vmatrix}$ and $\begin{vmatrix} 0 & 3 \\ 1 & 5 \end{vmatrix}$.

c) Use your formula to find $\begin{vmatrix} 0 & 0 & 3 \\ 0 & 1 & 2 \\ 4 & 3 & 5 \end{vmatrix}$ and $\begin{vmatrix} 2 & 1 & 3 \\ 3 & 5 & 0 \\ 2 & 0 & 0 \end{vmatrix}$

We can generalize Activity A6.14.1 beyond 3×3 matrices to obtain:

Proposition 12. *The determinant of an upper or lower triangular matrix will be the product of the entries along the diagonal.*

What about nontriangular matrices? A common strategy in mathematics is to reduce a problem to one we've already solved. Since we found a formula for the determinant of a triangular matrix, might we be able to transform an arbitrary matrix into a triangular matrix? Let's begin with a 3×3 matrix.

Activity 6.15: Determinant of a 3×3 Matrix

In the following, use **ONLY** the properties of determinants, and the formula for the determinant of a triangular matrix.

We will produce a formula for computing the determinant of a 3×3 matrix.

A6.15.1 Let $A = \begin{pmatrix} a & b & c \\ d & e & f \\ g & h & i \end{pmatrix}$

a) Show

$$\begin{vmatrix} a & b & c \\ d & e & f \\ g & h & i \end{vmatrix} = \begin{vmatrix} a & b & c \\ 0 & e & f \\ 0 & h & i \end{vmatrix} + \begin{vmatrix} 0 & b & c \\ d & e & f \\ 0 & h & i \end{vmatrix} + \begin{vmatrix} 0 & b & c \\ 0 & e & f \\ g & h & i \end{vmatrix}$$

b) Using the properties of the determinant, show

$$\begin{vmatrix} a & b & c \\ d & e & f \\ g & h & i \end{vmatrix} = \begin{vmatrix} a & b & c \\ 0 & e & f \\ 0 & h & i \end{vmatrix} - \begin{vmatrix} d & e & f \\ 0 & b & c \\ 0 & h & i \end{vmatrix} + \begin{vmatrix} g & h & i \\ 0 & b & c \\ 0 & e & f \end{vmatrix}$$

c) Using the properties of the determinant, show

$$\begin{vmatrix} a & b & c \\ d & e & f \\ g & h & i \end{vmatrix} = \begin{vmatrix} a & 0 & 0 \\ 0 & e & f \\ 0 & h & i \end{vmatrix} - \begin{vmatrix} d & 0 & 0 \\ 0 & b & c \\ 0 & h & i \end{vmatrix} + \begin{vmatrix} g & 0 & 0 \\ 0 & b & c \\ 0 & e & f \end{vmatrix}$$

A6.15.2 Consider the three matrices in the row expansion in Activity A6.15.1b.

a) Show $\begin{vmatrix} a & 0 & 0 \\ 0 & e & f \\ 0 & h & i \end{vmatrix} = \begin{vmatrix} a & 0 & 0 \\ 0 & e & f \\ 0 & 0 & i \end{vmatrix} - \begin{vmatrix} a & 0 & 0 \\ 0 & h & i \\ 0 & 0 & f \end{vmatrix}$

b) Find $\begin{vmatrix} a & 0 & 0 \\ 0 & e & f \\ 0 & h & i \end{vmatrix}$

c) Find $\begin{vmatrix} d & e & f \\ 0 & b & c \\ 0 & h & i \end{vmatrix}$ and $\begin{vmatrix} g & h & i \\ 0 & b & c \\ 0 & e & f \end{vmatrix}$

d) What do your results say about finding the determinant of $\begin{pmatrix} a & b & c \\ d & e & f \\ g & h & i \end{pmatrix}$ by expanding along the first column?

A6.15.3 In the preceding, you found the determinant by expanding along the first column.

a) Find the determinant of $\begin{pmatrix} a & b & c \\ d & e & f \\ g & h & i \end{pmatrix}$ by expanding along the first row.

b) Find the determinant of $\begin{pmatrix} a & b & c \\ d & e & f \\ g & h & i \end{pmatrix}$ by expanding along the second column.

c) Does the value of the determinant change, depending on which row or column is used to expand?

In Activity 6.15, you found a formula for the determinant of a 3×3 matrix when you expanded along the first column,

$$\begin{vmatrix} a & b & c \\ d & e & f \\ g & h & i \end{vmatrix} = aei - afh - dbi + dch + gbf - gce$$

Activity A6.15.3 suggests the determinant is the same regardless of which row or column we expand along, so we can simply use this formula to find the determinant of a 3×3 matrix. This appears to be an important result (and it is), but rather than memorizing it as a rather tedious and somewhat meaningless formula, let's take a closer look at it.

First, notice the coefficients a, d, g are the entries in the first column of the original matrix. In fact, we can rewrite our formulas as:

$$\begin{vmatrix} a & b & c \\ d & e & f \\ g & h & i \end{vmatrix} = a(ei - fh) - d(bi - ch) + g(bf - ce)$$

Consider the factors $ei - fh$, $bi - ch$, and $bf - ce$. These have the form of the difference of cross products, and we can rewrite them as determinants:

$$= a\begin{vmatrix} e & f \\ h & i \end{vmatrix} - d\begin{vmatrix} b & c \\ h & i \end{vmatrix} + g\begin{vmatrix} b & c \\ e & f \end{vmatrix}$$

If you look closely at the smaller determinants, you'll see that their entries are drawn from the rows and columns *other* than those that contained the coefficients. This leads to an important definition:

Definition 6.3 (Minor). *Let A be a matrix. The minor of a_{ij} (also called the ijth minor) is the determinant of the submatrix formed by deleting the ith row and jth column of A.*

We'll use A_{ij} to indicate the ijth minor of A, and $(A)_{ij}$ to refer to the submatrix.

Example 6.4. *Let* $A = \begin{pmatrix} 2 & 5 & 1 \\ -3 & 2 & 0 \\ 3 & -1 & 5 \end{pmatrix}$. *Find the minor of a_{32}, and a_{13}.*

Solution. *To find the minor of a_{32}, we find the determinant of the submatrix formed by deleting the third row and second column, leaving*

$$\begin{pmatrix} 2 & 5 & 1 \\ -3 & 2 & 0 \\ 3 & -1 & 5 \end{pmatrix} \rightarrow \begin{vmatrix} 2 & 1 \\ -3 & 0 \end{vmatrix}$$

The minor of a_{13} will be the determinant of the matrix formed by deleting the first row and third column:

$$\begin{pmatrix} 2 & 5 & 1 \\ -3 & 2 & 0 \\ 3 & -1 & 5 \end{pmatrix} \rightarrow \begin{vmatrix} -3 & 2 \\ 3 & -1 \end{vmatrix}$$

Before we can turn this into a formula for computing the determinant of other matrices, we have to settle an important question: Sometimes, that the product of the row or column entry and the minor is added, while at other times it is subtracted. How can we determine whether to add or subtract?

Activity 6.16: Cofactors

A6.16.1 In the following, a sequence of row and/or column switches are described. Determine whether the sign of the determinant of the final matrix will be the same or opposite the determinant of the original matrix.

a) Switching the second and third rows of a matrix, then switching the first and second rows.

b) Switching the fourth and fifth columns of a matrix; switching the third and fourth columns; then switching the second and third columns.

c) Moving the fourth row of a matrix to the top row. **Note**: This is *not* the same as switching the first and fourth rows; instead, all rows will move down one place:

$$\begin{pmatrix} \mathbf{r}_1 \\ \mathbf{r}_2 \\ \mathbf{r}_3 \\ \mathbf{r}_4 \\ \vdots \end{pmatrix} \rightarrow \begin{pmatrix} \mathbf{r}_4 \\ \mathbf{r}_1 \\ \mathbf{r}_2 \\ \mathbf{r}_3 \\ \vdots \end{pmatrix}$$

d) Moving the third column of a matrix to the first column. Again, this is not switching the first and third columns, but moving all the columns over on place to make room for the next first column:

$$\begin{pmatrix} \mathbf{c}_1 & \mathbf{c}_2 & \mathbf{c}_3 & \cdots \end{pmatrix} \rightarrow \begin{pmatrix} \mathbf{c}_3 & \mathbf{c}_1 & \mathbf{c}_2 & \cdots \end{pmatrix}$$

A6.16.2 In the following, let A be a $n \times n$ matrix, and let T_{ij} be the matrix formed by moving the ith row to the first row, and the jth column to the first column. As in the preceding problem, this moves all rows down and all columns over:

$$\underbrace{\begin{pmatrix} a_{11} & a_{12} & a_{13} & \cdots \\ a_{21} & a_{22} & a_{23} & \cdots \\ a_{31} & a_{32} & a_{33} & \cdots \\ \vdots & \vdots & \ddots & \vdots \end{pmatrix}}_{A} \rightarrow \underbrace{\begin{pmatrix} a_{ij} & a_{i1} & a_{i2} & \cdots \\ a_{1j} & a_{11} & a_{12} & \cdots \\ a_{2j} & a_{21} & a_{22} & \cdots \\ \vdots & \vdots & \ddots & \vdots \end{pmatrix}}_{T_{ij}}$$

Express $|T_{ij}|$ in terms of $|A|$.

a) T_{11}.

b) T_{23}.

c) T_{57}.

d) T_{42}.

A6.16.3 Let A be a $n \times n$ matrix, and suppose we find $|A|$ by expanding along the ith row.

a) Explain why one of the terms in the expansion will be

$$\begin{vmatrix} a_{11} & a_{12} & \cdots & 0 & \cdots & a_{1n} \\ a_{21} & a_{22} & \cdots & 0 & \cdots & a_{2n} \\ \vdots & \vdots & \ddots & \vdots & \ddots & \vdots \\ 0 & 0 & \cdots & a_{ij} & \cdots & 0 \\ \vdots & \vdots & \ddots & \vdots & \ddots & \vdots \\ a_{n1} & a_{n2} & \cdots & 0 & \cdots & a_{nn} \end{vmatrix}$$

b) Find c, so

$$\begin{vmatrix} a_{11} & a_{12} & \cdots & 0 & \cdots & a_{1n} \\ a_{21} & a_{22} & \cdots & 0 & \cdots & a_{2n} \\ \vdots & \vdots & \ddots & \vdots & \ddots & \vdots \\ 0 & 0 & \cdots & a_{ij} & \cdots & 0 \\ \vdots & \vdots & \ddots & \vdots & \ddots & \vdots \\ a_{n1} & a_{n2} & \cdots & 0 & \cdots & a_{nn} \end{vmatrix} = c \begin{vmatrix} a_{ij} & 0 & 0 & \cdots & 0 \\ 0 & a_{11} & a_{12} & \cdots & a_{1n} \\ 0 & a_{21} & a_{22} & \cdots & a_{2n} \\ \vdots & \vdots & \ddots & \vdots & \vdots \\ \vdots & \vdots & \ddots & \vdots & \vdots \\ 0 & a_{n1} & a_{n2} & \cdots & a_{nn} \end{vmatrix}$$

Suggestion: Switching adjacent rows or adjacent columns will multiply the determinant by -1, so c is some power of -1. How many row switches and column switches are necessary to move a_{ij} to the first row, first column position?

Activity 6.16 leads to the following conclusion: If we find the determinant of A by expanding along the ith row, the expression will have terms of the form $(-1)^{(i-1)+(j-1)}\begin{vmatrix} a_{ij} & \mathbf{0} \\ \mathbf{0} & (A)_{ij} \end{vmatrix}$ where the **0**s are the appropriately sized and oriented zero vectors and $(A)_{ij}$ is the submatrix formed by deleting the ith row and jth column of A.

We can simplify this expression slightly. Notice that $(-1)^{(i-1)+(j-1)}$ will be positive or negative, depending on whether the exponent is even or odd. But if $(i-1)+(j-1) = i+j-2$ is even, so is $i+j$; likewise, if $i+j-2$ is odd, so is $i-j$. This leads to the following result:

Proposition 13. *Let A be a square matrix, and $(A)_{ij}$ be the matrix formed by deleting the ith row and jth column of A.*

If we expand along the ith row of A,

$$|A| = (-1)^{i+1}\begin{vmatrix} a_{i1} & \mathbf{0} \\ \mathbf{0} & (A)_{i1} \end{vmatrix} + (-1)^{i+2}\begin{vmatrix} a_{i2} & \mathbf{0} \\ \mathbf{0} & (A)_{i2} \end{vmatrix} + \ldots + (-1)^{i+n}\begin{vmatrix} a_{in} & \mathbf{0} \\ \mathbf{0} & (A)_{in} \end{vmatrix}$$

Example 6.5. *Find an expression for*

$$\begin{vmatrix} 3 & 5 & 4 & -1 \\ 2 & 1 & 3 & 4 \\ 1 & -2 & 5 & 3 \\ -2 & 1 & 5 & 4 \end{vmatrix}$$

by expanding along the third row.

Solution. *Note that*

$$(A)_{31} = \begin{pmatrix} 5 & 4 & -1 \\ 1 & 3 & 4 \\ 1 & 5 & 4 \end{pmatrix} \qquad (A)_{32} = \begin{pmatrix} 3 & 4 & -1 \\ 2 & 3 & 4 \\ -2 & 5 & 4 \end{pmatrix}$$

$$(A)_{33} = \begin{pmatrix} 3 & 5 & -1 \\ 2 & 1 & 4 \\ -2 & 1 & 4 \end{pmatrix} \qquad (A)_{34} = \begin{pmatrix} 3 & 5 & 4 \\ 2 & 1 & 3 \\ -2 & 1 & 5 \end{pmatrix}$$

Proposition 13, gives us

$$\begin{vmatrix} 3 & 5 & 4 & -1 \\ 2 & 1 & 3 & 4 \\ 1 & -2 & 5 & 3 \\ -2 & 1 & 5 & 4 \end{vmatrix} = (-1)^{3+1}\begin{vmatrix} 1 & \mathbf{0} \\ \mathbf{0} & (A)_{31} \end{vmatrix} + (-1)^{3+2}\begin{vmatrix} -2 & \mathbf{0} \\ \mathbf{0} & (A)_{32} \end{vmatrix}$$
$$+ (-1)^{3+3}\begin{vmatrix} 5 & \mathbf{0} \\ \mathbf{0} & (A)_{33} \end{vmatrix} + (-1)^{3+4}\begin{vmatrix} 3 & \mathbf{0} \\ \mathbf{0} & (A)_{34} \end{vmatrix}$$

where the **0***s are appropriately sized and oriented zero vectors.*

Now that we know whether to add or subtract the determinants in the expansion, we can focus on finding the value of determinants of a particular type.

Activity 6.17: Cofactor Expansion

A6.17.1 Prove that the determinant of a $n \times n$ matrix can be expressed as the sum of the determinants of triangular matrices. Use the following steps.

a) Prove the base step: The determinant of any 2×2 matrix can be expressed as a sum of determinants of triangular matrices.

b) Prove the induction step: If the determinant of any $k \times k$ matrix can be expressed as the sum of determinants of triangular matrices, then the determinant of any $(k+1) \times (k+1)$ matrix can be expressed as the sum of determinants of triangular matrices.

A6.17.2 Let M be a square matrix where $|M| = |M_1| + |M_2|$ from the linearity property of the determinant. Show that $\begin{vmatrix} a & \mathbf{0} \\ \mathbf{0} & M \end{vmatrix} = \begin{vmatrix} a & \mathbf{0} \\ \mathbf{0} & M_1 \end{vmatrix} + \begin{vmatrix} a & \mathbf{0} \\ \mathbf{0} & M_2 \end{vmatrix}$

A6.17.3 In the following, you may use the result that the determinant of a triangular matrix is the product of the terms along its main diagonal.

a) Prove: $\begin{vmatrix} a_{11} & \mathbf{0} \\ \mathbf{0} & \Delta \end{vmatrix} = a_{11} |\Delta|$, where $\mathbf{0}$s are appropriately sized and oriented zero vectors, and Δ is a triangular matrix.

b) Prove $\begin{vmatrix} a_{11} & \mathbf{0} \\ \mathbf{0} & M \end{vmatrix} = a_{11} |M|$.

A6.17.4 According to Proposition 13, if we find the determinant of any matrix A by expanding along the ith row, the expression will have terms of the form

$$|A| = (-1)^{i+1} \begin{vmatrix} a_{i1} & \mathbf{0} \\ \mathbf{0} & A_{i,1} \end{vmatrix} + (-1)^{i+2} \begin{vmatrix} a_{i2} & \mathbf{0} \\ \mathbf{0} & A_{i,2} \end{vmatrix} + \ldots + (-1)^{i+n} \begin{vmatrix} a_{in} & \mathbf{0} \\ \mathbf{0} & A_{i,n} \end{vmatrix}$$

Rewrite this equation in terms of the minors of A.

Activity A6.17.3 proves:

Proposition 14. *Let a be a scalar, M a square matrix, and $\mathbf{0}$ appropriately sized and oriented zero vectors. Then* $\begin{vmatrix} a & \mathbf{0} \\ \mathbf{0} & M \end{vmatrix} = a|M|$

In Activity A6.17.4, we put this result together with Proposition 13 to obtain a general formula for the determinant. To summarize the results, it will be helpful to introduce the following term:

Definition 6.4 (Cofactor). *Let A be a square matrix. The ijth cofactor is $(-1)^{i+j} A_{ij}$, where A_{ij} is the ijth minor of A.*

Thus:

Proposition 15 (Cofactor Expansion). *To find $|A|$, choose any row or column.*

$|A|$ will be the sum of the products of the row or column entry with the corresponding cofactors.

Equivalently: If we choose to expand along the entries $a_{i1}, a_{i2}, \ldots, a_{in}$ of the ith row of A, then:

$$|A| = a_{i1}(-1)^{i+1}A_{i1} + a_{i2}(-1)^{i+2}A_{i2} + \ldots + a_{in}(-1)^{i+n}A_{in}$$

while if we choose to expand along the entries $a_{1j}, a_{2j}, \ldots, a_{nj}$ along the jth column of A, then:

$$|A| = a_{1j}(-1)^{1+j}A_{1j} + a_{2j}(-1)^{2+j}|A_{2j} + \ldots + a_{nj}(-1)^{n+j}A_{nj}$$

This formula is also referred to as the **Laplace expansion** for computing the determinant, after Pierre-Simon Laplàce (1749–1827).

Example 6.6. *Find the determinant by expanding along the third row.*
$\begin{vmatrix} 2 & 1 & 7 \\ 3 & 0 & 5 \\ -3 & 2 & 1 \end{vmatrix}$

Solution. *The entries of the third row are -3, 2, and 1.*

The cofactor of -3 (the entry in the 3rd row, 1st column) will be $(-1)^{3+1} = 1$ multiplied by the minor:

$$\begin{pmatrix} 2 & 1 & 7 \\ 3 & 0 & 5 \\ -3 & 2 & 1 \end{pmatrix} \rightarrow \begin{vmatrix} 1 & 7 \\ 0 & 5 \end{vmatrix}$$

The cofactor of 2 (the entry in the 3rd row, 2nd column) will be $(-1)^{3+2} = -1$ multiplied by the minor:

$$\begin{pmatrix} 2 & 1 & 7 \\ 3 & 0 & 5 \\ -3 & 2 & 1 \end{pmatrix} \rightarrow \begin{vmatrix} 2 & 7 \\ 3 & 5 \end{vmatrix}$$

Finally the cofactor of 1 (the entry in the 3rd row, 3rd column) will be $(-1)^{3+3} = 1$ multiplied by the minor

$$\begin{pmatrix} 2 & 1 & 7 \\ 3 & 0 & 5 \\ -3 & 2 & 1 \end{pmatrix} \rightarrow \begin{vmatrix} 2 & 1 \\ 3 & 0 \end{vmatrix}$$

The determinant will be the sum of the entries of the third row multiplied by the cofactors:

$$\begin{vmatrix} 2 & 1 & 7 \\ 3 & 0 & 5 \\ -3 & 2 & 1 \end{vmatrix} = (-3)(1)\begin{vmatrix} 1 & 7 \\ 0 & 5 \end{vmatrix} + (2)(-1)\begin{vmatrix} 2 & 7 \\ 3 & 5 \end{vmatrix} + (1)(1)\begin{vmatrix} 2 & 1 \\ 3 & 0 \end{vmatrix}$$

Using the formula for the determinant of a 2×2 *matrix:*

$$\begin{aligned} &= (-3)(1 \cdot 5 - 7 \cdot 0) + (-2)(2 \cdot 5 - 7 \cdot 3) + (1)(2 \cdot 0 - 1 \cdot 3) \\ &= (-3)(5) + (-2)(-11) + (1)(-3) \\ &= 4 \end{aligned}$$

Note that when computing determinants this way, we must repeatedly compute powers of the form $(-1)^{i+j}$. However, since these powers are either 1 or -1, we really only have to decide whether to add or subtract.

Activity 6.18: The Cofactor Checkerboard

In the following, let A be a square matrix, and let $A_{i,j}$ be the matrix formed by deleting the ith row and jth column of A.

A6.18.1 Suppose you're finding $|A|$ by expanding along the first row of A.

a) Write an expression for the cofactors of a_{11}, a_{12}, a_{13}.
b) Your expressions should have factors of $(-1)^{i+j}$. Simplify these powers.
c) What do you notice?

A6.18.2 Suppose you're finding $|A|$ by expanding along the second column.

a) Write an expression for the cofactors of a_{21}, a_{22}, a_{23}.
b) Your expressions should have factors of $(-1)^{i+j}$. Simplify these powers.
c) What do you notice?

Activity 6.18 leads to the "cofactor checkerboard," which is a visual reminder of when to add and when to subtract the minor:

$$\begin{pmatrix} + & - & + & - & \cdots \\ - & + & - & + & \cdots \\ + & - & + & - & \cdots \\ \vdots & \vdots & \vdots & \vdots & \ddots \end{pmatrix}$$

6.6 The Determinant Formula

A useful strategy:

Strategy. *Verify that a method works in a case where you know the answer.*

In Definition 6.2, we chose the properties the determinant should have; and in Propositions 9, 10, and others we established properties of the determinant. It's conceivable that all our work has produced a formula for something that isn't actually the determinant. So let's verify that the determinant computed by the formula has the same properties as the actual determinant.

Activity 6.19: Finding Determinants

In the following, do **NOT** use any properties of the determinant. Instead, **USE** the cofactor expansion to compute the determinant by finding its minors and cofactors.

A6.19.1 Let $A = \begin{pmatrix} 1 & -1 & 4 \\ 2 & 0 & -1 \\ 7 & 0 & 1 \end{pmatrix}$ Compute $|A|$ as indicated.

a) Find $|A|$ by expanding along the first row.
b) Find $|A|$ by expanding along the third column.
c) Find $|A|$ by expanding along the second row.
d) Find $|A|$ by expanding along the second column.
e) Do your results support the claim that the determinant is the same regardless of which row or column we use to expand along?
f) While you could find the determinant by expanding along any row or column, does your work suggest which row or column you might want to choose?

A6.19.2 Let $B = \begin{pmatrix} 2 & -3 & 1 \\ 1 & 2 & 5 \\ 2 & -3 & 1 \end{pmatrix}$ One of the properties of the determinant is that if two rows of a matrix are the same, the determinant is 0. Verify or refute that the determinant formula satisfies this property by computing $|B|$.

A6.19.3 Let $C = \begin{pmatrix} 1 & 2 & 0 \\ -1 & 5 & 1 \\ 1 & -1 & -1 \end{pmatrix}$ and $D = \begin{pmatrix} 1 & 2 & 0 \\ -1 & 5 & 2 \\ 1 & -1 & -2 \end{pmatrix}$ where D is formed by multiplying the third column of C by 2.

a) Find $|C|$ by expanding along the first row.
b) Find $|D|$ by expanding along the first column.
c) One of the required properties of the determinant is that if a row or column is multiplied by a constant, the determinant will also be multiplied by the same constant. Does the determinant formula appear to satisfy this property?

A6.19.4 Let $E = \begin{pmatrix} 1 & 2 & -3 \\ 3 & 0 & 5 \\ 1 & -2 & 4 \end{pmatrix}$ and $F = \begin{pmatrix} 1 & 2 & -3 \\ 1 & -2 & 4 \\ 3 & 0 & 5 \end{pmatrix}$ where F is formed by switching the second and third rows of E.

a) Find $|E|$. Use any row or column to expand.

b) Find $|F|$. Use any row or column to expand.

c) One of the required properties of the determinant is that switching adjacent rows or columns will change the sign of the determinant. Does the determinant formula satisfy this property?

A6.19.5 Let $T = \begin{pmatrix} 2 & 5 & 1 \\ 0 & 3 & 4 \\ 0 & 0 & -7 \end{pmatrix}$

a) Find $|T|$.

b) Does your result support the claim that the determinant of an upper triangular matrix is the product of the terms along the main diagonal?

A6.19.6 Let $G = \begin{pmatrix} 3 & 1 & 7 \\ 2 & 5 & 8 \\ 1 & 4 & 3 \end{pmatrix}$.

a) Choose matrices G_1, G_2, where (by the linearity property of the determinant) $|G| = |G_1| + |G_2|$.

b) Use the formula to compute $|G|$, $|G_1|$, and $|G_2|$.

c) Does the determinant formula satisfy the required linearity property of the determinant?

It appears that our formula for calculating the determinant does in fact give us the determinant.

There's one last problem. We should view the determinant as a function on the entries of a matrix. But we could easily imagine other formulas on the entries of a matrix, so it's possible that there might be more than one way to define a determinant on a matrix. In particular, *any* function that meets the requirements of Definition 6.2 is a determinant, so might there be more than one type of determinant?

Activity 6.20: Uniqueness of the Determinant

In the following, use **ONLY** the definition and properties of the determinant; in particular, **DO NOT** use any formula to compute the determinant.

A6.20.1 Consider the 2×2 matrix $\begin{pmatrix} a & b \\ c & d \end{pmatrix}$. Prove, using only the properties of the determinant, $\begin{vmatrix} a & b \\ c & d \end{vmatrix} = ad - bc$

Activity 6.20 shows that the determinant formula for a 2×2 matrix is unique: no other formula on the entries of a 2×2 matrix will meet all the requirements for being a determinant.

So what about larger matrices? Note that we proved Proposition 15 without using any preselected formula for the determinant. Thus, regardless of what formula we use to compute a determinant, we can always find the determinant of a matrix by choosing a row or column, then finding the sum of the product of the row or column entries with the corresponding cofactors.

This means the determinant of a 3×3 matrix will be based on whatever formula we have for the determinant of a 2×2 matrix. But since there's only one possible formula for the determinant of a 2×2 matrix, it follows there is only one possible formula for the determinant of a 3×3 matrix.

Similarly, the determinant of a 4×4 matrix will be based on whatever formula we have for the determinant of a 3×3 matrix. But since there's only one possible formula for the determinant of a 3×3 matrix, it follows there's only on possible formula for the determinant of a 4×4 matrix. It should be clear that this means that there is only one possible formula for the determinant of a $n \times n$ matrix, and so we claim:

Theorem 6.2 (Uniqueness of Determinant). *The determinant of a matrix is a unique value.*

Thus, we can speak of *the* determinant of a matrix.

Activity 6.21: Finding Determinants: Cross Products

A6.21.1 One way to visualize the determinant of a 2×2 matrix is that it is the product of the "down right diagonal" entries, minus the product of the "down left" diagonal entries.

a) Show that the difference of the product of the "down right" diagonal entries and the product of the "down left" diagonal entries does **NOT** give the correct value for the determinant of a 3×3 matrix.

b) Suppose we consider the double-wide matrix

$$\left(\begin{array}{ccc|ccc} a & b & c & a & b & c \\ d & e & f & d & e & f \\ g & h & j & g & h & j \end{array}\right)$$

There are six "diagonals" (more appropriately described as "bishop paths"): the "down right" diagonals aej, bfg, cdh; and the "down left" diagonals ceg, afh, bdj. What is the relationship of these diagonals to the added and subtracted quantities in the computation of the determinant of a 3×3 matrix?

A6.21.2 Suppose we try to extend the procedure in Activity A6.21.1 this computation to the determinant of a 4×4 matrix:

$$\left(\begin{array}{cccc|cccc} a & b & c & d & a & b & c & d \\ e & f & g & h & e & f & g & h \\ j & k & m & n & j & k & m & n \\ p & q & r & s & p & q & r & s \end{array}\right)$$

a) There are four distinct "down right" diagonals, and four distinct "down left" diagonals. Find them.

b) Explain why this method *cannot* be used to compute the determinant. Suggestion: How many terms should appear in the computation of the determinant?

Activity 6.21 leads to what is known as **Sarrus's Rule** for finding the determinant of a 3×3 matrix. Unfortunately, Sarrus's rule does not generalize to larger matrices, so it is an interesting curiosity of no real value.

6.7 More Properties of the Determinant

Proof is a central feature of higher mathematics. What might not be apparent is that *reproof* is also a feature of higher mathematics: proving a known result using a different approach. The value of a new proof is that it may offer insights into other problems, or draw attention to a different aspect of a concept. We might draw analogy: You might know one route to get from your home to your job. However, by taking different routes, you might discover a new favorite coffee shop or bookstore.

Earlier, you found several properties of determinants, based only on the definition of the determinant. This led to the Laplace expansion. We now reverse the process: suppose we *define* the determinant of a matrix based on the Laplace expansion. We can then try to prove the properties, and by doing so, gain further insight into the determinant itself.

Activity 6.22: The Laplace Expansion

In the following, use the Laplace expansion to find the determinant, but do **NOT** use or otherwise refer to any of the previously derived properties of the determinant.

A6.22.1 Prove $|I| = 1$. Use the following steps.

a) Prove the base case: $|I| = 1$, for a 2×2 identity matrix.

b) Prove that if the determinant of the $n \times n$ identity matrix is 1, then the determinant of the $(n+1) \times (n+1)$ identity matrix is also 1.

c) Explain why these two steps prove the general result.

A6.22.2 Prove: If B is a matrix formed by switching two adjacent rows or two adjacent columns of A, then $|B| = -|A|$. Suggestion: Find $|A|$ by expanding along one of the rows or columns that are switched.

A6.22.3 Prove: If $A = \left(\begin{array}{c|c} A' & \mathbf{u} \end{array} \right)$, $B = \left(\begin{array}{c|c} A' & \mathbf{v} \end{array} \right)$, and $C = \left(\begin{array}{c|c} A' & \mathbf{u}+\mathbf{v} \end{array} \right)$, then $|A| = |B| + |C|$.

Activity 6.22 shows that if we define the determinant by the Laplace expansion formula, it will be a function that meets all the requirements of Definition 6.2. Since the Laplace expansion formula was a consequence of Definition 6.2, we can summarize the two results as follows:

- If something meets the requirements of Definition 6.2, then the Laplace expansion formula holds.
- If the Laplace expansion formula holds for something, then it meets the requirements of Definition 6.2.

This gives us an if-and-only-if construction. Since one of the clauses is the definition of the determinant, it means that Definition 6.2, which defines the determinant; and the Laplace expansion formula, which computes the determinant, are equivalent definitions. In other words, we can use *either* Definition 6.2 *or* the Laplace expansion formula to define the determinant.

Suppose we used the Laplace expansion formula to define the determinant (many authors do). Could we derive all of our previous results? Certainly: since the Laplace expansion formula tells us the determinant meets all the requirements of Definition 6.2, the proofs we gave earlier still hold. However, while you might know the route from your favorite coffee shop to your home, and the route from your home to your favorite bookstore, this doesn't necessarily mean the best way to get from your favorite coffee shop to your favorite bookstore is to return home first. You might be able to find a more direct route.

Activity 6.23: Determinant of Triangular Matrices

Earlier, you showed that the determinant of an upper triangular matrix is the product of the terms along its diagonal. In the following, do **NOT** use or refer to this result. However, you may use the Laplace expansion formula for computing the determinant.

A6.23.1 We'll prove our result using induction.

a) Let T be the 2×2 upper triangular matrix

$$T = \begin{pmatrix} t_{11} & t_{12} \\ 0 & t_{22} \end{pmatrix}$$

Prove: The determinant of a 2×2 upper triangular matrix is the product of the terms along its main diagonal.

b) Explain why, if T is an upper triangular matrix, then $\left(\begin{array}{c|c} a_{11} & \mathbf{u} \\ \hline \mathbf{0} & T \end{array}\right)$ is a $(n+1) \times (n+1)$ upper triangular matrix.

c) Explain why, if T is an upper triangular matrix, $\left|\begin{array}{c|c} a_{11} & \mathbf{u} \\ \hline \mathbf{0} & T \end{array}\right| = a_{11}\,|T|$

d) Explain why, if $|T|$ is the product of the terms along its main diagonal, then $\left|\begin{array}{c|c} a_{11} & \mathbf{u} \\ \hline \mathbf{0} & T \end{array}\right| = a_{11}\,|T|$ will also be the product of the terms along the main diagonal of the block matrix.

e) Explain why this mean that the determinant of *any* upper triangular matrix will be the product of the terms along its main diagonal. Suggestion: Explain why the theorem is true for a 3×3 upper triangular matrix. Why does this tell you the theorem is also true for a 4×4 upper triangular matrix?

A6.23.2 Consider matrices like

$$A = \begin{pmatrix} a & b & c \\ d & e & 0 \\ f & 0 & 0 \end{pmatrix}, B = \begin{pmatrix} 0 & 0 & g \\ 0 & h & j \\ k & m & n \end{pmatrix}$$

a) Explain why these are *neither* upper triangular *nor* lower triangular matrices.

b) Find their determinants.

What other properties might the determinant have? Since a matrix corresponds to a linear transformation, we might draw inspiration from properties of linear transformations.

Activity 6.24: More Determinants, More Transformations

A6.24.1 Let P, Q be 2×2 transformation matrices. Base your answers to the following on the geometric interpretation of the determinant as a measure of how the orientation and content of a n-cube changes under a transformation.

a) What is the area of a unit square after the transformation P is applied to it?

b) What is the area of a unit square after the transformation Q is applied to it?

c) What is the area of a unit square after the transformation PQ is applied to it?

d) What does this suggest about the relationship between $|P|$, $|Q|$, and $|PQ|$?

A6.24.2 Let R be a $n \times n$ matrix corresponding to a rotation. Explain why $|R| = 1$.

A6.24.3 Suppose

$$A = \begin{pmatrix} a & b \\ c & d \end{pmatrix}, B = \begin{pmatrix} e & f \\ g & h \end{pmatrix}$$

Prove/disprove: $|A||B| = |AB|$.

Activity A6.24.3 proves:

Proposition 16. *Let A, B be 2×2 matrices. Then $|AB| = |A||B|$.*

We might try to generalize this. Unfortunately, when we multiply two matrices, the entries of the product are complicated expressions, so for larger matrices, the proof is an algebraic nightmare.

One strategy is to make the problem simpler: rather than trying to find the determinant of the product of *any* two matrices, what if we try to find the determinant of the product of two "simple" matrices?

Activity 6.25: Determinants of Diagonal and Triangular Matrices

A6.25.1 Let Δ_1 and Δ_2 be two diagonal matrices of the same size.

a) Prove: $\Delta_1\Delta_2$ is a diagonal matrix whose diagonal entries are the products of the corresponding entries of Δ_1 and Δ_2.

b) Prove: $|\Delta_1\Delta_2| = |\Delta_1||\Delta_2|$.

A6.25.2 Let L_1, L_2 be two lower triangular matrices of the same size.

a) Prove: L_1L_2 is a lower triangular matrix whose diagonal entries are the products of the corresponding entries of L_1 and L_2.

b) Prove: $|L_1 L_2| = |L_1||L_2|$.

When building a proof, mathematicians often rely on lemmas: results that minor in and of themselves, but are important because they lead to more general results. Activity 6.25 gives:

Lemma 2. *The determinant of a product of diagonal matrices is the product of the determinants.*

Likewise:

Lemma 3. *The determinant of a product of lower triangular matrices is the product of the determinants.*

It should be clear that

Lemma 4. *The determinant of a product of upper triangular matrices is the product of the determinants.*

This might seem useful but limited, since not all matrices are diagonal or triangular. However, there's an important set of triangular matrices: the elementary matrices (see Definition 4.8).

Activity 6.26: More Elementary Matrices

A6.26.1 Let A be a $n \times n$ coefficient matrix, and let T be the elementary matrix such that $A \xrightarrow{cR_i \to R_i} TA$.

a) Find $|T|$.
b) Based **ONLY** on the properties of the determinant, explain why $|TA| = c|A|$.
c) Prove: $|TA| = |T||A|$.

A6.26.2 Let A be a $n \times n$ coefficient matrix, and let U be the elementary matrix such that $A \xrightarrow{R_i \leftrightarrow R_{i+1}} UA$.

a) Find $|U|$.
b) Based **ONLY** on the properties of the determinant, explain why $|UA| = -|A|$.
c) Prove: $|UA| = |U||A|$.

A6.26.3 Let A be a $n \times n$ coefficient matrix, and let L be the elementary matrix such that $A \xrightarrow{cR_i + R_j \to R_j} LA$.

a) Find $|L|$.

b) Based **ONLY** on the properties of the determinant, explain why $|LA| = |A|$.

Activity 6.26 leads to the following:

Lemma 5. *Let T be an elementary matrix.* $|T| \neq 0$.

Indeed, in most cases $|T| = 1$ or $|T| = -1$.
Additionally:

Lemma 6. *Let A be any square matrix, and let T be an elementary matrix corresponding to a row operation performed on A. Then* $|TA| = |T||A|$.

Thus, while we don't know if the determinant of a product of two arbitrary matrices is the product of their determinants, we do know this holds if we're multiplying on the left by an elementary matrix. "Lather, rinse, repeat" to generalize this to:

Lemma 7. *Let T_1, T_2, ..., T_n be a sequence of elementary matrices where $T_n T_{n-1} \cdots T_2 T_1 A = B$. Then* $|T_n||T_{n-1}| \cdots |T_2||T_1||A| = |B|$.

While we can apply *any* sequence of elementary row operations to a matrix, we usually want to apply a sequence of elementary row operations that will allow us to solve the corresponding system of equations. This leads to an important result.

Activity 6.27: Determinants and Rank

In the following, suppose A is the coefficient matrix corresponding to a system of n equations in n unknowns, and let T_1, T_2, ..., T_k be elementary matrices where $T_k T_{k-1} \cdots T_2 T_1 A = B$, the row echelon form of A.

A6.27.1 Suppose the rank of A is n.

a) Explain why you know $|B| \neq 0$.
b) Why does this mean $|A| \neq 0$?

A6.27.2 Suppose $|A| = 0$.

a) Find $|B|$.
b) What does this say about the entries of B?

Activity 6.27 leads to an important result:

Theorem 6.3. *Suppose A is a $n \times n$ square matrix. The rank of A is n if and only if* $|A| = 0$.

In this theorem, we see one of the side effects of proof: We are *trying* to prove that the determinant of a product is the product of the determinants. But in the process of trying to prove that statement, we came upon an even greater discovery: the determinant of a matrix tells us whether a system of equations has a unique solution! This is sometimes known as the *serendipity effect*: finding something useful, interesting, or valuable when you're looking for something entirely unrelated.

Two more good strategies for mathematics are:

Strategy. *Look for corollaries and make conjectures.*

A *corollary* might be viewed as an "obvious" application of a main theorem, while a *conjecture* is a guess of what *might* be true.

Activity 6.28: Determinants and Inverses

In Activity 4.23 you found that the inverse of A could be found by row reducing the augmented coefficient matrix

$$\left(A \mid I \right) \rightarrow \left(I \mid A^{-1} \right)$$

A6.28.1 Suppose $|A| \neq 0$.

a) Explain why there must be a sequence of elementary row operations that transform $A \rightarrow I$.

b) Prove: If $|A| \neq 0$, then A^{-1} exists.

A6.28.2 Suppose $|A| = 0$.

a) Explain why there **CAN'T** be a sequence of elementary row operations that transform $A \rightarrow I$.

b) Prove: If $|A| = 0$, then A^{-1} does not exist.

A6.28.3 Remember one interpretation of $|A|$ for a 2×2 matrix is that it describes the area of a unit square transformed by A.

a) Suppose $|A| = c$. What would you expect $|A^{-1}|$ to be?

b) Let $A = \begin{pmatrix} a & b \\ c & d \end{pmatrix}$. Find A^{-1}.

c) Find $|A^{-1}|$. Does your result support your conjecture in Activity A6.28.3a?

Activity 6.28 gives us:

Corollary 1. A^{-1} *exists if and only if* $|A| \neq 0$.

Note that while a corollary is a minor result, it still requires a proof!

In Activity A6.28.3a, we returned to the idea that the determinant tells us something about how a linear transformation affects a region. This suggests the conjecture:

Conjecture 1. $|A^{-1}| = \dfrac{1}{|A|}$

Direct computation will show this is true for a 2×2 matrix. However, the formulas for determinants of larger matrices make this type of proof infeasible for larger matrices.

What about our quest to prove that the determinant of a product is the product of the determinants?

Activity 6.29: The Determinant of a Product

Let A be a $n \times n$ matrix.

A6.29.1 Suppose T is the elementary matrix where $A \xrightarrow{R_i \leftrightarrow R_{i+1}} TA$.

a) Find T^{-1}.

b) Suppose T^{-1} is applied to a $n \times n$ matrix B. If it exists, identify the elementary row operation $B \xrightarrow{?} T^{-1}B$.

A6.29.2 Suppose T is the elementary matrix where $A \xrightarrow{cR_i \to R_i} TA$.

a) Find T^{-1}.

b) Suppose T^{-1} is applied to a $n \times n$ matrix B. If it exists, identify the elementary row operation $B \xrightarrow{?} T^{-1}B$.

A6.29.3 Suppose T is the elementary matrix where $A \xrightarrow{cR_i + R_j \to R_j} TA$.

a) Find T^{-1}.

b) Suppose T^{-1} is applied to a $n \times n$ matrix B. If it exists, identify the elementary row operation $B \xrightarrow{?} T^{-1}B$.

A6.29.4 Suppose A, B are $n \times n$ matrices where $|A|$, $|B|$ are nonzero.

a) Prove: There is a sequence of elementary row operations R_1, R_2, ..., R_k where $A = R_1 R_2 \cdots R_{k-1} R_k$. Suggestion: Since $|A| \neq 0$, you know there is a sequence of elementary row operations T_1, T_2, ..., T_k where $T_k T_{k-1} \cdots T_2 T_1 A = I$.

b) Prove: There is a sequence of elementary row operations Q_1, Q_2, ..., Q_m where $B = Q_1 Q_2 \cdots Q_{m-1} Q_m$.

c) Express $|A|$, $|B|$ in terms of the $|R_i|$s and $|Q_j|$s.

d) Prove: $|AB| = |A||B|$.

A6.29.5 Suppose C is the reduced row echelon form of a $n \times n$ matrix, where the rank of C is less than n.

a) What is $|C|$?

b) Let M be any matrix of the appropriate size. Prove/disprove: CM must end in a row of 0s.

c) What does this say about $|CM|$? (Your answer should not refer to any computation of the determinant)

d) Prove: $|CM| = |C||M|$.

Activity 6.29 shows that if A, B are two $n \times n$ matrices with nonzero determinants, $|AB| = |A||B|$.

What if one of them has determinant 0? If $|B|$ has determinant zero, then AB will be

$$R_1 R_2 \cdots R_k Q_1 Q_2 \cdots Q_m C$$

where C is the reduced row echelon form of B. Since $|C| = 0$, and the remaining matrices are elementary matrices, the determinant of this product will be 0.

If $|A|$ has determinant 0, then AB will be

$$R_1 R_2 \cdots R_k C Q_1 Q_2 \cdots Q_m$$

where C is the reduced row echelon form of A. Consider the product $CQ_1Q_2 \cdots Q_m$. By Activity A6.29.5, this product will be a matrix with determinant 0. Since the remaining operations R_k, R_{k-1}, ... are elementary row operations, the determinant of the entire product will also be 0, and we again we will have $|AB| = |A||B|$. Thus:

Theorem 6.4 (Determinant of a Product). *If A, B are $n \times n$ matrices, $|AB| = |A||B|$.*

Again, it's helpful to look for corollaries.

Activity 6.30: Determinants and Inverses, Continued

A6.30.1 Suppose A is an invertible matrix.

a) Find $|AA^{-1}|$.

b) Prove: $|A^{-1}| = \dfrac{1}{|A|}$.

A6.30.2 Let A be a $n \times n$ matrix with integer entries.

a) Explain why $|A|$ must be an integer.

b) Suppose every term in a row of A has a factor c. Prove: $|A|$ also has a factor c.

c) Prove: If $|A|$ is prime, then no row or column has a common factor among its entries.

d) Suppose $|A|$ has a factor of c. Does this mean that a row or column of A must have a common factor of c among its entries? Prove or give a counterexample.

Activity A6.30.1 proves Conjecture 1. Since this is a simple result based off Theorem 6.4, we call it a corollary:

Corollary 2. *If A is invertible,* $|A^{-1}| = \frac{1}{|A|}$.

6.8 More Computations of the Determinant

An important concept in computer science is *computational complexity*. A complete introduction to computational complexity is beyond the scope of this book, so we'll focus on counting computations.

Activity 6.31: Computing the Determinant, Part One

In the following, assume that any multiplication counts as one operation; however, remember that a sequence of operations must be done pairwise, so an expression like $abcd$ requires three computations: ab; this product times c; and this product times d.

A6.31.1 Suppose you're finding the determinant of a matrix using the Laplace/cofactor expansion.

a) How many computations are needed to compute the determinant of a 2×2 matrix?

b) How many computations are needed to compute the determinant of a 3×3 matrix?

c) How many computations are needed to compute the determinant of a $n \times n$ matrix?

A6.31.2 Suppose A is a $n \times n$ invertible coefficient matrix.

a) How many computations are necessary to multiply a row of A by a constant c?

b) How many computations are necessary to multiply a row of A by c and add it to another row?

c) How many computations are necessary to eliminate the entries below the first row pivot?

d) How many computations are necessary to obtain the row echelon form of A?

Activity 6.31 shows that using the cofactor expansion to find the determinant of a $n \times n$ matrix requires performing $n!$ multiplications, where $n!$ is the product of the whole numbers from 1 to n. This value grows very rapidly: using the cofactor expansion to find the determinant of a 5×5 matrix will require $5! = 120$ multiplications, but for a 10×10 matrix, the number grows to $10! = 3,628,800$.

Can we do better?

Activity 6.32: Finding Determinants by Row Reduction

The following develops an alternative approach to finding the determinant of a matrix.

A6.32.1 Strictly speaking, elementary matrices are produced by performing a *single* elementary row operation on the identity matrix. However, we often perform several row operations at the same time. In each of the following, find $|T|$, where TA is the matrix produced by performing the given elementary row operations on A.

a) $\xrightarrow[]{\substack{cR_2 \to R_2 \\ cR_3 \to R_3 \\ cR_4 \to R_4}}$

b) $\xrightarrow[]{\substack{a_1R_1 + R_2 \to R_2 \\ a_2R_1 + R_3 \to R_3 \\ a_3R_1 + R_4 \to R_4}}$

c) $\xrightarrow[]{\substack{b_4R_5 + R_4 \to R_4 \\ b_3R_5 + R_3 \to R_3 \\ b_2R_5 + R_2 \to R_2 \\ b_1R_5 + R_1 \to R_1}}$

A6.32.2 Suppose $A = \begin{pmatrix} 2 & 3 & -1 \\ 5 & 7 & 3 \\ 3 & 5 & 4 \end{pmatrix}$.

a) Find $|A|$ using the cofactor expansion.

b) Suppose you row reduced A using the *fang cheng* algorithm. We can do this in four steps:

$$\begin{pmatrix} 2 & 3 & -1 \\ 5 & 7 & 3 \\ 3 & 5 & 4 \end{pmatrix} \xrightarrow[2R_3 \to R_3]{2R_2 \to R_2} \begin{pmatrix} 2 & 3 & -1 \\ 10 & 14 & 6 \\ 6 & 10 & 8 \end{pmatrix} \xrightarrow[-3R_1 + R_3 \to R_3]{-5R_1 + R_2 \to R_2} \begin{pmatrix} 2 & 3 & -1 \\ 0 & -1 & 11 \\ 0 & 1 & 11 \end{pmatrix}$$

$$\xrightarrow{-1R_2 \to R_2} \begin{pmatrix} 2 & 3 & -1 \\ 0 & 1 & -11 \\ 0 & 1 & 11 \end{pmatrix} \xrightarrow{R_3 + (-R_2) \to R_3} \begin{pmatrix} 2 & 3 & -1 \\ 0 & 1 & -11 \\ 0 & 0 & 22 \end{pmatrix}$$

Find the matrices T_1, T_2, T_3, T_4 that perform each of these steps. For example, find a matrix T_1 where $A \xrightarrow[2R_3 \to R_3]{2R_2 \to R_2} T_1 A$.

c) Explain why

$$T_4 T_3 T_2 T_1 A = \begin{pmatrix} 2 & 3 & -1 \\ 0 & 1 & -11 \\ 0 & 0 & 22 \end{pmatrix}$$

d) Use the result of Activity A6.32.2c to find $|A|$. Verify your answer by comparing it with the determinant found using the cofactor method.

A6.32.3 Suppose $C = \begin{pmatrix} 3 & 5 & 3 & 1 \\ 5 & 2 & 7 & 6 \\ 2 & 3 & 5 & 1 \\ 1 & 1 & 4 & 5 \end{pmatrix}$ To get C into row echelon form, our first step could be one of two things.

a) Suppose our first step is

$$\begin{pmatrix} 3 & 5 & 3 & 1 \\ 5 & 2 & 7 & 6 \\ 2 & 3 & 5 & 1 \\ 1 & 1 & 4 & 5 \end{pmatrix} \xrightarrow{R_1 \leftrightarrow R_4} \begin{pmatrix} 1 & 1 & 4 & 5 \\ 5 & 2 & 7 & 6 \\ 2 & 3 & 5 & 1 \\ 3 & 5 & 3 & 1 \end{pmatrix}$$

Find a matrix T that performs this step.

b) Alternatively, our first step could be

$$\begin{pmatrix} 3 & 5 & 3 & 1 \\ 5 & 2 & 7 & 6 \\ 2 & 3 & 5 & 1 \\ 1 & 1 & 4 & 5 \end{pmatrix} \xrightarrow{\substack{3R_2 \to R_2 \\ 3R_3 \to R_3 \\ 3R_4 \to R_4}} \begin{pmatrix} 3 & 5 & 3 & 1 \\ 15 & 6 & 21 & 18 \\ 6 & 9 & 15 & 3 \\ 3 & 3 & 12 & 15 \end{pmatrix}$$

Find a matrix R that performs this step.

c) Which of $|T|$ or $|R|$ is easier to find?

d) What does this suggest about the row operations you should use to row reduce a matrix A?

A6.32.4 Let $B = \begin{pmatrix} 2 & 7 & 3 & 5 \\ 1 & 4 & -1 & 2 \\ 3 & 1 & -5 & 2 \\ 5 & 2 & -6 & 3 \end{pmatrix}$.

a) Identify the row operations required to produce the row echelon form of B.

b) Find the corresponding matrices $T_1, T_2, \ldots, T_k$.

c) Use these elementary matrices and the row echelon form of B to find $|B|$.

d) Compute $|B|$ using the cofactor expansion.

Activity 6.32 develops a different approach to finding the determinant. Suppose $T_1, T_2, \ldots, T_k$ correspond to a sequence of elementary matrices that reduce A to row echelon form R:

$$T_k T_{k-1} \cdots T_2 T_1 A = R$$

The product property of the determinant gives us

$$|T_k||T_{k-1}| \cdots |T_2||T_1||A| = |R|$$

Since R is a matrix in row echelon form, it will be an upper triangular matrix, so computing $|R|$ will be trivial. And since (Activity 6.26) the determinant of an elementary matrix is either c (where the elementary matrix corresponds to multiplying a row by a constant), -1 (where the elementary matrix corresponds to switching adjacent rows) or 1 (where the elementary matrix corresponds to multiplying a row by a constant and adding it to another row), $|T_i|$ is easy to compute as well. Finally, while the determinant of the elementary matrix corresponding to switching nonadjacent rows will either be 1 or -1, it takes some effort to determine it; so it's probably best to avoid switching rows.

This gives us an equation we can solve for $|A|$. We'll refer to this as the *LU*-approach to finding the determinant (where we'll explain the "*LU*" part in Section 8.1).

Example 6.7. *Let* $A = \begin{pmatrix} 1 & 3 & 5 & 4 & 2 \\ 2 & 1 & 7 & -1 & 3 \\ 5 & 2 & 3 & 9 & 1 \\ 2 & 5 & 3 & 1 & 4 \\ 3 & 2 & -1 & -8 & 5 \end{pmatrix}$ *Find* $|A|$ *using the LU-approach.*

While we could specify the individual elementary matrices needed to row reduce A, we'll combine several row operations in each step.

Solution. *We produce the row echelon form of A. First, since our first row pivot is* 1*, we can just add multiples of the first row to the succeeding rows:*

$$\begin{pmatrix} 1 & 3 & 5 & 4 & 2 \\ 2 & 1 & 7 & -1 & 3 \\ 5 & 2 & 3 & 9 & 1 \\ 2 & 5 & 3 & 1 & 4 \\ 3 & 2 & -1 & -8 & 5 \end{pmatrix} \xrightarrow{\substack{-2R_1+R_2\to R_2 \\ -5R_1+R_3\to R_3 \\ -2R_1+R_4\to R_4 \\ -3R_1+R_5\to R_5}} \begin{pmatrix} 1 & 3 & 5 & 4 & 2 \\ 0 & -5 & -3 & -9 & -1 \\ 0 & -13 & -22 & -11 & -9 \\ 0 & -1 & -7 & -7 & 0 \\ 0 & -7 & -16 & -20 & -1 \end{pmatrix}$$

While we could switch the fourth row and second rows, it's easier if we don't switch rows (Activity A6.32.3), so our next step will be multiplying the remaining rows by -5*, then adding multiples of the second row to the third, fourth, and fifth rows:*

$$\xrightarrow{\substack{-5R_3\to R_3 \\ -5R_4\to R_4 \\ -5R_5\to R_5}} \begin{pmatrix} 1 & 3 & 5 & 4 & 2 \\ 0 & -5 & -3 & -9 & -1 \\ 0 & 65 & 110 & 55 & 45 \\ 0 & 5 & 35 & 35 & 0 \\ 0 & 35 & 80 & 100 & 5 \end{pmatrix} \xrightarrow{\substack{13R_2+R_3\to R_3 \\ R_2+R_4\to R_4 \\ 7R_2+R_5\to R_5}} \begin{pmatrix} 1 & 3 & 5 & 4 & 2 \\ 0 & -5 & -3 & -9 & -1 \\ 0 & 0 & 71 & -62 & 32 \\ 0 & 0 & 32 & 26 & -1 \\ 0 & 0 & 59 & 37 & -2 \end{pmatrix}$$

Completing our row reduction:

$$\xrightarrow{\substack{71R_4\to R_4 \\ 71R_5\to R_5}} \begin{pmatrix} 1 & 3 & 5 & 4 & 2 \\ 0 & -5 & -3 & -9 & -1 \\ 0 & 0 & 71 & -62 & 32 \\ 0 & 0 & 2272 & 1846 & -71 \\ 0 & 0 & 4189 & 2627 & -142 \end{pmatrix}$$

$$\xrightarrow{\substack{-32R_3+R_4\to R_4 \\ -59R_3+R_5\to R_5}} \begin{pmatrix} 1 & 3 & 5 & 4 & 2 \\ 0 & -5 & -3 & -9 & -1 \\ 0 & 0 & 71 & -62 & 32 \\ 0 & 0 & 0 & 3830 & -1095 \\ 0 & 0 & 0 & 6285 & -2030 \end{pmatrix}$$

$$\xrightarrow{3830R_5\to R_5} \begin{pmatrix} 1 & 3 & 5 & 4 & 2 \\ 0 & -5 & -3 & -9 & -1 \\ 0 & 0 & 71 & -62 & 32 \\ 0 & 0 & 0 & 3830 & -1095 \\ 0 & 0 & 0 & 24071550 & -7774900 \end{pmatrix}$$

$$\xrightarrow{-6285R_4+R_5\to R_5} \begin{pmatrix} 1 & 3 & 5 & 4 & 2 \\ 0 & -5 & -3 & -9 & -1 \\ 0 & 0 & 71 & -62 & 32 \\ 0 & 0 & 0 & 3830 & -1095 \\ 0 & 0 & 0 & 0 & -892825 \end{pmatrix}$$

If these row operations correspond to the matrices T_1, T_2, T_3, T_4, T_5, T_6, and T_7, we have

$$T_7T_6T_5T_4T_3T_2T_1A = \begin{pmatrix} 1 & 3 & 5 & 4 & 2 \\ 0 & -5 & -3 & -9 & -1 \\ 0 & 0 & 71 & -62 & 32 \\ 0 & 0 & 0 & 3830 & -1095 \\ 0 & 0 & 0 & 0 & -892825 \end{pmatrix}$$

Consequently

$$|T_7||T_6||T_5||T_4||T_3||T_2||T_1||A| = \begin{vmatrix} 1 & 3 & 5 & 4 & 2 \\ 0 & -5 & -3 & -9 & -1 \\ 0 & 0 & 71 & -62 & 32 \\ 0 & 0 & 0 & 3830 & -1095 \\ 0 & 0 & 0 & 0 & -892825 \end{vmatrix}$$

Remember that the matrix corresponding to multiplying a row by a constant and adding it to another row will have determinant 1, *so* $|T_i| = 1$ *for* $i = 1, 3, 5, 7$.

Since T_2 *corresponds to multiplying three rows by* -5,$|T_2| = (-5)^3$. *Similarly,* T_4 *corresponds to multiplying two rows by* 71, *so* $|T_4| = 71^2$, *and* $|T_6| = 3830$. *Finally, the determinant of the triangular matrix is the product of the entries along its main diagonal, so we have*

$$\begin{aligned} (3830)(71^2)(-5)^3|A| &= 1213929511250 \\ |A| &= -503 \end{aligned}$$

This seems to be a roundabout way of finding the determinant, so we might wonder if it's actually worth doing.

Activity 6.33: The LU-Approach to Determinants

A6.33.1 Use the LU-approach to compute the following determinants.

a) $\begin{vmatrix} 3 & 7 \\ 5 & -2 \end{vmatrix}$

b) $\begin{vmatrix} 3 & 7 & 3 \\ 2 & -4 & 1 \\ 5 & -1 & 2 \end{vmatrix}$

c) $\begin{vmatrix} 2 & 3 & 1 & 5 \\ 1 & 4 & -3 & 8 \\ 3 & 7 & 5 & 6 \\ 5 & -4 & -2 & 9 \end{vmatrix}$

A6.33.2 Suppose A is a 6×6 invertible coefficient matrix. Assume row reduction using the *fang cheng* algorithm, and that no pivot is 0.

a) The first step will be multiplying the second through sixth rows by the first row pivot. How many multiplications will this require?
b) The second step will be multiplying the first row by each row pivot and adding. How many multiplications will be required to eliminate all entries below the first row pivot?
c) How many multiplications will be required to obtain the row echelon form of A? **Note**: For convenience, count multiplication by 0 as a multiplication.
d) How many more multiplications will be required to find the determinant?

A6.33.3 In Activity 6.31, you found that using the cofactor expansion to find the determinant of a $n \times n$ matrix required $n!$ multiplications.

a) How many computations would be required to find the determinant of a 6×6 matrix using the cofactor expansion?
b) How many computations did the LU-approach require?
c) Suppose you're finding the determinant of a 10×10 matrix. Compare the number of computations required using the cofactor method to the number of computations required using the LU-approach. Which requires fewer computations?

It should be clear that using the LU-approach to find $|A|$ will require far fewer computations than using the cofactor method.

6.9 Use(lesses) of the Determinant

As Activity 6.31 shows, finding the determinant using the cofactor method requires $n!$ computations, which is impractical for large matrices. The LU-method is much more efficient, requiring only n^3 computations. But in either case, finding the determinant is a lot of work, so we might try to motivate it by finding useful applications.

Activity 6.34: Cramer's Rule

A6.34.1 Consider the system of equations

$$\begin{aligned} a_{11}x + a_{12}y &= b_1 \\ a_{21}x + a_{22}y &= b_2 \end{aligned}$$

a) Explain why we can assume $a_{11} \neq 0$. (In particular, why are we guaranteed that at least one of the equations will have a nonzero x coefficient, and we can make this equation our first equation.)

b) Row reduce the augmented coefficient matrix to solve for x, y. What relationship among the coefficients is required to be able to solve the system of equations?

c) Let A be the coefficient matrix. Find $|A|$.

d) Let $\mathbf{b}$ be the column vector $\begin{pmatrix} b_1 \\ b_2 \end{pmatrix}$, and let A_x be the matrix formed when the first column of the coefficient matrix is replaced with $\mathbf{b}$. Find $|A_x|$.

e) Let A_y be the matrix formed when the first column of the coefficient matrix is replaced with $\mathbf{b}$. Find $|A_y|$.

f) Express your solutions to x_1 and x_2 in terms of $|A|$, $|A_x|$, and $|A_y|$. (This result is known as **Cramer's rule**.)

A6.34.2 Use Cramer's rule to solve the system:

$$\begin{aligned} 3x - 5y &= 23 \\ 7x - 11y &= 5 \end{aligned}$$

A6.34.3 Cramer's rule generalizes to systems of three equations in three unknowns, with A_z defined in the obvious manner. Use Cramer's rule to solve:

$$\begin{aligned} 3x - 2y + 5z &= 11 \\ 2x - 4y + 7z &= 8 \\ 5x - 3y + 12z &= 5 \end{aligned}$$

Activity 6.34 leads to:

Theorem 6.5 (Cramer's Rule). *Let* $A\mathbf{x} = \mathbf{b}$ *be a system of linear equations, and let* A_i *be the matrix where the* i*th column of* A *has been replaced by* $\mathbf{b}$. *Then* $x_i = \dfrac{|A_i|}{|A|}$.

Cramer's rule appears to offer a neat, formulaic solution to systems of linear equations. On the other hand, computing the determinant is a tedious task, so we might ask: When should we use Cramer's rule?

Activity 6.35: When to Use Cramer's Rule

In the following, assume that you are finding the determinant using the cofactor expansion, so computing the determinant of a $n \times n$ matrix requires $n!$ multiplications.

A6.35.1 Consider a system of two equations with two unknowns:

$$\begin{aligned} ax + by &= c \\ dx + ey &= f \end{aligned}$$

a) Suppose you solve the system by row reduction. *Explicitly*, list the multiplications required to transform the augmented coefficient matrix into reduced row echelon form (using the *fang cheng* method to avoid having to divide until the end of the problem). For example, in the first step

$$\left(\begin{array}{cc|c} a & b & c \\ d & e & f \end{array}\right) \rightarrow \left(\begin{array}{cc|c} a & b & c \\ ad & ae & af \end{array}\right)$$

you need to find the products ad, ae, af. What other products must be found?

b) How many operations (multiplications) are required to transform the augmented coefficient matrix into reduced row echelon form? **Note**: For convenience of calculation, count multiplication by 0 as a multiplication.

c) How many divisions are necessary to obtain the solution to the system of equations?

d) Suppose you solve the system by Cramer's rule. How many operations (multiplications) are required to find the determinant of a 2×2 matrix? How many divisions?

e) Which method requires fewer operations?

A6.35.2 Consider a system of three equations with three unknowns.

a) How many multiplications are required to reduce the augmented coefficient matrix to reduced row echelon form? **Note**: For convenience of calculation, count multiplication by 0 as a multiplication.

b) How many multiplications are required to solve the system using Cramer's rule? (Note that a product like abc must be considered to require *two* multiplications, since we must first find ab, and then multiply the result by c)

c) Which method requires fewer multiplications?

A6.35.3 Suppose you have a system of n equations in n unknowns.

a) How many multiplications will be needed to solve a system of n equations in n unknowns using Cramer's rule?

b) How many multiplications will be needed to solve a system of n equations in n unknowns using row reduction?

c) At approximately what size of a system will Cramer's Rule require more computations than row reduction?

Roughly speaking, row reducing a $n \times n$ matrix will require n^3 multiplications. In contrast, if we use the cofactor expansion to find the required determinants, Cramer's rule will require $(n+1)!$ multiplications. This means that for $n = 2$, Cramer's rule is computationally "cheaper," requiring fewer calculations. For $n = 3$, the two methods require roughly the same number of calculations. By $n = 4$, however, Cramer's rule requires about twice as many computations as Gaussian elimination, and the disparity rapidly worsens: for $n = 9$, Cramer's rule will require *more than a thousand times as many computations* as Gaussian elimination.

To be sure, this analysis is based on using the cofactor expansion to find the determinant of a matrix, and we've already established that this is the worst algorithm for finding the determinant. The LU-approach requires about the same number of calculations as Gaussian elimination. Does this make Cramer's rule practical?

Unfortunately, no. Remember that the LU-approach *begins* by row reducing the coefficient matrix. So if you're going to row reduce the coefficient matrix to find the determinant so you can use Cramer's rule, you've already done the work necessary to solve the system using row reduction. The correct answer to, "When should we use Cramer's rule?" is "Almost never."

Similarly, we can find the inverse of a square matrix by row reducing a "double wide" matrix. Or we can use determinants to find the inverse.

Activity 6.36: The Inverse of a 2×2 Matrix

A6.36.1 Consider the 2×2 matrix

$$A = \begin{pmatrix} a & b \\ c & d \end{pmatrix}$$

a) Find $|A|$.
b) **WITHOUT** actually finding A^{-1}, what can you say about $|A^{-1}|$?
c) Find A^{-1}.
d) Find M, where $|A^{-1}|M = A^{-1}$. (Essentially, "factor out" the value of the determinant from A^{-1}).
e) Based on your work, find a formula for A^{-1}.

A6.36.2 If possible, find the inverse of the given matrix. If not possible, explain why not.

a) $\begin{pmatrix} 2 & 3 \\ 1 & 7 \end{pmatrix}$

b) $\begin{pmatrix} -2 & 1 \\ 5 & -1 \end{pmatrix}$

c) $\begin{pmatrix} 1 & 7 \\ 2 & 14 \end{pmatrix}$

Activity 6.36 gives a formula for the inverse of a 2×2 matrix:

Lemma 8 (Inverse of a 2×2 Matrix). *Let* $A = \begin{pmatrix} a & b \\ c & d \end{pmatrix}$. *Provided* $ad - bc \neq 0$,

$$A^{-1} = \frac{1}{ad - bc} \begin{pmatrix} d & -b \\ -c & a \end{pmatrix}$$

Can we generalize this to larger matrices? A useful strategy in math is to identify *every* concept that applies to a particular object. For example, the entry d, which is in the 1st row, 1st column position, is:

- a_{11}, the entry in the 1st row, 1st column of A,
- A_{11}, the cofactor of the 1st row, 1st column of A.

These observations are more useful if we make several of them, so consider $-b$, the entry in the 1st row, 2nd column of the inverse. We notice that $-b$ is:

- $-a_{12}$, the entry in the 1st row, 2nd column of A,
- A_{21}, the cofactor of the term in the 2nd row, 1st column.

By looking at the other entries $-c$ and a, we notice that *all* the entries in the inverse are cofactors of the original. This suggests we might want to take a closer look at cofactors.

Activity 6.37: The Adjoint Method

In the following, let A_{ij} be the ijth minor of A, and let $C_{ij} = (-1)^{i+j} A_{ij}$, the ijth cofactor of A.

A6.37.1 Let $A = \begin{pmatrix} a_{11} & a_{12} \\ a_{21} & a_{22} \end{pmatrix}$, and suppose

$$\begin{pmatrix} a_{11} & a_{12} \\ a_{21} & a_{22} \end{pmatrix} \begin{pmatrix} x_1 & x_3 \\ x_2 & x_4 \end{pmatrix} = |A| \, I$$

a) Show that $a_{11}C_{11} + a_{12}C_{12} = |A|$.

b) Show that $a_{21}C_{11} + a_{22}C_{12} = 0$.

c) Show that $x_1 = C_{11}$ and $x_2 = C_{12}$. Suggestion: What equations must x_1, x_2 satisfy, and what is a solution to that equation?

d) Show that $x_3 = C_{21}$ and $x_4 = C_{22}$.

A6.37.2 Let $A = \begin{pmatrix} a_{11} & a_{12} & a_{13} \\ a_{21} & a_{22} & a_{23} \\ a_{31} & a_{32} & a_{33} \end{pmatrix}$

a) Explain why $a_{11}C_{11} + a_{12}C_{12} + a_{13}C_{13} = |A|$

b) Let B be the matrix formed by replacing the first row of A with the second row (and keeping the original second row the same). Explain why $a_{21}C_{11} + a_{22}C_{12} + a_{23}C_{13} = 0$

c) Show $a_{31}C_{11} + a_{32}C_{12} + a_{33}C_{13} = 0$

d) Find a solution to the system

$$\begin{aligned} a_{11}x_1 + a_{12}x_2 + a_{13}x_3 &= |A| \\ a_{21}x_1 + a_{22}x_2 + a_{23}x_3 &= 0 \\ a_{31}x_1 + a_{32}x_2 + a_{33}x_3 &= 0 \end{aligned}$$

e) What is the relationship of your solution x_1, x_2, x_3 to the entries in the first column of the inverse of A?

f) Write down and solve a system of equations to find the entries in the second column of the inverse of A. Suggestion: One of the equations will give you the determinant, and the others will be equal to 0. How do you find $|A|$?

g) Write down and solve a system of equations to find the entries in the third column of the inverse of A.

A6.37.3 Let A be a $n \times n$ matrix.

a) Explain why, for any i,

$$a_{i1}C_{i1} + a_{i2}C_{i2} + \ldots + a_{in}C_{in} = |A|$$

b) Explain why, for any $i \neq j$,

$$a_{j1}C_{i1} + a_{j2}C_{i2} + \ldots + a_{jn}C_{jn} = 0$$

c) What does this tell you about a solution $x_{1i}, x_{2i}, \ldots, x_{ni}$ to the system

$$\begin{aligned} a_{11}x_{1i} + a_{12}x_{2i} + \ldots + a_{1n}x_{ni} &= 0 \\ a_{21}x_{1i} + a_{22}x_{2i} + \ldots + a_{2n}x_{ni} &= 0 \\ &\vdots \\ a_{i1}x_{1i} + a_{i2}x_{2i} + \ldots + a_{in}x_{ni} &= |A| \\ &\vdots \\ a_{n1}x_{1i} + a_{n2}x_{2i} + \ldots + a_{nn}x_{ni} &= 0 \end{aligned}$$

d) Explain why the solution to this system of equations gives the entries in the ith column of the inverse of A.

Activity 6.37 leads to the following conclusion: Let A be a square matrix. If A^{-1} exists, then the ijth entry will be $\frac{1}{|A|}C_{ji}$. This allows us to find the inverse of a $n \times n$ matrix as follows:

- Form matrix M whose ijth entry is C_{ij},
- $A^{-1} = \frac{1}{|A|}M^T$

Example 6.8. *Find the inverse of* $\begin{pmatrix} 2 & 7 \\ 3 & 4 \end{pmatrix}$

Solution. *We find the cofactors:*

$$C_{11} = 4 \qquad C_{12} = -3$$
$$C_{21} = -7 \qquad C_{22} = 2$$

We note $|A| = -13$, *so*

$$\begin{aligned} A^{-1} &= \frac{1}{-13}\begin{pmatrix} 4 & -3 \\ -7 & 2 \end{pmatrix}^T \\ &= -\frac{1}{13}\begin{pmatrix} 4 & -7 \\ -3 & 2 \end{pmatrix} \\ &= \begin{pmatrix} -\frac{4}{13} & \frac{7}{13} \\ \frac{3}{13} & -\frac{2}{13} \end{pmatrix} \end{aligned}$$

Example 6.9. *Find the inverse of* $A = \begin{pmatrix} 1 & 2 & 0 \\ 2 & 5 & -4 \\ 3 & 1 & 2 \end{pmatrix}$

Solution. *We need to find all the cofactors*

$$C_{11} = (-1)^{1+1}\begin{vmatrix} 5 & -4 \\ 1 & 2 \end{vmatrix} = 14 \qquad C_{12} = (-1)^{1+2}\begin{vmatrix} 2 & -4 \\ 3 & 2 \end{vmatrix} = -16 \qquad C_{13} = (-1)^{1+3}\begin{vmatrix} 2 & 5 \\ 3 & 1 \end{vmatrix} = -13$$

$$C_{21} = (-1)^{2+1}\begin{vmatrix} 2 & 0 \\ 1 & 2 \end{vmatrix} = -4 \qquad C_{22} = (-1)^{2+2}\begin{vmatrix} 1 & 0 \\ 3 & 2 \end{vmatrix} = 2 \qquad C_{23} = (-1)^{2+3}\begin{vmatrix} 1 & 2 \\ 3 & 1 \end{vmatrix} = 5$$

$$C_{31} = (-1)^{3+1}\begin{vmatrix} 2 & 0 \\ 5 & -4 \end{vmatrix} = -8 \qquad C_{32} = (-1)^{3+2}\begin{vmatrix} 1 & 0 \\ 2 & -4 \end{vmatrix} = 4 \qquad C_{33} = (-1)^{3+3}\begin{vmatrix} 1 & 2 \\ 2 & 5 \end{vmatrix} = 1$$

We also need the determinant. If we expand along the first row, we get

$$\begin{aligned} |A| &= a_{11}C_{11} + a_{12}C_{12} + a_{13}C_{13} \\ &= 1(14) + 2(-16) + 0(-13) \\ &= -18 \end{aligned}$$

So

$$A^{-1} = \frac{1}{-18}\begin{pmatrix} 14 & -16 & -13 \\ -4 & 2 & 5 \\ -8 & 4 & 1 \end{pmatrix}^T = -\frac{1}{18}\begin{pmatrix} 14 & -4 & -8 \\ -16 & 2 & 4 \\ -13 & 5 & 1 \end{pmatrix}$$

As with Cramer's rule, we appear to have a formula that generates the inverse, so we might wonder when we should use it.

Activity 6.38: When to Use the Adjoint Method

As in Activity 6.35, assume that computing the determinant of a $n \times n$ matrix requires $n!$ multiplications.

A6.38.1 Suppose you're trying to find the inverse of a 3×3 matrix.

a) How many multiplications are needed to find the inverse by row reducing $(A \mid I) \to (I \mid A^{-1})$?
b) How many multiplications are needed to find the inverse using the adjoint method?
c) Which method requires fewer multiplications?

A6.38.2 Suppose you're trying to find the inverse of a $n \times n$ matrix.

a) How many multiplications are needed to find the inverse by row reducing $(A \mid I) \to (I \mid A^{-1})$?
b) How many multiplications are needed to find the inverse using the adjoint method?
c) Which method requires fewer multiplications?

A6.38.3 Based on the number of multiplications required, when should you use the adjoint method to find the inverse of a matrix?

As with Cramer's rule, if you find the determinant of a matrix using the cofactor expansion, finding the inverse of a matrix using the adjoint method will generally require far more computations than the row reduction algorithm. And if you find the determinant of the matrix using the LU-approach, you will have effectively done the row reduction algorithm. So the proper answer to "When should we use the adjoint method to find the inverse of a matrix?" is "Almost never."

6.10 Uses of the Determinant

Based on our observations on Cramer's rule and the adjoint method, it might seem that the answer to any question that begins "When should we use a method based on determinants to ..." should be "Almost never." So does the determinant have a use?

We might return to a geometric interpretation of the determinant. Remember that if T is a 2×2 matrix, the absolute value of the determinant $|T|$ corresponds to the area of a unit square that has been transformed by T; if the determinant is negative, then the transformation has reversed the orientation of the unit square.

Activity 6.39: More Transformations

A6.39.1 In the following, a transformation matrix is given. Describe the effect of the transformation on a unit square: What will the area of the transformed square be, and will it have the same or opposite orientation from the original.

a) $\begin{pmatrix} 1 & 5 \\ 2 & 3 \end{pmatrix}$

b) $\begin{pmatrix} 5 & 3 \\ -6 & -1 \end{pmatrix}$

c) $\begin{pmatrix} 0 & 3 \\ 0 & 1 \end{pmatrix}$

d) $\begin{pmatrix} 5 & 3 \\ 10 & 6 \end{pmatrix}$

A6.39.2 Find a transformation from $\mathbb{R}^2$ to $\mathbb{R}^2$ with the listed properties, where all entries of the transformation matrix are nonzero.

a) A transformation A that preserves area and orientation.

b) A transformation B that preserves area but reverses orientation.

c) A transformation C that transforms a unit square into a figure with area 4 and reversed orientation.

More generally, suppose $\mathcal{V}$ and $\mathcal{W}$ are two different bases for a vector space in $\mathbb{F}^n$. Then there is a matrix A that transforms the coordinates $(v_1, v_2, \ldots, v_n)$ of a vector with respect to $\mathcal{V}$ into the coordinates $(w_1, w_2, \ldots, w_n)$ with respect to $\mathcal{W}$.

Of particular interest are bases where integer coordinates are transformed into integer coordinates. The easiest way to guarantee that is to make all entries of A and A^{-1} integers. While we could choose the entries of A to be integers, there's no guarantee that the entries of A^{-1} will also be integers. Thus we face the problem: Find a matrix of integers whose inverse is also a matrix of integers.

Activity 6.40: Custom Made Determinants

A6.40.1 Let L, U be a lower and an upper triangular matrix, respectively.

a) Find a 3×3 matrix L with determinant 2.

b) Find a 3×3 matrix U with determinant 3.

c) Find the determinant of LU.

d) Find a 3×3 matrix with determinant 6, where **all entries** are nonzero.

A6.40.2 Let L and U be lower and upper triangular matrices with integer entries, whose diagonal entries are ± 1, and let $M = LU$.

a) Prove: M is a matrix with integer entries.

b) Prove: $|M| = \pm 1$.

c) Prove: $m_{11} = \pm 1$.

A6.40.3 Let R, L be "reversed triangular matrices" (see Activity 6.14) with diagonal entries ± 1, and let $N = RL$.

a) Prove: N is a matrix with integer entries.

b) Prove: $|N| = \pm 1$.

c) Prove: One of the "corner" entries of N is ± 1.

A6.40.4 Find a 3×3 matrix T with integer entries, all of which are *greater* than 1, where $|T| = 1$.

A6.40.5 Suppose A is a $n \times n$ matrix of integers with determinant ± 1. Prove: A^{-1} is also a matrix of integers. Suggestion: While you shouldn't *use* the adjoint method to find the determinant, what does it tell you about the relationship between the entries of A and the entries of A^{-1}?

This ability to transform integer coordinates into integer coordinates is helpful in constructing a suitably "bad" basis for lattice cryptography.

Activity 6.41: Bad Basis From Good

Let $\mathcal{V} = \{\mathbf{v}_1, \mathbf{v}_2, \ldots, \mathbf{v}_n\}$ be a set of independent vectors. The lattice $\mathcal{L}$ is the set of all linear combinations of the vectors in $\mathcal{V}$ where the coefficients of the linear combination are integers. We say that $\mathcal{V}$ is a basis for the lattice $\mathcal{L}$.

A6.41.1 Let $\mathbf{x}$, $\mathbf{y}$ be vectors in $\mathcal{L}$, and let vector addition and scalar multiplication be defined as usual.

a) Show that the vectors in $\mathcal{L}$ are closed under vector addition.
b) Show that the vectors in $\mathcal{L}$ are closed under scalar multiplication when the scalars are drawn from the set of integers.
c) What other properties of a vector space do the lattice vectors satisfy, when the scalars are drawn from the set of integers?
d) What properties of a vector space do lattice vectors *fail* to satisfy, when the scalars are drawn from the set of integers?

A6.41.2 Let

$$\begin{aligned}\mathbf{w}_1 &= a_{11}\mathbf{v}_1 + a_{12}\mathbf{v}_2 + \ldots + a_{1n}\mathbf{v}_n \\ \mathbf{w}_2 &= a_{21}\mathbf{v}_1 + a_{22}\mathbf{v}_2 + \ldots + a_{2n}\mathbf{v}_n \\ &\vdots \\ \mathbf{w}_n &= a_{n1}\mathbf{v}_1 + a_{n2}\mathbf{v}_2 + \ldots + a_{nn}\mathbf{v}_n\end{aligned}$$

where the a_{ij}s are integers.

a) Suppose

$$\mathbf{x} = x_1\mathbf{v}_1 + x_2\mathbf{v}_2 + \ldots + x_n\mathbf{v}_n$$

where the x_is are all integers. What must be true about the a_{ij}s in order for

$$\mathbf{x} = y_1\mathbf{w}_1 + y_2\mathbf{w}_2 + \ldots + y_n\mathbf{w}_n$$

where the y_is are all integers? Suggestion: Remember that there is some matrix A that transforms the coordinates of a vector in one basis into the coordinates of the vector in another basis.

A6.41.3 Lattices are the basis for a form of cryptography known as **lattice cryptography**. A "good" bad basis is one where the coordinates (x, y) of a vector with respect to the bad basis change dramatically with small changes in the vector components (see Activity 5.37).

a) Let $\langle 1, 91\rangle$, $\langle 89, 1\rangle$ be a basis for a lattice $\mathcal{L}$. Find a different set of basis vectors.
b) Is your new basis a "good" bad basis? In other words: Suppose $\mathbf{x}$ has coordinates (x, y) with respect to your bad basis. Will a small change from $\mathbf{x}$ to $\mathbf{x}'$ cause a dramatic change in the coordinates from (x, y) to (x', y')?
c) Find a "good" bad basis for the lattice.

Finally, the determinant can be used to decide whether *functions* are independent.

Activity 6.42: Function Spaces

A6.42.1 Let $\mathbf{V}$ be the set of all second degree polynomials.

a) Show that $\mathbf{V}$ forms a vector space, where we use scalar multiplication by a real number and standard polynomial addition.
b) Explain why $\{1, x, x^2\}$ *spans* $\mathbf{V}$.
c) Prove/disprove: $\{1, x, x^2\}$ is an independent set of vectors.

A6.42.2 Suppose $f(x) = x^2 + 1$, $g(x) = 2x + 5$, and $h(x) = x^2 + x + 1$.

a) Find a, b, c where $af(x) + bg(x) + ch(x) = x^2 + 5x + 1$.
b) Prove/disprove: There is a nontrivial linear combination of f, g, h that is equal to 0.

A6.42.3 "Calculus is easier than algebra." Finding whether or not a nontrivial linear combination of functions is equal to the 0 vector can be a tedious and difficult task. We can simplify it by using calculus. Let f, g, h be differentiable functions of x, and suppose $a_1 f(x) + a_2 g(x) + a_3 h(x) = 0$

a) Find a system of equations that a_1, a_2, a_3 must satisfy if $a_1 f(x) + a_2 g(x) + a_3 h(x) = 0$. Suggestion: Differentiate this equation to get a second equation involving a_1, a_2, a_3.
b) What must be true in order for the system of equations to have a unique solution?

Activity 6.42 leads to:

Definition 6.5 (Wronskian). *Let $f_1, f_2, \ldots, f_n$ be function of x. The* ***Wronskian*** *is*

$$\mathcal{W} = \begin{vmatrix} f_1(x) & f_2(x) & \cdots & f_n(x) \\ f_1'(x) & f_2'(x) & \cdots & f_n'(x) \\ \vdots & \vdots & \ddots & \vdots \\ f^{(n-1)}(x) & f^{(n-1)}(x) & \cdots & f^{(n-1)}(x) \end{vmatrix}$$

where $f^{(i)}(x)$ is the ith derivative of $f(x)$.

If we view the Wronskian as the determinant of a system of equations in the $f_i(x)$s, then:

Proposition 17. *The functions $f_1(x)$, $f_2(x)$, $\ldots$, $f_n(x)$ are independent if and only if $\mathcal{W} \neq 0$.*

6.11 Permutations

Thus far we've found two ways to introduce the determinants:

- Begin with the idea of a multilinear function, which leads to the Laplace expansion. This was our path.
- Begin with the Laplace expansion, and show the determinant is multilinear. This is what we did in Activity 6.22.

We close this chapter by finding a third path to the determinant.

First, a definition:

Definition 6.6 (Permutation). *Let A be a set of k objects $a, b, c, \ldots$. A permutation of A is an ordered k-tuple of the elements of A.*

For example, if $A = \{1, 2, 3\}$, then $(1, 3, 2)$ and $(3, 2, 1)$ are permutations of A. We typically omit the commas and simply speak of the permutations (132) and (321).

If the elements of the set A have some defined ordering, we can define:

Definition 6.7 (Inversion). *In a permutation, an* ***inversion*** *occurs whenever a higher indexed element precedes a lower indexed element.*

We can view an inversion as measure of how "out of order" the permutation is. For example, the permutation (321) has three inversions: 3 is greater than both of the entries that follow it; 2 is greater than the entry that follows it. Meanwhile (132) has one inversion: 3 is higher than 2. Note that the "out of order" terms need not be adjacent: as long as a term is higher than *any* of the following terms, it counts as an inversion.

Definition 6.8 (Even and Odd). *A permutation is* ***even*** *if the total number of inversions is even; and* ***odd*** *if the total number of inversions is odd.*

So (321) is an even permutation, while (132) is an odd one. Another way to refer to whether the permutation is even or odd is to talk about its *parity*.

Since we're dealing with determinants of matrices, we'll define:

Definition 6.9 (Row Permutation). *A* ***row permutation of matrix*** *A is a permutation of the rows of A.*

For example, if $A = \begin{pmatrix} a & b & c \\ d & e & f \\ g & h & i \end{pmatrix}$, then $\begin{pmatrix} g & h & i \\ a & b & c \\ d & e & f \end{pmatrix}$ is a row permutation of A. Since the first, second, and third rows of this matrix are the 3rd, 1st, and 2nd row of A, we'll refer to this as the matrix (312).

First, let's investigate the relationship between row permutations and the determinant. We'll use Definition 6.2 in the following.

Activity 6.43: Permutations of Matrices

A6.43.1 Let $A = \begin{pmatrix} a & b & c \\ d & e & f \\ g & h & i \end{pmatrix}$.

a) List all the permutations of the rows. Use the notation that (xyz) is the matrix whose first, second, and third rows are the xth, yth, and zth rows of A.

b) Determine which of the permutations are even, and which are odd.

c) Find the matrix corresponding to each of the permutations.

A6.43.2 Suppose a permutation has k inversions.

a) Consider two adjacent elements p, q. Explain why, if $p < q$, that switching the two adjacent elements will increase the number of inversions by *exactly* 1. In particular: Show that the *only* new inversion is q, p.

b) Similarly, if $p > q$, switching will decrease the number of inversions by exactly 1.

c) Explain how you could obtain *any* permutation by switching adjacent elements.

What about the Laplace expansion?

Activity 6.44: Permutations and the Laplace Expansion

A6.44.1 Prove that the Laplace expansion for a $n \times n$ matrix will have $n!$ terms, using the following steps.

a) Prove that the Laplace expansion for a 2×2 matrix will have $2!$ terms.

b) Prove that if the Laplace expansion for a $k \times k$ matrix has $k!$ terms, the Laplace expansion for a $(k+1) \times (k+1)$ matrix will have $(k+1)!$ terms.

c) Explain why these two statements prove that the Laplace expansion for a $n \times n$ matrix will have $n!$ terms.

A6.44.2 Prove that each term of the Laplace expansion is a product of matrix entries from different rows and columns, using the following steps.

a) Prove that each term in the Laplace expansion for a 2×2 matrix is a product of matrix entries from different rows and columns.

b) Prove that if each term in the Laplace expansion for a $k \times k$ matrix is a product of matrix entries from different rows and columns, then each term in the Laplace expansion for a $(k+1) \times (k+1)$ matrix will be a product of matrix entries from different rows and columns.

c) Explain why these two statements prove that each term in the Laplace expansion for a $n \times n$ matrix will be a product of matrix entries from different rows and columns.

Activity 6.44 proves two important results about the Laplace expansion: First, there are $n!$ terms in the Laplace expansion. Second, each of these terms is the product of entries from different rows and columns of the matrix. Since in a $n \times n$ matrix there can be only $n!$ such products, it follows that:

Theorem 6.6. *The terms of the Laplace expansion include all possible products of matrix entries from different rows and columns.*

For example, when computing the determinant of $\begin{pmatrix} a & b & c \\ d & e & f \\ g & h & i \end{pmatrix}$, one such product is cdh, where c, d, h are in different rows and columns of the matrix; by Theorem 6.6, we know that this product will be one of the terms in the Laplace expansion.

What about the sign of the term?

Activity 6.45: Signs of Permutations

A6.45.1 Prove that *regardless* of which row or column you expand along, the Laplace expansion for a $n \times n$ matrix will include a term equal to the product of the entries on the main diagonal.

A6.45.2 Let A be a square matrix. Find the relationship between $|A'|$ and $|A|$ if:

a) A' is an odd row permutation of A.

b) A' is an even row permutation of A.

A6.45.3 Let A be a $n \times n$ matrix, and suppose $p_1 p_2 \cdots p_n$ is a term in the Laplace expansion, where p_i is an entry in the ith row of A.

a) Prove that $p_1 p_2 \cdots p_n$ is the main diagonal of some matrix A', a permutation of the rows of A.

b) Suppose the permutation that produces A' is even. Prove: $p_1 p_2 \cdots p_n$ will be added in the Laplace expansion of A.

c) Prove: If the permutation that produces A' is odd, $p_1 p_2 \cdots p_n$ will be subtracted in the Laplace expansion of A.

Activity 6.45 gives us an important result about the Laplace expansion:

Theorem 6.7. *Let A' be a row permutation of A. The product of the entries along the main diagonal of A' will be added or subtracted in the expression for the determinant of A depending on whether the row permutation is even or odd.*

For example, let $A = \begin{pmatrix} a & b & c & d \\ e & f & g & h \\ i & j & k & l \\ m & n & o & p \end{pmatrix}$ be row permuted into $\begin{pmatrix} i & j & k & l \\ a & b & c & d \\ e & f & g & h \\ m & n & o & p \end{pmatrix}$ which, in our permutation notation, is (3124). The product of the terms on the main diagonal is $ibgp$, and the permutation itself has 2 inversions, so it is an even permutation; consequently in the Laplace expansion of the determinant of A, this term will be added.

Let's take one final step. We'll define:

Definition 6.10. *Let A be a $n \times n$ matrix; let A_i be a row permutation of A; and let π_i be the product of the entries on the main diagonal of A_i. We'll define $\Pi(A)$ to be the sum or difference of all such products, where π_i is added if the row permutation is even, and subtracted if the row permutation is odd.*

For example, in a 3×3 matrix $\begin{pmatrix} a & b & c \\ d & e & f \\ g & h & i \end{pmatrix}$, we have

Matrix	Parity	π_i
$\begin{pmatrix} a & b & c \\ d & e & f \\ g & h & i \end{pmatrix}$	Even	aei
$\begin{pmatrix} a & b & c \\ g & h & i \\ d & e & f \end{pmatrix}$	Odd	ahf
$\begin{pmatrix} d & e & f \\ a & b & c \\ g & h & i \end{pmatrix}$	Odd	dbi
$\begin{pmatrix} d & e & f \\ g & h & i \\ a & b & c \end{pmatrix}$	Even	dhc
$\begin{pmatrix} g & h & i \\ a & b & c \\ d & e & f \end{pmatrix}$	Even	gbf
$\begin{pmatrix} g & h & i \\ d & e & f \\ a & b & c \end{pmatrix}$	Odd	gec

Adding the products of even permutations and subtracting the products of odd permutations gives us

$$\Pi(A) = aef + dhc + gbf - ahf - dbi - gec$$

Activity 6.46: Properties of Permutations

A6.46.1 Show that $\Pi(I) = 1$.

A6.46.2 Let A be a $n \times n$ matrix.

a) Let B be formed by switching two adjacent rows of A. Show that $\Pi(B) = -\Pi(A)$.

b) Suppose

$$A = \left(\, C \mid \mathbf{u}+\mathbf{v} \, \right), U = \left(\, C \mid \mathbf{u} \, \right), V = \left(\, C \mid \mathbf{v} \, \right),$$

where $\mathbf{u}$, $\mathbf{v}$ are column vectors. Show that $\Pi(A) = \Pi(U) + \Pi(V)$.

Activity 6.46 shows that $\Pi(A)$ satisfies all the requirements of Definition 6.2. Since the determinant of a matrix is unique, then $\Pi(A)$ *is* the determinant of A. It follows that we can use any of three definitions of the determinant:

- The determinant as a linear function (Definition 6.2),
- The determinant as a value computed by the Laplace expansion,
- The determinant as a permutation (Definition 6.10).

We close this section by trying to find alternate routes to our favorite coffee shop.

Activity 6.47: The Permutation Definition of the Determinant

In the following, use **ONLY** Definition 6.10. Do **NOT** use any other formula for or property of the determinant.

Assume A is a square matrix.

A6.47.1 Prove: If A has a row of 0s, $|A| = 0$.

A6.47.2 Prove: If A is triangular, $|A|$ will be the product of terms along the main diagonal.

A6.47.3 Prove: If B is formed by multiplying a row or column of A by a constant c, then $|B| = c|A|$.

A6.47.4 Prove: If two rows of A are the same, $|A| = 0$.

7

Eigenvalues and Eigenvectors

7.1 More Transformations

One important type of geometric transformation is called a **scaling** or a **dilation**: we might "zoom in" on a picture, making the whole much larger; or "zoom out" to shrink the image down.

Activity 7.1: Scaling

In a **scaling** or a **dilation**, all lengths of a figure are increased by the same amount, known as the **scaling factor**.

A7.1.1 Let $\vec{v}$ be a vector in $\mathbb{R}^2$. Find a transformation matrix M that does the indicated scaling.

a) Scales $\vec{v}$ by a factor of 3.
b) Scales $\vec{v}$ by a factor of $\frac{1}{3}$.
c) Scales $\vec{v}$ by a factor of -2.
d) Scales $\vec{v}$ by a factor of 0.

A7.1.2 Suppose PQR is a triangle in $\mathbb{R}^2$, and $M : \vec{v} \rightarrow c\vec{v}$, where c is a scalar. Show that if M is applied to the points P, Q, R, it will scale the triangle by a factor of c. In particular, M must scale the lengths PQ, QR, RP by a factor of c.

A7.1.3 Suppose M scales vectors in $\mathbb{R}^2$ by a factor of c. How does M affect areas in $\mathbb{R}^2$? Base your answer on the properties of transformation matrices.

Based on your work in Activity A7.1.1, you found that scaling by a factor of c corresponds to the matrix cI, where I is the appropriately-sized identity matrix.

However, there are times when we want to distort the picture in a specific fashion: for example, we might want to stretch it vertically *without* changing it horizontally. This is usually referred to as a **stretch** or a **compression** along an axis.

Activity 7.2: Stretching

A7.2.1 Find the matrix that performs the indicated stretch on vectors in $\mathbb{R}^2$.

a) The matrix M, corresponding to a stretch that triples all distances to the x-axis, but leaves the distance to the y-axis unchanged.

b) The matrix N, corresponding to a stretch that halves all distances to the y-axis, but leaves the distance to the x-axis unchanged.

A7.2.2 Find the matrix that performs the indicated two-way stretch on vectors in $\mathbb{R}^2$.

a) Tripling the distance to the x-axis while halving the distance to the y-axis.

b) Doubling the distance to the x-axis, and tripling the distance to the y-axis.

A7.2.3 Let M, N be the matrices you found in Activity A7.2.1.

a) Explain why, in general, $M\vec{v} \neq 3\vec{v}$.

b) Suppose

$$M\begin{pmatrix} x_1 \\ x_2 \end{pmatrix} = 3\begin{pmatrix} x_1 \\ x_2 \end{pmatrix}$$

Set up and solve a system of equations for finding x_1, x_2.

c) In Activity A7.2.3b, you found a vector algebraically. Explain how you could have found the vector geometrically, using the idea that M triples the distance to the x-axis.

d) Find a vector $\vec{u}$ where $N\vec{u} = \frac{1}{2}\vec{u}$ by setting up and solving a system of equations to find the components of $\vec{u}$. Then explain why your solution makes sense from a *geometric* perspective.

7.2 The Eigenproblem

In Activity 7.2, you discovered that for some matrices A, there are some special vectors $\mathbf{v}$ for which $A\mathbf{v}$ is a scalar multiple of $\mathbf{v}$. Let's explore this concept a little more.

Activity 7.3: Eigenvectors

A7.3.1 Suppose A is a nonsquare matrix.

a) Find the domain and codomain of the transformation represented by A.

b) Suppose A is a nonsquare matrix. Is it possible to find a scalar λ and a vector $\vec{v}$ where $A\vec{v} = \lambda\vec{v}$? Why/why not?

A7.3.2 Suppose A is a square matrix. Show that if $\vec{v}$ is the zero vector, then any value of λ will satisfy $A\vec{v} = \lambda\vec{v}$.

A7.3.3 Suppose A is a square matrix consisting of all zeroes. What must λ be in order for there to be a solution to $A\vec{v} = \lambda\vec{v}$ for a nonzero vector $\vec{v}$?

A7.3.4 If possible, find a matrix and vector or vectors with the given properties. A general strategy: Choose the vector *first*.

a) A 2×2 matrix A and a nonzero vector $\vec{v}$ where $A\vec{v} = 5\vec{v}$.

b) A 2×2 matrix A with nonzero entries, and a nonzero vector $\vec{v}$, where $A\vec{v} = 0\vec{v}$.

c) A 2×2 matrix A and *two* nonzero vectors $\vec{u}$, $\vec{v}$, where $A\vec{u} = 3\vec{u}$, and $A\vec{v} = -2\vec{v}$.

Activity 7.3 shows that if A isn't a square matrix, it's impossible to have $A\mathbf{v} = \lambda\mathbf{v}$. But as we saw in Activity 7.2, if A is square, then there could be a vector $\mathbf{v}$ and scalar λ where $A\mathbf{v} = \lambda\mathbf{v}$.

Now if $\mathbf{v} = \mathbf{0}$, then any value of λ works. But a question with answer "Anything at all" isn't a good question. It's not that we object to the *answer*; it's that *if* the answer is "anything at all," then the question is one we already known the answer to.[1] Thus, we'll require that $\mathbf{v}$ be a nonzero vector.

This leads to the definition:

Definition 7.1. *Given a matrix A, an* ***eigenvalue-eigenvector*** *pair is a value λ and a nonzero vector $\mathbf{v}$ for which $A\mathbf{v} = \lambda\mathbf{v}$.*

As we saw, while we don't want an eigen*vector* to be $\mathbf{0}$, it's possible for an eigen*value* to be 0.

So how can we find eigenvalues and eigenvectors? A good strategy in mathematics (and life): If you're looking for something, find out as much as you can about its properties. So if you're looking for an elephant, find out where it lives and what it eats. Since we're looking for eigenvalues and eigenvectors, let's find out as much as we can about them first.

---

[1]Remember Shannon's definition of information: it's answers to questions, where we don't already know the answer.

Activity 7.4: Properties of Eigenvalues and Eigenvectors

In the following, assume all vector components and scalars are drawn from $\mathbb{F}^n$.

A7.4.1 In the following, assume A is a square matrix with eigenvalue/eigenvector pair λ, $\vec{v}$.

a) Prove/disprove: $-\vec{v}$ is also an eigenvector corresponding to eigenvalue λ.

b) Prove/disprove: $c\vec{v}$ is also an eigenvector corresponding to eigenvalue λ.

c) Prove/disprove: If $\vec{u}$ is another eigenvector corresponding to λ, then $a\vec{u} + b\vec{v}$ is also an eigenvector corresponding to λ.

A7.4.2 Suppose A is a square matrix, and $\lambda, \vec{v}$ an eigenvalue-eigenvector pair for A. Further, suppose A^{-1} also exists.

a) Find an eigenvalue-eigenvector pair for A^{-1}. Suggestion: Since $\lambda, \vec{v}$ is an eigenvalue-eigenvector pair for A, we know $A\vec{v} = \lambda\vec{v}$. What happens if we multiply by A^{-1}?

b) What is the relationship between the eigenvectors of A and the eigenvectors of A^{-1}?

c) What is the relationship between the eigenvalues of A and the eigenvalues of A^{-1}?

d) Prove/disprove: If A has eigenvalue $\lambda = 0$, then A^{-1} does not exist.

A7.4.3 Suppose A is a square matrix with eigenvalue/eigenvector pair λ, $\vec{v}$, and let c be a scalar. Find an eigenvalue/eigenvector pair for cA.

Activity A7.4.1 leads to an important conclusion: if $\mathbf{u}$, $\mathbf{v}$ are eigenvectors corresponding to an eigenvalue λ, then any linear combination of $\mathbf{u}$ and $\mathbf{v}$ will also be an eigenvector corresponding to eigenvalue λ. This suggests:

Definition 7.2 (Eigenspace). *Let A have eigenvalue λ. The* ***eigenspace of*** *λ is the span of the eigenvectors of A corresponding to λ.*

Since the additive inverse of an eigenvector is an eigenvector (Activity A7.4.1a), and the sum of two eigenvectors is an eigenvector (Activity A7.4.1a), this proves:

Proposition 18. *The eigenspace of λ is a subspace.*

How can we use this to find eigenvalues and eigenvectors? A useful strategy:

Strategy. *You can assume anything you want, as long as you accept the consequences.*

In practice, this means that we can often solve a difficult problem by assuming that we've solved a simpler problem. In this case, we might tackle the problem "How can we find eigenvalues and eigenvectors for a matrix?" by assuming that we've found the eigenvalues, and seek the eigenvectors; or by assuming that we've found the eigenvectors, and seek the eigenvalues.

Activity 7.5: Solving the Eigenproblem

A7.5.1 Remember "every problem in linear algebra begins with a system of linear equations." Let $A = \begin{pmatrix} a & b \\ c & d \end{pmatrix}$.

a) Suppose you know $\mathbf{v}$ is an eigenvector for A. What system of linear equations could you set up and solve to find the corresponding eigenvalue?

b) Suppose you know λ is an eigenvalue for A. What system of linear equations could you set up and solve to find the corresponding eigenvalue?

Activity 7.5 shows that if we know an eigenvalue, we can find an eigenvector; and if we know an eigenvector, we can find the eigenvalues. So, which should we look for first?

We might reason as follows. If we want to find an eigen*vector* for a $n \times n$ matrix, we have to find the n components of a vector in $\mathbb{F}^n$, so we need to find n things all at once. In contrast, if we want to find an eigen*value*, we only need to find one thing, namely the eigenvalue itself. Since it's generally easier to find *one* thing than it is to find multiple things, it follows that we should begin by looking for the eigenvalues.

So our first task will be to find the eigenvalues. While we could develop an algorithm for finding the eigenvalues, let's make sure we can actually get to our destination: It's easy to convince yourself that you can solve a problem, up until the point you actually *have* to solve the problem. So let's try to find some eigenvectors, given some eigenvalues.

Activity 7.6: Finding Eigenvectors

Suppose you find that $A = \begin{pmatrix} 8 & -12 & -30 \\ -3 & 8 & 15 \\ 3 & -6 & -13 \end{pmatrix}$ has eigenvalues $\lambda = -1$ and $\lambda = 2$.

A7.6.1 Consider the eigenvalue $\lambda = -1$.

a) Set up and solve a system of linear equations for finding a corresponding eigenvector.

b) Find a basis for the eigenspace of $\lambda = -1$.

A7.6.2 Consider the eigenvalue $\lambda = 2$.

a) Set up and solve a system of linear equations for finding a corresponding eigenvector.

b) Find a basis for the eigenspace of $\lambda = 2$.

A7.6.3 Could $\lambda = -2$ be an eigenvalue? Why/why not?

Activity 7.6 leads to two conclusions. First, once we find the eigen*values*, we can find the eigen*vectors* easily enough.

Example 7.1. *Suppose* $A = \begin{pmatrix} 3 & 14 & -14 & -6 \\ -6 & 27 & -22 & -10 \\ 12 & 50 & -51 & -22 \\ -42 & -56 & 70 & 29 \end{pmatrix}$ *has eigenvalue* $\lambda =$ *3. Find a basis for the corresponding eigenspace.*

Solution. *Since* $\lambda = 3$ *is an eigenvalue, we know an eigenvector* $\langle x_1, x_2, x_3, x_4 \rangle$ *satisfies the matrix equation*

$$\begin{pmatrix} 3 & 14 & -14 & -6 \\ -6 & 27 & -22 & -10 \\ 12 & 50 & -51 & -22 \\ -42 & -56 & 70 & 29 \end{pmatrix} \begin{pmatrix} x_1 \\ x_2 \\ x_3 \\ x_4 \end{pmatrix} = 3 \begin{pmatrix} x_1 \\ x_2 \\ x_3 \\ x_4 \end{pmatrix}$$

Performing the matrix multiplication:

$$\begin{pmatrix} 3x_1 + 14x_2 - 14x_3 - 6x_4 \\ -6x_1 + 27x_2 - 22x_3 - 10x_4 \\ 12x_1 + 50x_2 - 51x_3 - 22x_4 \\ -42x_1 - 56x_2 + 70x_3 + 29x_4 \end{pmatrix} = 3 \begin{pmatrix} x_1 \\ x_2 \\ x_3 \\ x_4 \end{pmatrix}$$

which gives us the system of equations

$$\begin{aligned} 3x_1 + 14x_2 - 14x_3 - 6x_4 &= 3x_1 \\ -6x_1 + 27x_2 - 22x_3 - 10x_4 &= 3x_2 \\ 12x_1 + 50x_2 - 51x_3 - 22x_4 &= 3x_3 \\ -42x_1 - 56x_2 + 70x_3 + 29x_4 &= 3x_4 \end{aligned}$$

Rearranging

$$\begin{aligned} 0x_1 + 14x_2 - 14x_3 - 6x_4 &= 0 \\ -6x_1 + 24x_2 - 22x_3 - 10x_4 &= 0 \\ 12x_1 + 50x_2 - 54x_3 - 22x_4 &= 0 \\ -42x_1 - 56x_2 + 70x_3 + 26x_4 &= 0 \end{aligned}$$

Row reducing

$$\begin{pmatrix} 0 & 14 & -14 & -6 \\ -6 & 24 & -22 & -10 \\ 12 & 50 & -54 & -22 \\ -42 & -56 & 70 & 26 \end{pmatrix} \rightarrow \begin{pmatrix} 3 & -12 & 11 & 5 \\ 0 & 7 & -7 & -3 \\ 0 & 0 & 0 & 0 \\ 0 & 0 & 0 & 0 \end{pmatrix}$$

so the eigenspace will be

$$s\langle 1, 9, 0, 21\rangle + t\langle 7, 21, 21, 0\rangle$$

and the basis will be

$$\mathcal{V} = \{\langle 1, 9, 0, 21\rangle, \langle 7, 21, 21, 0\rangle\}$$

(where we can simplify the second basis vector to $\langle 1, 3, 3, 0\rangle$ *if we wish).*

This also motivates the definition:

Definition 7.3 (Geometric Multiplicity). *Suppose* λ *is an eigenvalue for matrix* M*. The* ***geometric multiplicity of*** λ *is the dimension of the eigenspace of* λ*.*

Since the dimension is also the maximum number of linearly independent vectors in a set, we could use an equivalent definition:

Definition 7.4 (Geometric Multiplicity, Equivalent Definition). *The* ***geometric multiplicity of*** λ *is the number of linearly independent eigenvectors corresponding to* λ*.*

We'll do one last thing before trying to find the eigenvalues: Let's see how *many* eigenvalues we'll have to find. To proceed, we'll make one important assumption: If A is a $n \times n$ matrix, then A will have a *finite* number of eigenvalues. We'll justify this assumption later, but for now we'll take the viewpiont that since we're trying to figure out when we've found all the eigenvalues, we'd best hope that A has a finite number.

Activity 7.7: Independence of Eigenvectors

In the following, suppose A has k (not necessarily distinct) eigenvalues λ_1, $\lambda_2, \ldots, \lambda_k$ with corresponding eigenvectors $\vec{v}_1, \vec{v}_2, \ldots, \vec{v}_k$, where the vectors corresponding to the same eigenvalues are linearly independent; we make no assumptions about vectors corresponding to different eigenvalues.

A7.7.1 Suppose one of the eigenvectors can be expressed as a linear combination of the others, say $\vec{v}_1 = a_2\vec{v}_2 + a_3\vec{v}_3 + \ldots + a_k\vec{v}_k$

a) Explain why at least one of the $\vec{v}_i$s corresponds to eigenvalue $\lambda_i \neq \lambda_1$.

b) Explain why it is true that $A\vec{v}_1 = a_2\lambda_2\vec{v}_2 + a_3\lambda_3\vec{v}_3 + \ldots + a_k\lambda_k\vec{v}_k$

c) Explain why it is also true that $A\vec{v}_1 = a_2\lambda_1\vec{v}_2 + a_3\lambda_1\vec{v}_3 + \ldots + a_k\lambda_1\vec{v}_k$

d) Why does this mean we can write $\vec{0}$ as a nontrivial linear combination of the eigenvectors $\vec{v}_i$, where $\lambda_i \neq \lambda_1$? (Remember to allow for the possibility that some of the λ_is are equal to λ_1.)

e) Explain why this proves the set of eigenvectors for eigenvalues not equal to λ_1 is dependent.

Activity 7.7 leads us to the following result. Suppose we begin with a set of eigenvectors $\mathcal{V}$, where the eigenvectors are *either* linearly independent eigenvectors for the same eigenvalue, or eigenvectors for different eigenvalues. If $\mathcal{V}$ is a dependent set, we can write one of the eigenvectors (say $\vec{v}_1$) as a linear combination of the remaining vectors of $\mathcal{V}$ and produce a dependent subset $\mathcal{V}'$, which omits *all* eigenvectors with eigenvalue λ_1.

But now we have a dependent set of eigenvectors $\mathcal{V}'$, which are *either* linearly independent eigenvectors for the same eigenvalue, or eigenvectors for different eigenvalues. So we're back at our starting point and we can "lather, rinse, repeat" to produce another dependent set of eigenvectors $\mathcal{V}''$, which will omit all eigenvectors corresponding to another eigenvalue.

We can continue to do this. But since we've assumed A has a finite number of eigenvalues, this means we'll eventually end with a set corresponding to the linearly independent eigenvectors for a single eigenvalue. But (by our reasoning) this set will also be dependent, which is impossible.

Thus our original assumption must be flawed, and our set of eigenvectors *can't* be dependent. Consequently:

Proposition 19 (Independence of Eigenvectors). *Let $\mathcal{V}$ be a set of eigenvectors. If the vectors corresponding to the same eigenvalue are linearly independent, then $\mathcal{V}$ is an independent set of vectors.*

Since each eigen*value* corresponds to at least one eigen*vector*, it follows that a $n \times n$ matrix can have at most n eigenvalues.

We can go one step further. Since the geometric multiplicity of an eigenvalue is the number of linearly independent eigenvectors it corresponds to, we have:

Proposition 20 (Geometric Multiplicity). *The sum of the geometric multiplicities of the eigenvalues of a $n \times n$ matrix is at most n.*

7.3 Finding Eigenvalues: Numerical Methods

Now that we know how many eigenvalues we can expect to find, it's worth expending the effort to find them. One approach is to try and find them numerically, using the properties of eigenvalues and eigenvectors.

Activity 7.8: Finding Eigenvalues Numerically

Suppose A is a matrix with eigenvalues λ_1, λ_2 with corresponding eigenvectors $\vec{v}_1$, $\vec{v}_2$. We'll assume λ_1, λ_2 are real, and $|\lambda_1| > |\lambda_2| > 0$.

A7.8.1 Suppose $\vec{x} = a\vec{v}_1 + b\vec{v}_2$. Let $\vec{x}_n = A^n\vec{x}$.

a) Express $\vec{x}_1$ in terms of vectors v_1, v_2 only. (Your answer should **NOT** include the matrix factor A)

b) Express $\vec{x}_n$ in terms of vectors v_1, v_2 only. (Your answer should **NOT** include the matrix factor A)

c) Explain why, if n is a very large number, $|\lambda_1| \approx \dfrac{||\vec{x}_{n+1}||}{||\vec{x}_n||}$.

A7.8.2 Suppose A is a 2×2 matrix with two real and distinct eigenvalues λ_1, λ_2, with $|\lambda_1| > |\lambda_2|$.

a) Suppose $\vec{v}$ is any vector in $\mathbb{R}^2$. Explain why $\vec{v}$ can be expressed as a linear combination of the two eigenvectors $\vec{v}_1$, $\vec{v}_2$.

b) Let

$$A = \begin{pmatrix} 2 & 1 \\ 6 & 1 \end{pmatrix}$$

Find A^{16} and A^{17}.

c) Choose any nonzero vector $\vec{x}$ and find $A^{17}\vec{x}$ and $A^{16}\vec{x}$.

d) Find $\dfrac{||A^{17}\vec{x}||}{||A^{16}\vec{x}||}$.

e) What *two* values does this suggest for λ_1, the larger of the two eigenvalues for A?

f) Which value leads to an actual eigenvalue-eigenvector pair?

A7.8.3 One problem with this approach is that it doesn't give us the eigenvalue or the eigenvector directly. Let's consider a variation.

a) Let $\vec{x}$ be any nonzero vector, and n some large positive integer. Explain why $A^n\vec{x}$ is *approximately* an eigenvector.

b) Since in general $A^n\overrightarrow{x}$ will have very large components, find a unit vector $\overrightarrow{y}$ in the same direction as $A^n\overrightarrow{x}$. Explain why $\overrightarrow{y}$ will also be an approximate eigenvector.

c) If $\overrightarrow{y}$ is an approximate eigenvalue, $A\overrightarrow{y} \approx \lambda\overrightarrow{y}$. Explain why we **CAN'T** find λ by dividing by $\overrightarrow{y}$.

d) "If you can't divide, multiply." What happens if we take the dot product of both sides with $\overrightarrow{y}$?

A7.8.4 Find an approximate eigenvalue-eigenvector pair for the given matrix. Verify that your eigenvalue/eigenvector pair is an approximate eigenvalue/eigenvector pair.

a) $\begin{pmatrix} 0 & 1 \\ 1 & 1 \end{pmatrix}$

b) $\begin{pmatrix} 1 & 2 \\ 3 & 5 \end{pmatrix}$

c) $\begin{pmatrix} 2 & -4 \\ -4 & 3 \end{pmatrix}$

d) $\begin{pmatrix} 2 & 3 & -5 \\ 2 & 2 & 8 \\ 1 & 1 & 5 \end{pmatrix}$

Numerical methods are attractive, because they often lead to solutions that might be difficult or impossible to find analytically.

Example 7.2. *Find an eigenvalue-eigenvector pair for* $\begin{pmatrix} 2 & 2 & -3 \\ 5 & 3 & 1 \\ 1 & -1 & 2 \end{pmatrix}$

Solution. *We'll pick* $\overrightarrow{x} = \langle 1, 0, 0\rangle$ *and choose* $n = 16$. *We find*

$$A^{16}\overrightarrow{x} = \langle 786597608269, 1319059822164, -137318992489\rangle$$

which we normalize to $\overrightarrow{y} \approx \langle 0.5101, 0.8555, -0.0891\rangle$, *which will be an approximate eigenvector.*

Next, we find

$$A\overrightarrow{y} = \langle 2.9984, 5.0281, -0.5234\rangle \approx 5.9\overrightarrow{y}$$

so the corresponding eigenvalue is $\lambda = 5.9$.

However, there are risks with numerical methods: it's often not clear what the limits of the numerical methods are. It's useful to test numerical methods to their breaking point.

Activity 7.9: Numerical Methods: To the Breaking Point

A7.9.1 Let $A = \begin{pmatrix} 5 & 2 \\ 4 & 3 \end{pmatrix}$. A has two real eigenvalues λ_1, λ_2, where $|\lambda_1| > |\lambda_2|$, so the method of Activity 7.8 can be used.

a) Use the method of Activity 7.8 to find an eigenvalue-eigenvector pair for A. (Call this eigenvector $\vec{v}_1$)
b) From Activity 7.8, the eigenvectors $\vec{v}_1$ and $\vec{v}_2$ must span $\mathbb{R}^2$. Why can't you use this fact to find the eigenvector $\vec{v}_2$?
c) Show that choosing a *different* starting vector will still give you the same eigenvalue.
d) Which eigenvalue will be found using this method?

A7.9.2 Let $B = \begin{pmatrix} 0 & 1 \\ -1 & 0 \end{pmatrix}$.

a) Suppose you try to find an eigenvalue-eigenvector pair using the method of Activity 7.8. What happens?
b) What does this suggest about the eigenvalues of B? (Remember that *if* the eigenvalues meet certain conditions, we can use the method of Activity 7.8 to find the largest eigenvalue.)

Activity A7.9.1 reveals one problem with the numerical approach: We can only use this method to find *one* eigenvalue-eigenvector pair.

Since the eigenvalue we find is the largest (in absolute value), and we're often interested in this largest eigenvalue, the numerical method can be used in many cases.

However, Activity A7.9.2 reveals a more fundamental problem: Sometimes, our approach doesn't work at all!

Clearly something unusual is going on in these cases. But what? To answer this question, let's return to the geometric interpretation of eigenvectors and eigenvalues.

Activity 7.10: Complex Eigenvalues

A7.10.1 Suppose λ, $\vec{x}$ is an eigenvalue-eigenvector pair for matrix A, and suppose λ is a real number. What is the geometric significance of $A\vec{x}$?

A7.10.2 Let

$$C = \begin{pmatrix} 0 & 1 \\ -1 & 0 \end{pmatrix}$$

a) Use the approach of Activity 7.8 to show $|\lambda_1| = 1$.

b) Show that neither $\lambda = 1$ nor $\lambda = -1$ is an eigenvalue for C.

c) By interpreting C as a geometric transformation, explain why you shouldn't be surprised by this result.

A7.10.3 Show that $D = \begin{pmatrix} 0 & 8 \\ -2 & 0 \end{pmatrix}$ has eigenvalue $\lambda = 4i$. What is an eigenvector corresponding to this eigenvalue?

Thus the problem is our underlying assumption that the eigenvalues are real.

7.4 Eigenvalues and Eigenvectors for a 2×2 Matrix

The preceding shows that while a numerical approach can work to find eigenvalues under *some* conditions, there are many common situations where the numerical methods will fail to produce results. Thus, we want to be able to find eigenvalues analytically. To do that, we'll employ our standard strategy: We'll set down and solve a system of linear equations.

Activity 7.11: Finding Eigenvectors

A7.11.1 Suppose

$$\begin{pmatrix} 2 & 4 \\ 3 & 1 \end{pmatrix} \begin{pmatrix} x_1 \\ x_2 \end{pmatrix} = \lambda \begin{pmatrix} x_1 \\ x_2 \end{pmatrix}$$

a) Write down the corresponding system of equations to find x_1, x_2.

b) Find and row reduce the coefficient matrix for the system.

c) In Activity 7.3, you found the eigenvector is not unique. Consequently this system must have an infinite number of solutions. What does this tell you about the entries in the row reduced coefficient matrix?

d) Find the eigenvalues and corresponding eigenvectors for the matrix.

A7.11.2 Suppose

$$\begin{pmatrix} 0 & 1 \\ -1 & 0 \end{pmatrix} \begin{pmatrix} x_1 \\ x_2 \end{pmatrix} = \lambda \begin{pmatrix} x_1 \\ x_2 \end{pmatrix}$$

a) Write down a system of equations to find x_1, x_2.

b) Find and row reduce the coefficient matrix for the system.

c) Remember this matrix must have an infinite number of solutions. What does this say about the eigenvalues λ?

d) Find the eigenvalues and eigenvectors for this matrix.

A7.11.3 Suppose

$$A = \begin{pmatrix} a & b \\ c & d \end{pmatrix}, \overrightarrow{x} = \begin{pmatrix} x_1 \\ x_2 \end{pmatrix}$$

a) Write down the coefficient matrix corresponding to the system $A\overrightarrow{x} = \lambda\overrightarrow{x}$, where x_1, x_2 are the unknowns to be solved for.

b) Show that this coefficient matrix is the same as $A - \lambda I$, where I is the identity matrix of the appropriate size.

c) What relationship must be true about the values a, b, c, d, λ for there to be an infinite number of solutions x_1, x_2?

7.5 The Characteristic Equation

In Activity 7.11, you found that in order for λ, $\mathbf{v}$ to be an eigenvalue-eigenvector pair for a 2×2 matrix $\begin{pmatrix} a & b \\ c & d \end{pmatrix}$, we require

$$(a - \lambda)(d - \lambda) - bc = 0$$

This is known as the **characteristic equation** for finding the eigenvalues of A. More specifically, it's the characteristic equation for 2×2 matrices. Notice that, since a, b, c, d are the entries of the matrix, the only actual unknown in this equation is λ. Thus the characteristic equation is a polynomial in λ.

Can we find the characteristic equation for matrices of larger dimension?

Activity 7.12: The Characteristic Equation

In the following, assume A is a $n \times n$ matrix.

A7.12.1 Show that if $\lambda, \overrightarrow{x}$ is an eigenvalue-eigenvector pair, then $(A - \lambda I)\overrightarrow{x} = \overrightarrow{0}$.

A7.12.2 Show that if $(A - \lambda I)\overrightarrow{x} = \overrightarrow{0}$ (with $\overrightarrow{x} \neq \overrightarrow{0}$), then $\lambda, \overrightarrow{x}$ is an eigenvalue-eigenvector pair.

A7.12.3 Suppose $|A - \lambda I| \neq 0$. What does this say about $\overrightarrow{x}$, a solution to $(A - \lambda I)\overrightarrow{x} = \overrightarrow{0}$?

A7.12.4 In order for $(A - \lambda I)\overrightarrow{x} = \overrightarrow{0}$ to have a nontrivial solution, what must be true about the determinant of $A - \lambda I$?

Activity 7.12 leads to the following:

Lemma 9. *Let A be a matrix. $\lambda, \mathbf{x}$, with $\mathbf{x} \neq \mathbf{0}$, is an eigenvector-eigenvalue pair if and only if $(A - \lambda I)\mathbf{x} = \mathbf{0}$*

In order for the equation $(A - \lambda I)\mathbf{x} = \mathbf{0}$ to have a nontrivial solution, we require $|A - \lambda I| = 0$. This leads to the following definition:

Definition 7.5 (Characteristic Equation). *Let A be a square matrix. The* ***characteristic equation*** *is*

$$|A - \lambda I| = 0$$

Since $|A - \lambda I|$ will be a polynomial in λ, we call it the **characteristic polynomial**. We can state the main result of Activity 7.12 as:

Theorem 7.1. *The eigenvalues λ of A are solutions to the characteristic equation $|A - \lambda I| = 0$.*

Because the characteristic polynomial *is* a polynomial, it's possible for it to have a repeated root. We define:

Definition 7.6 (Algebraic Multiplicity). *The algebraic multiplicity of an eigenvalue is number of times it appears as a root of the characteristic polynomial.*

Example 7.3. *Use the characteristic equation to find the eigenvalues of $A = \begin{pmatrix} 1 & 5 \\ 3 & 3 \end{pmatrix}$. Then find the eigenvectors.*

Solution. *We find the characteristic equation $|A - \lambda I| = 0$:*

$$\begin{aligned}
\left| \begin{pmatrix} 1 & 5 \\ 3 & 3 \end{pmatrix} - \lambda \begin{pmatrix} 1 & 0 \\ 0 & 1 \end{pmatrix} \right| &= 0 \\
\left| \begin{pmatrix} 1 & 5 \\ 3 & 3 \end{pmatrix} - \begin{pmatrix} \lambda & 0 \\ 0 & \lambda \end{pmatrix} \right| &= 0 \\
\left| \begin{pmatrix} 1 - \lambda & 5 \\ 3 & 3 - \lambda \end{pmatrix} \right| &= 0 \\
(1 - \lambda)(3 - \lambda) - 5 \cdot 3 &= 0 \\
\lambda^2 - 4\lambda + 3 - 15 &= 0 \\
\lambda^2 - 4\lambda - 12 &= 0 \\
(\lambda - 6)(\lambda + 2) &= 0
\end{aligned}$$

Solving this equation gives us $\lambda = 6$, $\lambda = -2$; note that since the factors corresponding to these solutions only appear once, both of these eigenvalues have algebraic multiplicity 1.

To find the eigenvector corresponding to $\lambda = 6$, remember that this will be a vector $\mathbf{x}$ that solves $(A - 6I)\mathbf{x} = \mathbf{0}$, so its components will be a nontrivial solution to the homogeneous system with coefficient matrix $A - 6I$. Finding, then row reducing this matrix:

$$\begin{pmatrix} 1-6 & 5 \\ 3 & 3-6 \end{pmatrix} = \begin{pmatrix} -5 & 5 \\ 3 & -3 \end{pmatrix} \rightarrow \begin{pmatrix} 1 & -1 \\ 0 & 0 \end{pmatrix}$$

which has solution $t\langle 1, 1\rangle$.

Similarly the eigenvector corresponding to $\lambda = -2$ will be a vector $\mathbf{x}$ that solves $(A-(-2)I)\mathbf{x} = \mathbf{0}$, so it will be a nontrivial solution to the homogeneous system with coefficient matrix $A - (-2)I$. Finding, then row reducing this matrix:

$$\begin{pmatrix} 1-(-2) & 5 \\ 3 & 3-(-2) \end{pmatrix} = \begin{pmatrix} 3 & 5 \\ 3 & 5 \end{pmatrix} \rightarrow \begin{pmatrix} 3 & 5 \\ 0 & 0 \end{pmatrix}$$

which has solution $s\langle -5, 3\rangle$.

Thus the eigenvalue $\lambda = 6$ has eigenvector $\langle 1, 1\rangle$, and $\lambda = -2$ has eigenvector $\langle -5, 3\rangle$.

Example 7.4. *Find the eigenvalues and eigenvectors of* $\begin{pmatrix} 1 & 3 \\ 0 & 1 \end{pmatrix}$

Solution. *We find the characteristic equation:*

$$\begin{vmatrix} 1-\lambda & 3 \\ 0 & 1-\lambda \end{vmatrix} = 0$$
$$(1-\lambda)(1-\lambda) - 3\cdot 0 = 0$$
$$(1-\lambda)(1-\lambda) = 0$$

which has solution $\lambda = 2$. Since the factor giving this solution is raised to the 2nd power, then $\lambda = 1$ has algebraic multiplicity 2.

The eigenvector will be a nontrivial solution to $(A - I)\mathbf{x} = \mathbf{0}$, so finding and reducing the coefficient matrix $A - I$ gives us:

$$\begin{pmatrix} 1-1 & 3 \\ 0 & 1-1 \end{pmatrix} = \begin{pmatrix} 0 & 3 \\ 0 & 0 \end{pmatrix}$$

which has solution $t\langle 1, 0\rangle$, so $\lambda = 1$ has eigenvalue $\langle 1, 0\rangle$.

The characteristic equation relies on finding the determinant of a matrix, and as we've seen, methods that rely on finding the determinant are good for theoretical results.

Activity 7.13: Eigenvalues and the Characteristic Equation

In the following, let A be a $n \times n$ square matrix.

A7.13.1 Suppose B is formed by multiplying a row of A by a constant c.

a) Explain why $|B| = c|A|$.

b) Explain why $|B - \lambda I| \neq c|A - \lambda I|$.

c) Why does this mean the eigenvalues of A and the eigenvalues of B are unrelated?

A7.13.2 Suppose C is formed by switching adjacent rows of A.

a) Explain why $|C| = -|A|$.

b) Explain why $|C - \lambda I| \neq -|A - \lambda I|$.

c) Why does this mean the eigenvalues of A and the eigenvalues of C are unrelated?

A7.13.3 Suppose A is an upper triangular matrix, a lower triangular matrix, or a diagonal matrix. What are the eigenvalues of A?

A7.13.4 Consider the matrix A^T.

a) Express the characteristic equation (in terms of the determinant) for finding the eigenvalues of A^T.

b) How does the characteristic equation for finding the eigenvalues of A^T compare to the characteristic equation for finding the eigenvalues of A?

c) What is the relationship between the eigenvalues of A and the eigenvalues of A^T? Why?

A7.13.5 Find the eigenvalues and eigenvectors for the given matrices.

a) $\begin{pmatrix} 7 & -5 \\ 10 & -8 \end{pmatrix}$

b) $\begin{pmatrix} 3 & -2 \\ 6 & -4 \end{pmatrix}$

c) $\begin{pmatrix} 8 & 5 \\ -10 & -7 \end{pmatrix}$

A7.13.6 Find the characteristic polynomial for the given matrices. **DO NOT** try to solve them for the eigenvalues.

a) $\begin{pmatrix} 1 & 0 & 0 \\ 3 & 0 & 5 \\ 1 & 1 & 3 \end{pmatrix}$

b) $\begin{pmatrix} 2 & 0 & 0 & 0 \\ 0 & 1 & 3 & 5 \\ 1 & 2 & 0 & 1 \\ 0 & 1 & 0 & 3 \end{pmatrix}$

A7.13.7 Consider your characteristic equations from Activity A7.13.6.

a) If the entries of A are real, what can you say about the coefficients of the characteristic polynomial for A?

b) What does your work suggest about the degree of the characteristic polynomial for a $n \times n$ matrix?

c) What does this suggest about the number of eigenvalues for a $n \times n$ matrix?

One of the most important results of Activity 7.13 is:

Theorem 7.2. *A and A^T have the same eigenvalues.*

Moreover:

Theorem 7.3. *The characteristic polynomial for a $n \times n$ matrix will be a nth degree polynomial, and there will be n eigenvalues.*

If A has real entries, the corresponding characteristic polynomial will have real coefficients. However, it's possible for a polynomial with real coefficients to have complex roots. Consequently, it's possible for our eigenvalues to be complex numbers.

Activity 7.14: Complex Eigenvalues and Eigenvectors

A7.14.1 Suppose A is a $n \times n$ matrix.

a) What is the degree of the characteristic polynomial?

b) Suppose n is odd. Why does this mean that A always has at least one real eigenvalue?

c) Suppose the entries of A are real numbers. Why does this mean that if $\lambda = a + bi$ is an eigenvalue for A, so is $a - bi$?

A7.14.2 Find the eigenvalues and eigenvectors of the matrices given.

a) $\begin{pmatrix} 0 & 2 \\ -2 & 0 \end{pmatrix}$

b) $\begin{pmatrix} 3 & 1 \\ -9 & 3 \end{pmatrix}$

c) $\begin{pmatrix} 2 & 0 & 0 \\ 0 & 1 & -2 \\ 0 & 8 & 1 \end{pmatrix}$

A7.14.3 Suppose A is a matrix with real entries, and let $\lambda, \mathbf{u}$ be an eigenvalue/eigenvector pair for A.

a) Suppose λ is complex. Prove/disprove: $\mathbf{u}$ is a real vector.
b) Prove: $\overline{\lambda}$ is an eigenvalue of A. If it is an eigenvalue, what is the corresponding eigenvector?
c) Prove: $\overline{\mathbf{u}}$ is an eigenvector of A. If it is an eigenvector, what is the corresponding eigenvalue?

What if the entries of our matrix are complex numbers? While the roots of a polynomial with complex coefficients are complex numbers, we don't have the guarantee they will occur in conjugate pairs. However, we can still determine other properties of the eigenvalues and eigenvectors of such a matrix.

Activity 7.15: Hermitian Matrices

In the following, let A be a square matrix with complex entries.

A7.15.1 Suppose $\lambda = a + bi$ is a complex eigenvalue for A.

a) Find an eigenvalue for $\overline{A}$.
b) Find an eigenvalue of $\overline{(A^T)}$.
c) What does this say about the eigenvalues of Hermitian matrices? Why?

A7.15.2 Actually, we can strengthen our statement about the eigenvalues of Hermitian matrices from Activity A7.15.1c as follows. Suppose A is Hermitian and $\lambda, \mathbf{u}$ is an eigenvalue/eigenvector pair for A. Let $c = \overline{\mathbf{u}} \cdot A\mathbf{u}$.

a) Prove: $c = \lambda ||\mathbf{u}||^2$.
b) One objection to the notation $\overline{\mathbf{u}} \cdot A\mathbf{u}$ is that there are two types multiplication present: the dot product $\cdot$, and the matrix multiplication $A\mathbf{u}$. Rewrite the definition of c using only matrix multiplication. (Assume $\mathbf{u}$ is a column vector, so $A\mathbf{u}$ is a legitimate expression of a matrix product).
c) From your expression of c as a matrix product, find an expression for $\overline{c}$.
d) Show that $\overline{c} = \lambda ||\mathbf{u}||^2$. Suggestion: Remember $(MN)^T = N^T M^T$, and that we're assuming A is Hermitian.
e) Why does your work prove λ must be a real number?

Activity A7.15.2 proves:

Theorem 7.4. *The eigenvalues of Hermitian matrices are real.*

The H notation and the term Hermitian are in honor of Hermite's proof of this result in 1855.

One of the interesting features about mathematics is that problems work in both directions: On the one hand, if we can solve the characteristic equation $|A - \lambda I| = 0$ to find the eigenvalues of A, we can find the roots of a polynomial equation by finding the eigenvalues of some matrix A.

Activity 7.16: Solving Polynomial Equations

A useful feature of mathematics is that problems are generally reversible: If we solve a polynomial equation to find eigenvalues, then we can find eigenvalues to solve polynomial equations.

A7.16.1 Suppose you want to find a matrix A where $|A - \lambda I| = \lambda^2 + p\lambda + q$.

a) If $|A - \lambda I| = \lambda^2 + p\lambda + q$, what are the dimensions of A?

b) Find A where $|A - \lambda I| = \lambda^2 + p\lambda + q$. Suggestion: Rewrite the polynomial as $\lambda^2 + p\lambda + q = -\lambda(-\lambda - p) - (-q)(1)$.

A7.16.2 Let A be a matrix where $|A - \lambda I| = \lambda^2 + p\lambda + q$, and consider the matrix $A' = \begin{pmatrix} 0 & \mathbf{0} \\ \mathbf{0} & A \end{pmatrix}$ where $\mathbf{0}$ is the zero vector of the appropriate size and orientation, and 0 is 0.

a) Find $|A' - \lambda I|$.

b) Find row vector $\mathbf{u}$ and column vector $\mathbf{v}$ where $B = \begin{pmatrix} 0 & \mathbf{u} \\ \mathbf{v} & A \end{pmatrix}$ has $|B - \lambda I| = -(\lambda^3 + p\lambda^2 + q\lambda + r)$. Suggestion: The first term of the expansion along the first row of $B - \lambda I$ will be $-\lambda|A - \lambda I|$. What other cofactors are necessary to give the desired determinant?

A7.16.3 Let $f(x) = x^n + c_{n-1}x^{n-1} + c_{n-2}x^{n-2} + \ldots + c_2x^2 + c_1x + c_0$.

a) Explain why the roots of $f(x)$ are the same as the roots of $-f(x)$.

b) Find M where $|M - \lambda I| = f(\lambda)$ or $|M - \lambda I| = -f(\lambda)$.

c) Explain why the eigenvalues of M correspond to the roots of $f(x)$.

A7.16.4 Consider $x^4 - 4x^3 + 9x^2 - 10x + 40$.

a) Find a matrix M whose eigenvalues are the roots of the polynomial.

b) What happens if you apply the method from Activity 7.8 for finding approximate eigenvalues?

c) What does this say about the roots of $x^4 - 4x^3 + 9x^2 - 10x + 40$?

A7.16.5 For each polynomial, find a matrix M whose eigenvalues are the roots of the polynomial, then use the method of approximating eigenvalues from Activity 7.8 to find an approximate root of the given polynomial. Verify your solution is an approximate root.

a) $x^2 + 5x + 3$

b) $x^2 - 12x + 10$

c) $x^3 - 4x^2 + 15x - 110$

7.6 Stochastic Matrices

Theorem 7.2 leads to several very important result for stochastic matrices.

Activity 7.17: Eigenvalues and Stochastic Matrices

In the following, let M be a $n \times n$ stochastic matrix.

A7.17.1 Consider $\vec{1}$, the vector with n components all equal to 1.

a) Show that $\vec{1}$ is an eigenvector for M^T. What is the corresponding eigenvalue?

b) Prove: M has an eigenvalue of 1.

c) Why does this mean that every stochastic matrix must have a fixed point?

A7.17.2 Suppose λ is an eigenvalue of M^T, with corresponding eigenvector $\langle v_1, v_2, \ldots, v_n \rangle$. Without loss of generality, we'll assume v_1 is the greatest component of the eigenvector (in other words, $|v_1| \geq |v_i|$ for all i).

a) Prove: $|m_{11}v_1 + m_{21}v_2 + \ldots m_{n1}v_n| = |\lambda||v_1|$. (Note: The values m_{ij} are the entries in the first *column* of the matrix M, while the values v_i are the components of the eigenvector for M^T.)

b) Prove: $|m_{11}v_1 + m_{21}v_2 + \ldots m_{n1}v_n| \leq |v_1|$. Suggestion: What do you know about v_1 and the m_{i1}s?

c) Why was it reasonable to say "Without loss of generality?"

d) Why does this mean that if λ is an eigenvalue of M, then $|\lambda| \leq 1$?

A7.17.3 Suppose the stochastic matrix M represents the movement rule for how people move around a park with n locations (as in Activity 3.12, for example). Let $\vec{x}_0 = \langle x_1, x_2, \ldots, \rangle$ be the initial placement of the parkgoers, where x_1 is the number of parkgoers in location 1; x_2 is the number of parkgoers in location 2; and so on.

a) Explain why $x_i \geq 0$ for all i.

b) What is the significance of the sum of the components of $\vec{x}_0$?

c) Let $\vec{x}_k = M^k \vec{x}_0$. What do you know about the components of x_k, and the sum of the components?

d) Prove: If $|\lambda| < 1$, $\vec{x}_0$ **CANNOT** be an eigenvalue of M corresponding to λ.

Activity 7.17 proves:

Theorem 7.5. *Let M be a stochastic matrix. Then:*

- *M always has a fixed point,*
- *M always has eigenvalue $\lambda = 1$,*
- *All other eigenvalues of M have absolute value less than or equal to 1.*
- *If $\mathbf{v}$ is an eigenvector for M corresponding to eigenvalue λ, where $|\lambda| < 1$, then $\mathbf{v}$ has both positive and negative components.*

7.7 A Determinant-Free Approach

Finding eigenvalues and eigenvectors using the characteristic equation relies on computing the determinant of a matrix, and for any matrix of reasonable size, this is an arduous task. Thus, like Cramer's rule and the adjoint method, finding the eigenvalues through the characteristic equation is a useful theoretical result, but it is challenging to implement it practically.

So how can we find eigenvalues and eigenvectors? To begin with, remember that a square matrix A is a linear transformation that acts on vectors in $\mathbb{F}^n$ and produces vectors in $\mathbb{F}^n$. This means we can apply A repeatedly to some randomly chosen vector $\mathbf{v}$, known as a **seed vector**.

Activity 7.18: More Equations for Eigenvalues

A7.18.1 Explain why the set consisting of just the seed vector $\mathbf{v}$ must be independent.

A7.18.2 Suppose $\{\mathbf{v}, A\mathbf{v}\}$ is *dependent*. Explain why this means that $\mathbf{v}$ is an eigenvector.

A7.18.3 Prove/disprove: If A is a 2×2 matrix and $\{\mathbf{v}, A\mathbf{v}\}$ is independent, $\{\mathbf{v}, A\mathbf{v}, A^2\mathbf{v}\}$ must be *dependent*.

A7.18.4 Suppose $\{\mathbf{v}, A\mathbf{v}\}$ is independent, but there exist nonzero values a_1 and a_0 where

$$A^2\mathbf{v} + a_1 A\mathbf{v} + a_0\mathbf{0} = \mathbf{0}$$

Further, suppose $x^2 + a_1 x + a_0 = (x - \lambda_1)(x - \lambda_2)$.

a) Prove

$$A^2\mathbf{v} + a_1 A\mathbf{v} + a_0\mathbf{v} = (A - \lambda_1 I)(A - \lambda_2 I)\mathbf{v}$$

where I is the identity matrix of the appropriate size.

b) Also prove

$$A^2\mathbf{v} + a_1 A\mathbf{v} + a_0\mathbf{v} = (A - \lambda_2 I)(A - \lambda_1 I)\mathbf{v}$$

(in other words, the factors can be rearranged)

c) Explain why neither $(A - \lambda_1 I)\mathbf{v}$ nor $(A - \lambda_2 I)\mathbf{v}$ can be $\mathbf{0}$.

d) Explain why both $(A - \lambda_1 I)\mathbf{v}$ and $(A - \lambda_2 I)\mathbf{v}$ are eigenvectors. What are their corresponding eigenvalues?

Activity 7.18 suggests a way to find the eigenvectors and eigenvalues for any 2×2 matrix A and (if we're lucky) some larger matrices.

Example 7.5. *Find eigenvalues and eigenvectors for*

$$A = \begin{pmatrix} 9 & -12 \\ 8 & -11 \end{pmatrix}$$

Solution. *We'll start with the simplest possible seed vector and choose* $\mathbf{v} = \langle 1, 0 \rangle$. *We find:*

$$\mathbf{v} = \begin{pmatrix} 1 \\ 0 \end{pmatrix}, \qquad A\mathbf{v} = \begin{pmatrix} 9 \\ 8 \end{pmatrix}, \qquad A^2\mathbf{v} = \begin{pmatrix} -15 \\ -16 \end{pmatrix}$$

We note that $\{\mathbf{v}, A\mathbf{v}\}$ *is an independent set, but* $\{\mathbf{v}, A\mathbf{v}, A^2\mathbf{v}\}$ *is not, so there is a nontrivial solution to*

$$x_1\mathbf{v} + x_2 A\mathbf{v} + x_3 A^2\mathbf{v} = \mathbf{0}$$

Note that finding a solution follows the same process as trying to determine whether a set of vectors is independent. Remember that we checked for independence by letting our vectors be the columns of a matrix, then row reducing. Thus:

$$\begin{pmatrix} 1 & 9 & -15 \\ 0 & 8 & -16 \end{pmatrix} \longrightarrow \begin{pmatrix} 1 & 9 & -15 \\ 0 & 1 & -2 \end{pmatrix}$$

One nontrivial solution will be

$$x_1 = -3, x_2 = 2, x_3 = 1$$

and so

$$-3\mathbf{v} + 2A\mathbf{v} + A^2\mathbf{v} = \mathbf{0}$$

or, in a more standard form

$$A^2\mathbf{v} + 2A\mathbf{v} - 3\mathbf{v} = \mathbf{0}$$

which we can write as

$$(A^2 + 2A - 3I)\mathbf{v} = \mathbf{0}$$

which has factors $A + 3I$ and $A - I$.

To find the eigenvectors, we note that since

$$(A + 3I)(A - I)\mathbf{v} = \mathbf{0}$$

then $(A - I)\mathbf{v}$ is an eigenvector corresponding to $\lambda = -3$, so the eigenvector is

$$\left(\begin{pmatrix} 9 & -12 \\ 8 & -11 \end{pmatrix} - \begin{pmatrix} 1 & 0 \\ 0 & 1 \end{pmatrix}\right)\begin{pmatrix} 1 \\ 0 \end{pmatrix} = \begin{pmatrix} 8 & -12 \\ 8 & -12 \end{pmatrix}\begin{pmatrix} 1 \\ 0 \end{pmatrix}$$
$$= \begin{pmatrix} 8 \\ 8 \end{pmatrix}$$

While we could use $\langle 8, 8\rangle$ as an eigenvector, remember any scalar multiple of an eigenvector is also an eigenvector; conversely, we can remove any common factor of the components, so we'll scale this to $\langle 1, 1\rangle$

Next, since

$$(A - I)(A + 3I)\mathbf{v} = \mathbf{0}$$

then $(A+3I)\mathbf{v}$ will be an eigenvector corresponding to $\lambda = 1$, so the eigenvector is

$$\left(\begin{pmatrix} 9 & -12 \\ 8 & -11 \end{pmatrix} + 3\begin{pmatrix} 1 & 0 \\ 0 & 1 \end{pmatrix}\right)\begin{pmatrix} 1 \\ 0 \end{pmatrix} = \begin{pmatrix} 12 & -12 \\ 8 & -8 \end{pmatrix}\begin{pmatrix} 1 \\ 0 \end{pmatrix}$$
$$= \begin{pmatrix} 12 \\ 8 \end{pmatrix}$$

and again while we could use $\langle 12, 8\rangle$ as an eigenvector, we can also scale it to $\langle 3, 2\rangle$.

Can we extend this to larger matrices?

Activity 7.19: Higher Dimensional Matrices

In the following, assume A is a $n \times n$ square matrix, and $\mathbf{v}$ is a randomly chosen (nonzero) vector in $\mathbb{F}^n$.

A7.19.1 Explain why there must be a greatest value $k-1$ for which the set

$$\left\{\mathbf{v}, A\mathbf{v}, A^2\mathbf{v}, \ldots, A^{k-1}\mathbf{v}\right\}$$

is independent.

A7.19.2 Why does this mean there are nontrivial values $a_1, a_2, \ldots a_k$ for which

$$A^k\mathbf{v} + a_k A^{k-1}\mathbf{v} + \ldots + a_2 A\mathbf{v} + a_1\mathbf{v} = \mathbf{0}$$

A7.19.3 Suppose

$$A^k\mathbf{v} + a_k A^{k-1}\mathbf{v} + \ldots + a_2 A\mathbf{v} + a_1\mathbf{v} = (A - \lambda_1 I)(A - \lambda_2 I)\cdots(A - \lambda_k I)$$

Prove: the factors $(A - \lambda_i I)$ can be rearranged in any order.

A7.19.4 Let k be the least value for which there are nontrivial solutions to

$$A^k\mathbf{v} + a_{k-1} A^{k-1}\mathbf{v} + \ldots + a_1 A\mathbf{v} + a_0\mathbf{v} = \mathbf{0}$$

where

$$A^k + a_{k-1}A^k + \ldots + a_1 A + a_0 I = (A - \lambda_1 I)(A - \lambda_2 I)\cdots(A - \lambda_k I)$$

a) Prove: $\mathbf{v}_1 = (A - \lambda_2 I)\cdots(A - \lambda_k I)\mathbf{x}$ is an eigenvector. Note: Since $(A - \lambda_1 I)\mathbf{v}_1 = \mathbf{0}$, you only have to prove that $\mathbf{v}_1$ is not $\mathbf{0}$.

b) Prove: Each of the λ_is is an eigenvalue. Also find the corresponding eigenvectors.

A7.19.5 Let k be the least value for which there are nontrivial solutions to

$$x_{k+1}A^k\mathbf{v} + x_k A^{k-1}\mathbf{v} + \ldots + x_2 A\mathbf{v} + x_1\mathbf{v} = \mathbf{0}$$

a) Explain why you can find the x_is by row reducing

$$\begin{pmatrix} \mathbf{v} & A\mathbf{v} & A^2\mathbf{v} & \ldots & A^k\mathbf{v} \end{pmatrix}$$

(in other words, by row reducing the matrix of column vectors)

b) Prove: x_{k+1} is a free variable. Suggestion: If it's not a free variable, what must it be equal to?

c) Prove: The system has no other free variables. Suggestion: Suppose x_i, with $i < k+1$, is another free variable. What happens if you let $x_{k+1} = 0$?

d) Prove: There is a unique solution $x_1, x_2, \ldots, x_k$ to

$$A^k\mathbf{v} + x_k A^{k-1}\mathbf{v} + \ldots + x_2 A\mathbf{v} + x_1\mathbf{v} = \mathbf{0}$$

To talk about the result from Activity 7.19, we introduce the following terms. Let $\mathbf{v}$ be a nonzero vector, and let k be the least value for which there are nontrivial solutions to

$$A^k\mathbf{v} + a_{k-1}A^{k-1}\mathbf{v} + a_{k-2}A^{k-2}\mathbf{v} + \ldots + a_1A\mathbf{v} + a_0\mathbf{v} = \mathbf{0}$$

The **minimal polynomial of A with respect to $\mathbf{v}$** (or simply the **minimal polynomial** if $\mathbf{v}$ is understood) is

$$x^k + a_{k-1}x^{k-1} + a_{k-2}x^{k-2} + \ldots + a_1x + a_0$$

(note the coefficient of the leading term is 1). Activity A7.19.4 gives us:

Theorem 7.6. *All roots of the minimal polynomial are eigenvalues.*

This means the approach of Activity 7.18 can be generalized to arbitrary matrices.

We could find the minimal polynomial by checking to see if $\{\mathbf{v}, A\mathbf{v}\}$ is independent; if it is, we can then check to see if $\{\mathbf{v}, A\mathbf{v}, A^2\mathbf{v}\}$ is independent; if it is, we can then check to see if $\{\mathbf{v}, A\mathbf{v}, A^2\mathbf{v}, A^3\mathbf{v}\}$ is independent; and so on. But is there a more efficient approach?

Activity 7.20: The Minimal Polynomial

A7.20.1 Let $A = \begin{pmatrix} 3 & -2 & -4 \\ -8 & 5 & 12 \\ 4 & -2 & -5 \end{pmatrix}$ and $\mathbf{v} = \begin{pmatrix} 1 \\ 0 \\ 0 \end{pmatrix}$.

a) Find $\mathbf{v}$, $A\mathbf{v}$, $A^2\mathbf{v}$, $A^3\mathbf{v}$, and $A^4\mathbf{v}$.

b) Find the solution(s) to the equation

$$x_4A^4\mathbf{v} + x^3A^3\mathbf{v} + x_2A^2\mathbf{v} + x_1A\mathbf{v} + x_0\mathbf{v} = \mathbf{0}$$

c) Find the least value k, and the corresponding values of x_i, for which there are nontrivial solutions to

$$A^k\mathbf{v} + x_{k-1}A^{k-1}\mathbf{v} + \ldots + x_1A\mathbf{v} + x_0\mathbf{v} = \mathbf{0}$$

A7.20.2 Let $B = \begin{pmatrix} 3 & 14 & -14 & -6 \\ -6 & 27 & -22 & -10 \\ 12 & 50 & -51 & -22 \\ -42 & -56 & 70 & 29 \end{pmatrix}$ and $\mathbf{v} = \begin{pmatrix} 1 \\ 0 \\ 0 \\ 0 \end{pmatrix}$. Find the least value k, and the corresponding values x_i, for which there are nontrivial solutions to

$$A^k\mathbf{v} + x_{k-1}A^{k-1}\mathbf{v} + \ldots + x_1A\mathbf{v} + x_0\mathbf{v} = \mathbf{0}$$

If A is a $n \times n$ matrix, then the set

$$\{\mathbf{v}, A\mathbf{v}, A^2\mathbf{v}, \ldots, A^n\mathbf{v}\}$$

is necessarily dependent. Activity 7.20 suggests that we can produce the minimal polynomial from this set alone. Then once we have the eigenvalues, we can find the eigenvectors by solving the eigenequation $A\mathbf{v} = \lambda\mathbf{v}$.

Example 7.6. *Find the eigenvalues and eigenvectors for*

$$A = \begin{pmatrix} 11 & 50 & 30 & 10 \\ 2 & 11 & 6 & 2 \\ -30 & -150 & -89 & -30 \\ 72 & 360 & 216 & 73 \end{pmatrix}$$

Solution. *We'll pick a random vector, say* $\mathbf{v} = \langle 1, 0, 0, 0 \rangle$ *and compute*

$$\mathbf{v} = \begin{pmatrix} 1 \\ 0 \\ 0 \\ 0 \end{pmatrix}, \qquad A\mathbf{v} = \begin{pmatrix} 11 \\ 2 \\ -30 \\ 72 \end{pmatrix}, \qquad A^2\mathbf{v} = \begin{pmatrix} 41 \\ 8 \\ -120 \\ 288 \end{pmatrix}$$

$$A^3\mathbf{v} = \begin{pmatrix} 131 \\ 26 \\ -390 \\ 936 \end{pmatrix}, \qquad A^4\mathbf{v} = \begin{pmatrix} 401 \\ 80 \\ -1200 \\ 2880 \end{pmatrix}$$

Letting these be column vectors and row reducing, we find

$$\begin{pmatrix} 1 & 11 & 41 & 131 & 401 \\ 0 & 2 & 8 & 26 & 80 \\ 0 & -30 & -120 & -390 & -1200 \\ 0 & 72 & 288 & 936 & 2880 \end{pmatrix} \longrightarrow \begin{pmatrix} 1 & 0 & -3 & -12 & -39 \\ 0 & 1 & 4 & 13 & 40 \\ 0 & 0 & 0 & 0 & 0 \\ 0 & 0 & 0 & 0 & 0 \end{pmatrix}$$

We have x_5, x_4, x_3 *as independent variables. Since we want a minimal polynomial, we'll use the solution*

$$x_1 = 3, x_2 = -4, x_3 = 1, x_4 = 0, x_5 = 0$$

which gives minimal polynomial $x^2 - 4x + 3 = (x - 3)(x - 1)$, *giving us eigenvalues* $\lambda = 3$, $\lambda = 1$.

For $\lambda = 3$, *we have*

$$\begin{pmatrix} 11 & 50 & 30 & 10 \\ 2 & 11 & 6 & 2 \\ -30 & -150 & -89 & -30 \\ 72 & 360 & 216 & 73 \end{pmatrix} \begin{pmatrix} x_1 \\ x_2 \\ x_3 \\ x_4 \end{pmatrix} = 3 \begin{pmatrix} x_1 \\ x_2 \\ x_3 \\ x_4 \end{pmatrix}$$

which reduces to

$$\begin{pmatrix} 8 & 50 & 30 & 10 \\ 2 & 8 & 6 & 2 \\ -30 & -150 & -92 & -30 \\ 72 & 360 & 216 & 70 \end{pmatrix} \begin{pmatrix} x_1 \\ x_2 \\ x_3 \\ x_4 \end{pmatrix} = \begin{pmatrix} 0 \\ 0 \\ 0 \\ 0 \end{pmatrix}$$

Row reducing

$$\begin{pmatrix} 8 & 50 & 30 & 10 \\ 2 & 8 & 6 & 2 \\ -30 & -150 & -92 & -30 \\ 72 & 360 & 216 & 70 \end{pmatrix} \longrightarrow \begin{pmatrix} 1 & 4 & 3 & 1 \\ 0 & 9 & 3 & 1 \\ 0 & 0 & -12 & -5 \\ 0 & 0 & 0 & 0 \end{pmatrix}$$

which has solution

$$\langle x_1, x_2, x_3, x_4 \rangle = t\langle 5, 1, -15, 36 \rangle$$

as an eigenvector with geometric multiplicity 1.

For $\lambda = 1$*, we have:*

$$\begin{pmatrix} 11 & 50 & 30 & 10 \\ 2 & 11 & 6 & 2 \\ -30 & -150 & -89 & -30 \\ 72 & 360 & 216 & 73 \end{pmatrix} \begin{pmatrix} x_1 \\ x_2 \\ x_3 \\ x_4 \end{pmatrix} = 1 \begin{pmatrix} x_1 \\ x_2 \\ x_3 \\ x_4 \end{pmatrix}$$

which reduces to

$$\begin{pmatrix} 10 & 50 & 30 & 10 \\ 2 & 10 & 6 & 2 \\ -30 & -150 & -90 & -30 \\ 72 & 360 & 216 & 72 \end{pmatrix} \begin{pmatrix} x_1 \\ x_2 \\ x_3 \\ x_4 \end{pmatrix} = \begin{pmatrix} 0 \\ 0 \\ 0 \\ 0 \end{pmatrix}$$

Row reducing

$$\begin{pmatrix} 10 & 50 & 30 & 10 \\ 2 & 10 & 6 & 2 \\ -30 & -150 & -90 & -30 \\ 72 & 360 & 216 & 72 \end{pmatrix} \longrightarrow \begin{pmatrix} 1 & 5 & 3 & 1 \\ 0 & 0 & 0 & 0 \\ 0 & 0 & 0 & 0 \\ 0 & 0 & 0 & 0 \end{pmatrix}$$

which gives solution

$$\langle x_1, x_2, x_3, x_4 \rangle = t\langle -1, 0, 0, 1 \rangle + s\langle -5, 1, 0, 0 \rangle + r\langle -3, 0, 1, 0 \rangle$$

So $\lambda = 1$ *is an eigenvalue with geometric multiplicity* 3.

Since this gives us a total of four eigenvectors in $\mathbb{R}^4$*, we have found all of the eigenvectors.*

While the roots of the minimal polynomial will be eigenvalues, and we'll be able to find linearly independent eigenvectors once we know the eigenvalues, there's no guarantee as yet that we'll find *all* the eigenvalues this way. So what happens if our minimal polynomial doesn't allow us to find all the eigenvalues? We'll consider two different cases.

Remember that a $n \times n$ matrix can have at most n linearly independent eigenvectors. If it does:

Definition 7.7 (Nondefective). *A $n \times n$ matrix A is* ***nondefective*** *if it has n linearly independent eigenvectors; it is* ***defective*** *otherwise.*

Suppose we try to find the eigenvectors and eigenvalues for a nondefective $n \times n$ matrix using the approach in Activity 7.20. Our first seed vector will give us a minimal polynomial, and the roots of the minimal polynomial will give us some of the eigenvalues. We can then find the associated eigenvectors.

If we don't have enough eigenvectors, the obvious strategy is to pick another seed vector. But which one? In some sense, if our seed vector simply reproduces the known eigenvalues, then our effort will be wasted; what we want is a seed vector that gives us *new* eigenvalues. Is there some way to choose such a seed vector?

Activity 7.21: Seedling Vectors

In the following, assume A is a nondefective $n \times n$ matrix, with n linearly independent eigenvectors

$$\mathcal{V} = \{\mathbf{v}_1, \mathbf{v}_2, \dots \mathbf{v}_n, \}$$

Assume that you've found some of eigenvalues λ_1, λ_2, ..., λ_k. Let

$$H = (A - \lambda_1 I)(A - \lambda_2 I) \cdots (A - \lambda_k I)$$

A7.21.1 Explain why $\mathcal{V}$ is a basis for $\mathbb{F}^n$.

A7.21.2 Suppose $\mathbf{x}$ is a linear combination of the eigenvectors corresponding to the eigenvalues you've already found. Prove: $H\mathbf{x} = \mathbf{0}$.

A7.21.3 Suppose $H\mathbf{z} \neq \mathbf{0}$. Prove: $H\mathbf{z}$ is a linear combination of eigenvectors corresponding to eigenvalues you don't already know.

A7.21.4 Consider the minimal polynomial of A with respect to $H\mathbf{z}$, where $H\mathbf{z} \neq \mathbf{0}$. Explain why this minimal polynomial *must* have a degree less than or equal to $n - k$, where k is the number of eigenvalues you've already found.

A7.21.5 Let $A = \begin{pmatrix} 2 & 14 & -10 & -7 \\ 0 & -3 & 2 & 2 \\ 0 & 6 & 1 & -2 \\ 0 & -18 & 6 & 9 \end{pmatrix}$

a) Use seed vector $\langle 1, 0, 0, 0\rangle$ to find a minimal polynomial. Then find a basis for the corresponding eigenspace.
b) Choose another seed vector and (if possible) find additional eigenvalues and eigenvectors.

Activity 7.21 suggests the following approach. Suppose we've found some of the eigenvalues, say λ_1 through λ_k. Let

$$H = (A - \lambda_1 I)(A - \lambda_2 I)\cdots(A - \lambda_k I)$$

If $\mathbf{x}$ is a linear combination of the eigenvectors associated with the λ_is, then $H\mathbf{x} = \mathbf{0}$. Consequently, if $H\mathbf{z} \neq \mathbf{0}$, then $\mathbf{z}$ must include some of the eigenvectors for eigenvalues we *haven't* found. Even better, $H\mathbf{z}$ can be expressed *entirely* in terms of these unknown eigenvectors. Thus, if we want to find our eigenvectors efficiently, our next seed vector should in fact be the "seedling" vector $H\mathbf{z}$. In effect, we choose a seed vector, then "grow" it into a more suitable form.

A further advantage is the following: The minimal polynomial for a $n \times n$ matrix could have degree n. But suppose we've found some of the eigenvalues and their corresponding eigenvectors. $H\mathbf{z}$ must be a linear combination of the *remaining* eigenvectors, which means the minimal polynomial will have a lesser degree.

Example 7.7. *Find all eigenvectors of*

$$A = \begin{pmatrix} 35 & 88 & 46 & 12 \\ -8 & -25 & -14 & -4 \\ -52 & -124 & -63 & -16 \\ 162 & 414 & 216 & 57 \end{pmatrix}$$

Solution. *If we use our standard seed vector, we'll find all the eigenvalues and eigenvectors in a single step, so for illustrative purposes we'll choose* $\mathbf{v} = \begin{pmatrix} 0 \\ 4 \\ -10 \\ 9 \end{pmatrix}$ *and compute*

$$\mathbf{v} = \begin{pmatrix} 0 \\ 4 \\ -10 \\ 9 \end{pmatrix} \quad A\mathbf{v} = \begin{pmatrix} 0 \\ 4 \\ -10 \\ 9 \end{pmatrix} \quad A^2\mathbf{v} = \begin{pmatrix} 0 \\ 4 \\ -10 \\ 9 \end{pmatrix} \quad A^3\mathbf{v} = \begin{pmatrix} 0 \\ 4 \\ -10 \\ 9 \end{pmatrix} \quad A^4\mathbf{v} = \begin{pmatrix} 0 \\ 4 \\ -10 \\ 9 \end{pmatrix}$$

Using these as column vectors and row reducing

$$\begin{pmatrix} 0 & 0 & 0 & 0 & 0 \\ 4 & 4 & 4 & 4 & 4 \\ -10 & -10 & -10 & -10 & -10 \\ 9 & 9 & 9 & 9 & 9 \end{pmatrix} \rightarrow \begin{pmatrix} 1 & 1 & 1 & 1 & 1 \\ 0 & 0 & 0 & 0 & 0 \\ 0 & 0 & 0 & 0 & 0 \\ 0 & 0 & 0 & 0 & 0 \end{pmatrix}$$

with corresponding minimal polynomial $f(A) = A - I$. Thus $\lambda = 1$ is an eigenvalue, and we find the corresponding set of linearly independent eigenvectors $\begin{pmatrix} 2 \\ -2 \\ 0 \\ 9 \end{pmatrix}$ *and* $\begin{pmatrix} 1 \\ -3 \\ 5 \\ 0 \end{pmatrix}$. *Since A is a 4×4 matrix, we can find two more linearly independent eigenvectors and/or generalized eigenvectors.*

To find them, we'll pick another seed vector, then apply our minimal polynomial $f(A) = A - I$ to grow it into a seedling. This time we'll use the standard choice, and compute:

$$\mathbf{x} = \begin{pmatrix} 1 \\ 0 \\ 0 \\ 0 \end{pmatrix} \qquad (A - I)\mathbf{x} = \begin{pmatrix} 34 \\ -8 \\ -52 \\ 162 \end{pmatrix}$$

Since we've already found two linearly independent eigenvectors, and there are at most 4, then the minimal polynomial with respect to $\mathbf{v} = (A - I)\mathbf{x}$ will have degree at most $4 - 2 = 2$, so we only need to compute

$$\mathbf{v} = \begin{pmatrix} 34 \\ -8 \\ -52 \\ 162 \end{pmatrix} \qquad A\mathbf{v} = \begin{pmatrix} 38 \\ 8 \\ -92 \\ 198 \end{pmatrix} \qquad A^2\mathbf{v} = \begin{pmatrix} 178 \\ -8 \\ -340 \\ 882 \end{pmatrix}$$

Using these as column vectors and row reducing

$$\begin{pmatrix} 34 & 38 & 178 \\ -8 & 8 & -8 \\ -52 & -92 & -340 \\ 162 & 198 & 882 \end{pmatrix} \to \begin{pmatrix} 1 & 0 & 3 \\ 0 & 1 & 2 \\ 0 & 0 & 0 \\ 0 & 0 & 0 \end{pmatrix}$$

with corresponding minimal polynomial $f(A) = A^2 - 2A - 3I$, giving us the remaining eigenvalues $\lambda = 3, -1$; these in turn give us the remaining eigenvectors (which we'll leave for the reader to find).

7.8 Generalized Eigenvalues

What if we have a defective matrix?

Activity 7.22: Defective Matrices

A7.22.1 Consider the matrix $A = \begin{pmatrix} 1 & 0 \\ 2 & 1 \end{pmatrix}$

a) Using seed vector $\vec{x} = \langle 1, 0 \rangle$, find a minimal polynomial. Then find the eigenvalue(s) and corresponding eigenvector(s).

b) Choose a different seed vector and find a minimal polynomial. Find the eigenvalue(s) and corresponding eigenvector(s).

c) Choose a third seed vector and find a minimal polynomial. Find the eigenvalue(s)and corresponding eigenvector(s).

d) Since A is a 2×2 matrix, how many independent eigenvectors *could* it have, and why is your work on Activities A7.22.1a through A7.22.1c troubling?

Activity 7.22 suggests that we might not be able to find all eigenvalues of a matrix this way.

But let's take a closer look. When we choose seed vector $\langle 1, 0 \rangle$, we found minimal polynomial $x^2 - 2x + 1 = (x-1)^2$, so $\lambda = 1$ was an eigenvalue, and we found eigenvector $\langle 0, 1 \rangle$.

Since A is a 2×2 matrix, it could have up to two eigenvectors, so we choose another seed vector. We find the seed vector $\mathbf{z} = \langle 1, 1 \rangle$ will grow into seedling vector

$$\begin{aligned} \mathbf{v} &= (A - I)\mathbf{z} \\ &= \begin{pmatrix} 0 \\ 2 \end{pmatrix} \end{aligned}$$

and we find

$$\mathbf{v} = \begin{pmatrix} 0 \\ 2 \end{pmatrix} \qquad A\mathbf{v} = \begin{pmatrix} 0 \\ 2 \end{pmatrix}$$

which gives us a minimal polynomial $x - 1$, and again returns the eigenvalue $\lambda = 1$. In fact, we find that no matter what we choose for our seed vector, as long as it grows into a nonzero seedling vector, we will not be able to find any other eigenvalues!

What happened? Remember that our procedure assumes we're working with a *non*defective matrix, which has a full set of eigenvectors. The fact that we can't seem to find any other eigenvalues or eigenvectors should cause us to suspect that our matrix is *defective.*

However, suspicion isn't proof: Just because we can't find a second eigenvector doesn't mean a second eigenvector doesn't exist. Perhaps we're not clever enough; perhaps some bizarre seed vector will produce just the right seedling vector to give us a second linearly independent eigenvector.

To answer this question, let's take a closer look at our work. We obtained the minimal polynomial

$$(A - I) \begin{pmatrix} 0 \\ 2 \end{pmatrix} = \mathbf{0}$$

and so we know that $\langle 0,2\rangle$ is an eigenvector. But remember we found $\langle 0,2\rangle$ by applying $(A-I)$ to $\langle 1,1\rangle$, so we can write this as:

$$(A-I)(A-I)\begin{pmatrix}1\\1\end{pmatrix}=\mathbf{0}$$

We can interpret this equation as follows: The seed $\langle 1,1\rangle$ "grew into" the eigenvector $\langle 0,2\rangle$. This leads to the following idea:

Definition 7.8 (Generalized Eigenvector). *A nonzero vector* $\mathbf{u}$ *is a* ***generalized eigenvector for*** λ ***of rank*** k *if* k *is the least whole number for which* $(A-\lambda I)^k\mathbf{u}=\mathbf{0}$.

Thus, $\langle 1,1\rangle$ is a generalized eigenvector of rank 2.

Activity 7.23: Generalized Eigenvectors

In the following, let A be a matrix and λ be one of its eigenvalues. Let $\vec{v}$ be a generalized eigenvector of rank k corresponding to λ , where $k>1$.

A7.23.1 Consider $(A-\lambda I)\vec{v}$.

a) Explain why $(A-\lambda I)\vec{v}$ must be a nonzero vector.

b) Why does this show that $(A-\lambda I)\vec{v}$ is a generalized eigenvector for λ of rank $k-1$?

c) Find a generalized eigenvector for λ of rank $k-2$. (Assume $k>2$)

A7.23.2 Suppose $\vec{v}\neq\vec{0}$, and $(A-\lambda I)^k\vec{v}=\vec{0}$. Can you conclude $\vec{v}$ is a generalized eigenvector of rank k? Why/why not?

As Activity 7.23 shows, once we find a generalized eigenvector of rank k, we can immediately find a generalized eigenvector of rank $k-1$, $k-2$, and so on. This leads to the definition:

Definition 7.9 (Chain of Generalized Eigenvectors). *Suppose* $\mathbf{v}$ *is a generalized eigenvector of rank* k. *The sequence*

$$\mathbf{v},(A-\lambda I)\mathbf{v},(A-\lambda I)^2\mathbf{v},\ldots,(A-\lambda I)^{k-1}\mathbf{v}$$

is a ***chain of generalized eigenvectors***,

Earlier we introduced the idea of an eigenspace, which consists of all eigenvectors corresponding to an eigenvalue λ. Now that we have generalized eigenvectors, we also have

Definition 7.10 (Generalized Eigenspaces). *Let Λ be a set of eigenvalues for A. The* ***generalized eigenspace*** *includes all eigenvectors and generalized eigenvectors corresponding to an eigenvalue in Λ.*

Given a set of vectors, the natural question to ask (in linear algebra) is: Do we have an independent set of vectors?

Activity 7.24: Independence of Generalized Eigenvectors

In the following, assume $\vec{v}$ is a generalized eigenvector for A of rank k corresponding to eigenvalue λ.

A7.24.1 Consider the sequence of vectors

$$\vec{v}, (A - \lambda I)\vec{v}, (A - \lambda I)^2\vec{v}, \ldots, (A - \lambda I)^{k-1}\vec{v}$$

and remember that since $\vec{v}$ has rank k, $(A - \lambda I)^k\vec{v} = \vec{0}$.

a) Show that if there is a nontrivial linear combination

$$a_0\vec{v} + a_1(A - \lambda I)\vec{v} + a_2(A - \lambda I)^2\vec{v} + \ldots + a_{k-1}(A - \lambda I)^{k-1}\vec{v} = \vec{0}$$

then there is a nontrivial linear combination

$$b_1(A - \lambda I)\vec{v} + b_2(A - \lambda I)^2\vec{v} + \ldots + b_{k-1}(A - \lambda I)^{k-1}\vec{v} = \vec{0}$$

Suggestion: What happens when you multiply the first linear combination by $(A - \lambda I)$?

b) "Lather, rinse, repeat." Why does this mean $m(A - \lambda I)^{k-1}\vec{v} = \vec{0}$?

c) Explain why this last is impossible, and so the vectors

$$\{\vec{v}, (A - \lambda I)\vec{v}, (A - \lambda I)^2\vec{v}, \ldots, (A - \lambda I)^{k-1}\vec{v}\}$$

must be independent.

d) Consider the set that includes an (ordinary) eigenvector for λ, and some or all of the generalized eigenvectors for λ. Why must such a set be independent?

A7.24.2 Suppose $\vec{v}$ is a generalized eigenvector of rank k corresponding to λ. Prove $\vec{v}$ cannot be an eigenvector for any eigenvalue $\mu \neq \lambda$. Suggestion: If $\vec{v}$ *is* an eigenvalue for μ, what is $(A - \lambda I)\vec{v}$?

In Activity 7.23, you showed that each of the vectors in the chain is a generalized eigenvector for λ, and in Activity 7.24, you showed that the chain was a set of independent vectors.

Let's return to the eigenproblem. Suppose A is a $n \times n$ matrix, and we've found a set of $k < n$ eigenvectors. If A is a nondefective matrix, then we can find additional eigenvectors by choosing seed vectors $\mathbf{u}$ that are linearly independent of the known eigenvectors; the resulting minimal polynomial must contain at least one new eigenvalue.

However, suppose we obtain a minimal polynomial that does not contain any new eigenvalues. How should we proceed then? One possibility is to look for generalized eigenvectors corresponding to our known eigenvalues.

Since this requires going through *every* eigenvalue we've found and trying to see there are any generalized eigenvectors, it might be easier to proceed as follows:

- Pick a seed vector $\mathbf{v}$, and find the corresponding minimal polynomial.
- For each eigenvalue, find the linearly independent eigenvectors *and* generalized eigenvectors.

At this point, we will either have an eigenbasis for $\mathbb{F}^n$ or not. If we do, we're done, since there can be no more eigenvectors. If not, we'll choose a seed vector that is linearly independent and repeat the process, finding new eigenvalues and eigenvectors.

We should view the fact that this process must end with mixed emotions. On the one hand, it means we will eventually be done with the problem of finding an eigenbasis, since every time we pick a new seed vector, we will find at least one more vector in our eigenbasis.

But is it possible that, through our choice of seed vectors, we'll find enough generalized eigenvectors to halt the process *before* we've found all of the ordinary eigenvectors?

Activity 7.25: Finding Generalized Eigenvectors

Suppose A is a $n \times n$ matrix, where $\mathcal{V} = \{\vec{v}_1, \vec{v}_2, \ldots, \vec{v}_n\}$ is an eigenbasis for $\mathbb{F}^n$.

A7.25.1 Let $\lambda, \vec{v}$ be an eigenvalue-eigenvector pair. Prove: For any eigenvalue $\mu \neq \lambda$, $(A - \mu I)\vec{v} = c\vec{v}$, where $c \neq 0$.

A7.25.2 Suppose $\vec{v}_1, \vec{v}_2, \ldots \vec{v}_m$ are either eigenvectors or generalized eigenvectors for eigenvalue λ of A. Let k be the greatest rank among the generalized eigenvectors. Prove: If $\vec{x}$ is a linear combination of the $\vec{v}_i$s, then $(A - \lambda I)^k \vec{x} = \vec{0}$.

A7.25.3 Suppose A is a nondefective matrix, so all of the $\vec{v}_i$s are ordinary eigenvectors.

a) Suppose $\lambda_i = \lambda_j$ for all i, j. Prove/disprove: Every vector in $\mathbb{F}^n$ is an eigenvector corresponding to eigenvalue λ_i.

b) Suppose there exists a vector $\vec{x}$ in $\mathbb{F}^n$, which is not an eigenvector. Why does this mean that there are at least two distinct eigenvalues?

A7.25.4 Suppose A is a nondefective matrix, and let λ be any of its eigenvalues.

a) Prove: $(A-\lambda I)\vec{x}$ is a (possibly trivial) linear combination of eigenvectors for eigenvalues $\lambda_i \neq \lambda$.

b) Prove: If $(A-\lambda I)\vec{x} \neq \vec{0}$, then for all k, $(A-\lambda I)^k\vec{x} \neq \vec{0}$.

c) What does this say about the existence of generalized eigenvectors for nondefective matrices?

Activity 7.25 leads to an important conclusion:

Proposition 21. *Suppose A is a nondefective matrix. Then A has no generalized eigenvectors.*

Thus, we don't have to worry that finding generalized eigenvectors will cause us to miss actual eigenvalues.

The preceding activities give us a general strategy for finding all eigenvalues and eigenvectors for a $n \times n$ matrix A:

- Pick a random seed vector $\mathbf{u}$, and find the minimal polynomial for A,
- Use the roots of the minimal polynomial to find some of the eigenvalues.
- Find the corresponding eigenvectors and generalized eigenvectors.
- If you haven't found a n eigenvectors, let

$$H = (A - \lambda_1 I)(A - \lambda_2 I)\cdots(A - \lambda_k I)$$

  where λ_i are the eigenvalues you've already found (included as many times as their rank warrants), and choose a new seed vector $\mathbf{z}$, which you can grow into a nonzero seedling vector $\mathbf{v} = H\mathbf{z}$.

- Use $\mathbf{v}$ as a new seed vector and find a new minimal polynomial, which will either give you new eigenvalues (and eigenvectors), or generalized eigenvectors.
- As needed, replace chains of generalized eigenvectors.

Example 7.8. *Find all eigenvalues and eigenvectors for*

$$A = \begin{pmatrix} 9 & -3 & -1 \\ 10 & -3 & 0 \\ 14 & -5 & -2 \end{pmatrix}$$

Solution. *We start with seed vector* $\mathbf{v} = \langle 1, 0, 0 \rangle$ *and find*

$$\mathbf{v} = \begin{pmatrix} 1 \\ 0 \\ 0 \end{pmatrix} \qquad A\mathbf{v} = \begin{pmatrix} 9 \\ 10 \\ 14 \end{pmatrix} \qquad A^2\mathbf{v} = \begin{pmatrix} 37 \\ 60 \\ 48 \end{pmatrix} \qquad A^3\mathbf{v} = \begin{pmatrix} 105 \\ 190 \\ 122 \end{pmatrix}$$

Row reducing the matrix of column vectors

$$\begin{pmatrix} 1 & 9 & 37 & 105 \\ 0 & 10 & 60 & 190 \\ 0 & 14 & 48 & 122 \end{pmatrix} \longrightarrow \begin{pmatrix} 1 & 9 & 37 & 105 \\ 0 & 1 & 6 & 19 \\ 0 & 0 & 1 & 4 \end{pmatrix}$$

which gives solution

$$x_1 = -2, x_2 = 5, x_3 = -4, x_4 = 1$$

and corresponding minimal polynomial $x^3 - 4x^2 + 5x - 2 = (x-1)(x-2)(x-1)$, *so we find eigenvalues* $\lambda = 1$, *with eigenvector* $\langle 2, 5, 1 \rangle$*; and* $\lambda = 2$, *with eigenvector* $\langle 1, 2, 1 \rangle$.

Since we've only found two eigenvectors, we might have a third. We'll use our eigenvalues $\lambda = 1$, $\lambda = 2$, *and construct* $H = (A - I)(A - 2I)$, *then pick a new seed vector, say* $\langle 0, 1, 0 \rangle$. *Growing this into a seedling* $\mathbf{u}$*:*

$$\begin{aligned} \mathbf{u} &= (A - I)(A - 2I) \begin{pmatrix} 0 \\ 1 \\ 0 \end{pmatrix} \\ &= (A - I) \begin{pmatrix} -3 \\ -5 \\ -5 \end{pmatrix} \\ &= \begin{pmatrix} -4 \\ -10 \\ -2 \end{pmatrix} \end{aligned}$$

and we compute

$$\mathbf{u} = \begin{pmatrix} -4 \\ -10 \\ -2 \end{pmatrix} \qquad A\mathbf{u} = \begin{pmatrix} -4 \\ -10 \\ -2 \end{pmatrix}$$

Consequently

$$(A - I)\mathbf{u} = \mathbf{0}$$

But remember $\mathbf{u} = (A - I)(A - 2I) \begin{pmatrix} 0 \\ 1 \\ 0 \end{pmatrix}$, *so we can rewrite this as:*

$$(A - I)(A - I)(A - 2I) \begin{pmatrix} 0 \\ 1 \\ 0 \end{pmatrix} = (A - I)^2 \begin{pmatrix} -3 \\ -5 \\ -5 \end{pmatrix}$$

Thus $\mathbf{x} = \langle -3, -5, -5 \rangle$ *is a generalized eigenvector of rank* 2, *and the seed vector in the chain* $\mathbf{x}$, $(A - I)\mathbf{x}$; *note that this second vector is our seedling vector* $\langle -4, -10, -2 \rangle$, *and so our chain of generalized eigenvectors for* $\lambda = 1$ *will be*

$$\langle -3, -5, -5 \rangle, \langle -4, -10, -2 \rangle$$

This chain replaces our original eigenvector for $\lambda = 1$, *so all together we have*

- $\lambda = 2$, *with eigenvector* $\langle 1, 2, 1 \rangle$,
- $\lambda = 1$, *with chain of generalized eigenvectors* $\langle -3, -5, -5 \rangle$ *(rank* 2*) and* $\langle -4, -10, -2 \rangle$ *(a "normal" eigenvector).*

Finally, we remind the reader that while the characteristic polynomial is not a particularly good way to *find* eigenvalues and eigenvectors, it nevertheless remains a powerful tool for investigating their theoretical properties.

Activity 7.26: The Trace

A7.26.1 Suppose A is an upper or lower triangular matrix.

a) What are the eigenvalues of A?

b) What is the determinant of A?

c) What seems to be the relationship between the eigenvalues and the determinant?

A7.26.2 For this Activity, you may want to review the properties of polynomials (Activity 9.1). Let A be a $n \times n$ square matrix with characteristic polynomial $|A - \lambda I|$.

a) Suppose the characteristic polynomial of A can be written as

$$(\lambda - \lambda_1)(\lambda - \lambda_2) \cdots (\lambda - \lambda_n)$$

In the expanded form of the characteristic polynomial, what is the sign of λ^n and the sign of $\lambda_1 \lambda_2 \cdots \lambda_n$?

b) Show that the constant term in the cofactor expansion of $|A - \lambda I|$ is the product of the eigenvalues.

c) Use the properties of the determinant to show

$$|A - \lambda I| = |A| - \lambda \left(|A_1| + |A_2| + \ldots \right)$$

where the $A$is are $(n - 1) \times (n - 1)$ matrices.

d) What does this say about the relationship between $|A|$ and the eigenvalues of A?

A7.26.3 Consider the cofactor expansion of $|A - \lambda I|$.

a) Explain why, in the cofactor expansion of $|A - \lambda I|$, there will be a term corresponding to the product of

$$(a_{11} - \lambda)(a_{22} - \lambda)\cdots(a_{nn} - \lambda)$$

(in other words, the product of all the diagonal entries of $A - \lambda I$).

b) Explain why it is *not* possible for a term in the cofactor expansion to consist of exactly $n - 1$ of the diagonal entries.

c) Why does this mean all the remaining terms in the cofactor expansion will have degree λ^{n-2} or less?

d) What will be the coefficient of the λ^{n-1} term in the cofactor expansion of $|A - \lambda I|$?

e) Suppose the characteristic polynomial of A can be written as

$$(\lambda - \lambda_1)(\lambda - \lambda_2)\cdots(\lambda - \lambda_n)$$

What will be the coefficient of the λ^{n-1} term in the characteristic polynomial?

f) By comparing the two coefficients of λ^{n-1}, what does this tell you about the relationship between the eigenvalues of A and its diagonal entries?

Activity A7.26.2 gives us:

Theorem 7.7. *The determinant of A is the product of its eigenvalues.*

It also points to the importance of something we might have overlooked before. We define:

Definition 7.11 (Trace)**.** *The trace of a square matrix, designated Tr A, is the sum of the entries along its main diagonal.*

Activity A7.26.3 proves:

Theorem 7.8. *The trace of a matrix A is the sum of its eigenvalues.*

This gives us a new way to find eigenvalues for some matrices.

Activity 7.27: Eigenvalues for $n \times n$ matrices

Suppose $A = \begin{pmatrix} a & b \\ c & d \end{pmatrix}$ with eigenvalues λ_1, λ_2.

A7.27.1 Refer to Theorems 7.7 and 7.8.

a) Find $\lambda_1\lambda_2$ and $\lambda_1 + \lambda_2$.

b) Suppose a, b, c, d are real. If λ_1 is the complex number $p + iq$, what must λ_2 be?

A7.27.2 Suppose you have the sum $x + y$ and product xy of two numbers.

a) Show: $(x + y)^2 - 4xy = (x - y)^2$.

b) Explain how you can use the result in Activity A7.27.2a to find two numbers, given their sum and product.

c) Solve: The sum of two numbers is 24 and the product is 11. Find the numbers.

d) Solve: The sum of two numbers is 30 and the product is -25. Find the numbers.

e) Solve: The sum of two numbers is 10 and the product is 40. Find the numbers.

A7.27.3 Find the eigenvalues of the following matrices.

a) M, where $|M| = 10$ and Tr $M = -7$.

b) P, where $|P| = -1$ and Tr $P = 0$.

c) Q, where $|Q| = 1$ and Tr $Q = 0$.

7.9 Symmetric Matrices

As a general rule, we won't know if a matrix is defective or nondefective until we find all the eigenvalues. However, in some cases, the structure of the matrix gives us a hint.

Activity 7.28: Symmetric Matrices

In the following, suppose A is a symmetric matrix.

A7.28.1 Prove: The eigenvalues and eigenvectors of A and A^T are the same.

A7.28.2 Suppose $\lambda \neq \mu$ are two eigenvalues of A, whose eigenvectors $\vec{u}$, $\vec{v}$ can be expressed as the column vectors $\mathbf{u}$, $\mathbf{v}$.

a) Prove: $\mathbf{u}^T A = \lambda \mathbf{u}^T$.

b) Use the fact that $\vec{v}$ is an eigenvector for A corresponding to eigenvalue μ to find an expression for $\mathbf{u}^T A \mathbf{v}$.

c) Use the fact that $\mathbf{u}^T A = \lambda \mathbf{u}^T$ to find a different expression for $\mathbf{u}^T A\mathbf{v}$.

d) Compare your expressions from Activities A7.28.2b and A7.28.2c. Why does this show $\mathbf{u}^T\mathbf{v} = \mathbf{0}$, the matrix of the appropriate size whose entries are all 0?

e) What does this say about $\vec{u}$, $\vec{v}$?

Activity 7.28 shows:

Proposition 22. *The eigenvectors of a symmetric matrix are orthogonal.*

An even more remarkable result occurs if we consider the eigenvalues of a symmetric matrix.

Activity 7.29: Eigenvalues of Symmetric Matrices

Suppose A is a symmetric matrix with real entries. Further, suppose λ, $\vec{v}$ is an eigenvalue/eigenvector pair; and that λ is complex.

A7.29.1 Prove: $\vec{v}$ has complex components.

A7.29.2 Explain why $\vec{v} \cdot \overline{\vec{v}} \neq 0$.

A7.29.3 Prove: A *can't* have complex eigenvalues.

Activity 7.29 leads to:

Theorem 7.9. *All eigenvalues of a real symmetric matrix A are real.*

It's important to note that this result relies on the entries of A being real: if some entries of A are complex, there's no guarantee that the eigenvalues occur on conjugate pairs.

While this is a useful result, it might appear to be rather restrictive: after all, it would seem that *most* matrices are not symmetric. However, there are some important matrices that are symmetric: in particular, the adjacency matrix for a graph (see Activity 4.9) is symmetric, so they will always have real eigenvalues; this means that the numerical approach of Activity 7.8 will always determine the largest eigenvalue. Consequently, we can always find the eigenvalue centrality of a graph numerically (see Activity 7.34).

In fact, we can go further.

Activity 7.30: Eigenvalues of Symmetric Matrices, Continued

A7.30.1 Let A be a symmetric matrix.

a) Explain why the method of Activity 7.8 will always find an eigenvalue/eigenvector pair for A.

b) Suppose λ_1, $\vec{x}_1$ is an eigenvalue/eigenvector pair for A. Prove: If $\vec{x}$ is a linear combination of the eigenvectors of A *exclusive of* $\vec{x}_1$, then the method of Activity 7.8 will find the *second* largest eigenvalue of A.

c) How can you find a vector $\vec{x}$ that is a linear combination of the *other* eigenvectors of A *without* knowing what the other eigenvectors are? Suggestion: Suppose you chose a linear combination that included $\vec{x}_1$. How could you eliminate $\vec{x}_1$ from the linear combination *without* knowing the other eigenvectors?

A7.30.2 Find all eigenvalues and eigenvectors of the matrix given. Note: Because of roundoff error, you might want to limit the powers of the matrix used.

a) $\begin{pmatrix} 3 & 5 & 1 \\ 5 & 1 & 9 \\ 1 & 9 & 4 \end{pmatrix}$

b) $\begin{pmatrix} 2 & 3 & 5 & 1 \\ 3 & 4 & 7 & 1 \\ 5 & 7 & 2 & 6 \\ 1 & 1 & 6 & 3 \end{pmatrix}$

Finally:

Activity 7.31: Can Symmetric Matrices Be Defective?

A7.31.1 Let M be a square matrix and $\vec{v}$ a vector that M can act on. Find a matrix expression corresponding to the dot product of $M\vec{v}$ with itself.

A7.31.2 Suppose A is a symmetric matrix.

a) Prove, for any scalar c, $(A - cI)^T = A - cI$.

b) Suppose λ is an eigenvalue of A, and $\vec{x}$ satisfies $(A - \lambda I)^2\vec{x} = \vec{0}$. Prove: $\vec{x} = \vec{0}$. Suggestion: Find the dot product of $(A - \lambda I)\vec{x}$ with itself.

c) Why does this mean that A can't have generalized eigenvectors of any rank?

By Activity 7.31, a symmetric matrix can't have generalized eigenvectors. This means:

Theorem 7.10. *Symmetric matrices are nondefective.*

Some symmetric matrices have an even more interesting property.

Activity 7.32: Positive Definite Matrices

In the following, assume A is a $n \times n$ symmetric matrix that acts on vectors in $\mathbb{R}^n$.

A7.32.1 Let $\overrightarrow{x}$ be a vector in $\mathbb{R}^n$. What is the significance of the sign $\overrightarrow{x} \cdot A\overrightarrow{x}$ (where $\cdot$ is the dot product)?

A7.32.2 Suppose A has a negative eigenvalue. Prove: There exists some $\overrightarrow{x}$ for which $\overrightarrow{x} \cdot A\overrightarrow{x} < 0$.

A7.32.3 Suppose $\overrightarrow{x} \cdot A\overrightarrow{x} > 0$ for all vectors $\overrightarrow{x}$. Prove/disprove: All eigenvalues of A are positive.

A7.32.4 Suppose all eigenvalues of A are positive.

a) Consider any vector $\overrightarrow{x}$. Explain *geometrically* why $\overrightarrow{x} \cdot A\overrightarrow{x} > 0$.
b) Prove your result algebraically.

Activity 7.32 motivates the following definition:

Definition 7.12 (Positive Definite). *A symmetric matrix A is* ***positive definite*** *if, for all $\overrightarrow{x}$, $\overrightarrow{x} \cdot A\overrightarrow{x} > 0$.*

Since we can also express the dot product using a matrix transpose, this can also be expressed as $\mathbf{u}^T A\mathbf{u} > 0$. However, since the left hand side is a product of matrices, the 0 on the right hand side must be viewed as the 0 matrix. As you showed in Activity 7.32:

Proposition 23. *All eigenvalues of a positive definite matrix are positive.*

7.10 Graphs

One particularly important use of linear algebra is in the analysis of graphs.

Activity 7.33: More Graphs

You may want to refer to Activity 4.9 for a review of some basic graph properties.

A7.33.1 Let A be the adjacency matrix for a graph.

a) What do you know about the eigenvalues of A?

b) Could A be a positive definite matrix?

One potential use of eigenvalues and eigenvectors has to do with so called **centrality measures** of graphs.

Activity 7.34: Centrality Measures

Suppose A is the adjacency matrix for a graph (see Activity 4.9). We can interpret the vertices and edges of the graph as a description of a social network, where the vertices represent people and the edges represent a "friend" relationship between them. An important question regarding such social networks is known as the **key player problem**: Of the people in the network, who is the most important?

For the following activities, use the graph shown:

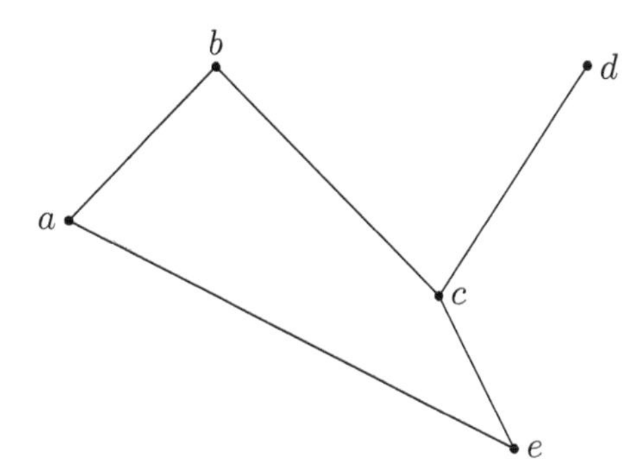

A7.34.1 Find the adjacency matrix A for the graph.

A7.34.2 Suppose at $t = k$, each person has some amount x_i of information (you might think of this as "gossip"). Assume that at time $t = k + 1$, each person communicates this amount to *each* of their friends. How can the adjacency matrix be used to compute the amount of gossip each person has at $t = k + 1$?

A7.34.3 Suppose λ is the largest eigenvalue of A, with corresponding eigenvector $\vec{v}$. Let $\vec{x}_0$ be some nonzero vector, and let $\vec{x}_{n+1} = A\vec{x}_n$.

a) If we treat the components of a vector $\vec{x}_0$ as the amount of “gossip” possessed by a given node at a starting time, what is the significance of $\vec{x}_1 = A\vec{x}_0$?

b) What would be the significance of the eigenvector $\vec{v}$, corresponding to the largest eigenvalue λ of matrix A?

c) Suppose you wanted to spread some information (“gossip”) as rapidly as possible. How could you use the information about the eigenvalues and eigenvectors of the adjacency matrix?

8

Decomposition

Our goal in this chapter is to factor a matrix by writing it as a product of (hopefully) simpler matrices. A useful strategy to keep in mind, in mathematics and in life:

Strategy. *Things that do the same thing are the same thing.*

8.1 LU-Decomposition

Let's consider elementary row operations in more detail.

Activity 8.1: Row Reduction, Revisited

In the following, let A_{R} be the row echelon form of matrix A, and A_{redR} be the reduced row echelon form of a matrix. Let F_i be an elementary matrix corresponding to an elementary row operation required to produce A_{R} from A; and B_i be an elementary matrix corresponding to an elementary row operation required to produce A_{redR} from A_{R}, so we have

$$\begin{aligned} A_{\mathsf{R}} &= F_i F_{i-1} \cdots F_2 F_1 A \\ A_{\mathsf{redR}} &= B_j B_{j-1} \cdots B_2 B_1 A_{\mathsf{R}} \end{aligned}$$

Finally, let

$$\begin{aligned} \mathcal{F} &= F_i F_{i-1} \cdots F_2 F_1 \\ \mathcal{B} &= B_j B_{j-1} \cdots B_2 B_1 \end{aligned}$$

A8.1.1 For each of the following matrices, find the F_is and the B_is. Also find $\mathcal{F}$, the product of the F_is; and $\mathcal{B}$, the product of the B_is. If possible, avoid switching rows.

a) $A = \begin{pmatrix} 2 & 3 \\ 4 & 9 \end{pmatrix}$

b) $B = \begin{pmatrix} 4 & 9 \\ 2 & 3 \end{pmatrix}$

c) $C = \begin{pmatrix} 2 & 1 & 5 \\ 6 & -3 & 7 \\ -8 & -10 & -27 \end{pmatrix}$

d) $D = \begin{pmatrix} 6 & -3 & 7 \\ 2 & 1 & 5 \\ -8 & -10 & -27 \end{pmatrix}$

A8.1.2 What do you notice about $\mathcal{F}$ and $\mathcal{B}$? Prove your results.

From Activity 8.1, you found that if A could be put into reduced row echelon form using a sequence of elementary row operations that *didn't* require row interchanges, there was a lower triangular matrix $\mathcal{F}$ and an upper triangular matrix $\mathcal{B}$ where $\mathcal{B}\mathcal{F}A = I$.

Example 8.1. *Find $\mathcal{B}$, $\mathcal{D}$, $\mathcal{F}$ for* $A = \begin{pmatrix} 2 & 5 & 3 \\ 4 & 8 & 11 \\ -6 & -19 & -1 \end{pmatrix}$

Solution. *Producing the row echelon form and then the reduced row echelon form of A:*

$$\begin{pmatrix} 2 & 5 & 3 \\ 4 & 8 & 11 \\ -6 & -19 & -1 \end{pmatrix} \xrightarrow{-2R_1+R_2 \to R_2} \begin{pmatrix} 2 & 5 & 3 \\ 0 & -2 & 5 \\ -6 & -19 & -1 \end{pmatrix}$$

$$\xrightarrow{3R_1+R_3 \to R_3} \begin{pmatrix} 2 & 5 & 3 \\ 0 & -2 & 5 \\ 0 & -4 & 8 \end{pmatrix} \xrightarrow{-2R_2+R_3 \to R_3} \begin{pmatrix} 2 & 5 & 3 \\ 0 & -2 & 5 \\ 0 & 0 & -2 \end{pmatrix}$$

The elementary matrices corresponding to these elementary row operations are

$$F_1 = \begin{pmatrix} 1 & 0 & 0 \\ -2 & 1 & 0 \\ 0 & 0 & 1 \end{pmatrix} \qquad F_2 = \begin{pmatrix} 1 & 0 & 0 \\ -2 & 1 & 0 \\ 7 & -2 & 1 \end{pmatrix}$$

$$F_3 = \begin{pmatrix} 1 & 0 & 0 \\ 0 & 1 & 0 \\ 3 & 0 & 1 \end{pmatrix} \qquad F_4 = \begin{pmatrix} 1 & 0 & 0 \\ 0 & 1 & 0 \\ 0 & -2 & 1 \end{pmatrix}$$

Multiplying these together (remember the order must be reversed)

$$\begin{aligned} \mathcal{F} &= F_4F_3F_2F_1 \\ &= \begin{pmatrix} 1 & 0 & 0 \\ -2 & 1 & 0 \\ 7 & -2 & 1 \end{pmatrix} \end{aligned}$$

If we continue the row reduction to produce the reduced row echelon form:

$$\begin{pmatrix} 2 & 5 & 3 \\ 0 & -2 & 5 \\ 0 & 0 & -2 \end{pmatrix} \xrightarrow{2R_2 \to R_2} \begin{pmatrix} 4 & 10 & 6 \\ 0 & -4 & 10 \\ 0 & 0 & -2 \end{pmatrix}$$

$$\xrightarrow{2R_1 \to R_1} \begin{pmatrix} 4 & 10 & 6 \\ 0 & -4 & 10 \\ 0 & 0 & -2 \end{pmatrix} \xrightarrow{5R_3 + R_2 \to R_2} \begin{pmatrix} 4 & 10 & 6 \\ 0 & -4 & 0 \\ 0 & 0 & -2 \end{pmatrix}$$

$$\xrightarrow{3R_3 + R_1 \to R_1} \begin{pmatrix} 4 & 10 & 0 \\ 0 & -4 & 0 \\ 0 & 0 & -2 \end{pmatrix} \xrightarrow{2R_1 \to R_1} \begin{pmatrix} 8 & 20 & 0 \\ 0 & -4 & 0 \\ 0 & 0 & -2 \end{pmatrix}$$

$$\xrightarrow{5R_2 + R_1 \to R_1} \begin{pmatrix} 8 & 0 & 0 \\ 0 & -4 & 0 \\ 0 & 0 & -2 \end{pmatrix} \xrightarrow{\frac{1}{8}R_1 \to R_1} \begin{pmatrix} 1 & 0 & 0 \\ 0 & -4 & 0 \\ 0 & 0 & -2 \end{pmatrix}$$

$$\xrightarrow{-\frac{1}{4}R_2 \to R_2} \begin{pmatrix} 1 & 0 & 0 \\ 0 & 1 & 0 \\ 0 & 0 & -2 \end{pmatrix} \xrightarrow{-\frac{1}{2}R_3 \to R_3} \begin{pmatrix} 1 & 0 & 0 \\ 0 & 1 & 0 \\ 0 & 0 & 1 \end{pmatrix}$$

These steps correspond to the elementary matrices

$$B_1 = \begin{pmatrix} 1 & 0 & 0 \\ 0 & 2 & 0 \\ 0 & 0 & 1 \end{pmatrix} \qquad B_2 = \begin{pmatrix} 2 & 0 & 0 \\ 0 & 1 & 0 \\ 0 & 0 & 1 \end{pmatrix} \qquad B_3 = \begin{pmatrix} 1 & 0 & 0 \\ 0 & 1 & 5 \\ 0 & 0 & 1 \end{pmatrix}$$

$$B_4 = \begin{pmatrix} 1 & 0 & 3 \\ 0 & 1 & 0 \\ 0 & 0 & 1 \end{pmatrix} \qquad B_5 = \begin{pmatrix} 2 & 0 & 0 \\ 0 & 1 & 0 \\ 0 & 0 & 1 \end{pmatrix} \qquad B_6 = \begin{pmatrix} 1 & 5 & 0 \\ 0 & 1 & 0 \\ 0 & 0 & 0 \end{pmatrix}$$

$$B_7 = \begin{pmatrix} \frac{1}{8} & 0 & 0 \\ 0 & 1 & 0 \\ 0 & 0 & 1 \end{pmatrix} \qquad B_8 = \begin{pmatrix} 1 & 0 & 0 \\ 0 & -\frac{1}{4} & 0 \\ 0 & 0 & 1 \end{pmatrix} \qquad B_9 = \begin{pmatrix} 1 & 0 & 0 \\ 0 & 1 & 0 \\ 0 & 0 & -\frac{1}{2} \end{pmatrix}$$

So

$$\begin{aligned} \mathcal{B} &= B_9 B_8 B_7 B_6 B_5 B_4 B_3 B_2 B_1 \\ &= \begin{pmatrix} \frac{1}{2} & \frac{5}{4} & \frac{31}{8} \\ 0 & -\frac{1}{2} & -\frac{5}{4} \\ 0 & 0 & -\frac{1}{2} \end{pmatrix} \end{aligned}$$

Consequently

$$\begin{pmatrix} \frac{1}{2} & \frac{5}{4} & \frac{31}{8} \\ 0 & -\frac{1}{2} & -\frac{5}{4} \\ 0 & 0 & -\frac{1}{2} \end{pmatrix} \begin{pmatrix} 1 & 0 & 0 \\ -2 & 1 & 0 \\ 7 & -2 & 1 \end{pmatrix} \begin{pmatrix} 2 & 5 & 3 \\ 4 & 8 & 11 \\ -6 & -19 & -1 \end{pmatrix} = \begin{pmatrix} 1 & 0 & 0 \\ 0 & 1 & 0 \\ 0 & 0 & 1 \end{pmatrix}$$

Our ability to express the row reduction process by a product of elementary matrices leads to a factorization of A.

Activity 8.2: More Row Reduction

A8.2.1 Suppose A can be reduced to the identity matrix through a sequence of elementary operations that don't involve switching rows. By Activity 8.1, we have $\mathcal{BF}A = I$, where $\mathcal{B}$ is an upper triangular matrix and $\mathcal{F}$ is a lower triangular matrix.

a) Prove: $\mathcal{F}^{-1}$ and $\mathcal{B}^{-1}$ exist.

b) What type of matrices are $\mathcal{F}^{-1}$ and $\mathcal{B}^{-1}$?

c) Why does this mean A can be written as a product LU, where L is a lower triangular matrix and U is an upper triangular matrix?

A8.2.2 If possible, express each of the following as a product LU, where L is a lower triangular matrix and U is an upper triangular matrix.

a) $A = \begin{pmatrix} 2 & 3 \\ 4 & 9 \end{pmatrix}$

b) $B = \begin{pmatrix} 4 & 9 \\ 2 & 3 \end{pmatrix}$

c) $C = \begin{pmatrix} 2 & 1 & 5 \\ 6 & -3 & 7 \\ -8 & -10 & -27 \end{pmatrix}$

d) $D = \begin{pmatrix} 6 & -3 & 7 \\ 2 & 1 & 5 \\ -8 & -10 & -27 \end{pmatrix}$

A8.2.3 Suppose you wanted to write $A = UL$, the product of an upper triangular matrix with a lower triangular matrix.

a) Assuming such matrices exist, how could you find U and L?

b) Write $C = \begin{pmatrix} 2 & 1 & 5 \\ 6 & -3 & 7 \\ -8 & -10 & -27 \end{pmatrix}$ as a product UL, an upper triangular matrix multiplied by a lower triangular matrix.

Activity 8.2 leads to:

Proposition 24. *Suppose A is a nonsingular matrix whose reduced row echelon form can be produced without switching rows. Then $A = LU$, the product of a lower triangular matrix L and an upper triangular matrix U.*

Example 8.2. *Find an LU-decomposition of* $A = \begin{pmatrix} 2 & 5 & 3 \\ 4 & 8 & 11 \\ -6 & -19 & -1 \end{pmatrix}$.

Solution. *We found*

$$\begin{pmatrix} \frac{1}{2} & \frac{5}{4} & \frac{31}{8} \\ 0 & -\frac{1}{2} & -\frac{5}{4} \\ 0 & 0 & -\frac{1}{2} \end{pmatrix} \begin{pmatrix} 1 & 0 & 0 \\ -2 & 1 & 0 \\ 7 & -2 & 1 \end{pmatrix} \begin{pmatrix} 2 & 5 & 3 \\ 4 & 8 & 11 \\ -6 & -19 & -1 \end{pmatrix} = \begin{pmatrix} 1 & 0 & 0 \\ 0 & 1 & 0 \\ 0 & 0 & 1 \end{pmatrix}$$

We find

$$\begin{pmatrix} \frac{1}{2} & \frac{5}{4} & \frac{31}{8} \\ 0 & -\frac{1}{2} & -\frac{5}{4} \\ 0 & 0 & -\frac{1}{2} \end{pmatrix}^{-1} = \begin{pmatrix} 2 & 5 & 3 \\ 0 & -2 & 5 \\ 0 & 0 & -2 \end{pmatrix}$$

$$\begin{pmatrix} 1 & 0 & 0 \\ -2 & 1 & 0 \\ 7 & -2 & 1 \end{pmatrix}^{-1} = \begin{pmatrix} 1 & 0 & 0 \\ 2 & 1 & 0 \\ -3 & 2 & 1 \end{pmatrix}$$

Multiplying by the inverses of the upper and lower triangular matrices:

$$\begin{pmatrix} 1 & 0 & 0 \\ -2 & 1 & 0 \\ 7 & -2 & 1 \end{pmatrix} \begin{pmatrix} 2 & 5 & 3 \\ 4 & 8 & 11 \\ -6 & -19 & -1 \end{pmatrix} = \begin{pmatrix} 2 & 5 & 3 \\ 0 & -2 & 5 \\ 0 & 0 & -2 \end{pmatrix}$$

$$\begin{pmatrix} 2 & 5 & 3 \\ 4 & 8 & 11 \\ -6 & -19 & -1 \end{pmatrix} = \begin{pmatrix} 1 & 0 & 0 \\ 2 & 1 & 0 \\ -3 & 2 & 1 \end{pmatrix} \begin{pmatrix} 2 & 5 & 3 \\ 0 & -2 & 5 \\ 0 & 0 & -2 \end{pmatrix}$$

Activity A8.2.3 suggests we could also write $A = U'L'$, the product of an upper triangular matrix and a lower triangular matrix.

What if row reduction requires a row interchange?

Activity 8.3: Required Row Interchanges

As in Activity 8.1, let F_i be an elementary matrix corresponding to an elementary row operation required to produce the row echelon form of a matrix, and B_i be an elementary matrix corresponding to an elementary row operation required to produce the reduced row echelon form of a matrix from the row echelon form; let $\mathcal{F}$ and $\mathcal{B}$ be the product of the F_is and the B_is, respectively.

A8.3.1 If possible, find a 4×4 matrix A that *cannot* be put in row echelon form without switching rows.

A8.3.2 We'll say that a matrix is in **quasi-row echelon form** if a sequence of row interchanges (and no other elementary operation) can put it into row echelon form. Suppose A can be put in quasi-row echelon form A_{qR}, and let D_i be an elementary matrix corresponding to a row interchange necessary to produce A_{R} from A_{qR}.

a) Let $\mathcal{D}$ be the product of all the D_is. Describe $\mathcal{D}$.

b) Prove/disprove: To produce A_{redR} from A_{R}, no row interchanges are necessary.

A8.3.3 For each of the following matrices, find the F_is, the B_is, and if necessary the D_is. Also find $\mathcal{F}$, the product of the F_is; $\mathcal{B}$, the product of the B_is; and $\mathcal{D}$, the product of the D_is.

a) $M = \begin{pmatrix} 2 & 1 & 5 \\ 6 & 3 & 17 \\ -8 & -7 & -22 \end{pmatrix}$

b) $N = \begin{pmatrix} 2 & 4 & -6 \\ 1 & 2 & 2 \\ -3 & -7 & 9 \end{pmatrix}$

If a matrix can be reduced to the identity matrix, we know it is invertible. Thus:

Theorem 8.1 (LDU-decomposition). *Any invertible matrix can be written as LDU, the product of a lower triangular matrix L, an upper triangular matrix U, and a permuted elementary matrix D. If the matrix can be reduced to row echelon form without using row interchanges, it can be written as LU.*

Example 8.3. *Find a LDU or LU-decomposition of* $A = \begin{pmatrix} 1 & 1 & 2 \\ 2 & 2 & 5 \\ -3 & -4 & 0 \end{pmatrix}$.

Solution. *Row reducing A:*

$$\begin{pmatrix} 1 & 1 & 2 \\ 2 & 2 & 5 \\ -3 & -4 & 0 \end{pmatrix} \xrightarrow[R_3 + 3R_1 \to R_3]{R_2 - 2R_1 \to R_2} \begin{pmatrix} 1 & 1 & 2 \\ 0 & 0 & 1 \\ 0 & -1 & 6 \end{pmatrix} \xrightarrow{R_2 \leftrightarrow R_3} \begin{pmatrix} 1 & 1 & 2 \\ 0 & -1 & 6 \\ 0 & 0 & 1 \end{pmatrix}$$

$$\xrightarrow{-R_2 \to R_2} \begin{pmatrix} 1 & 1 & 2 \\ 0 & 1 & -6 \\ 0 & 0 & 1 \end{pmatrix} \xrightarrow[R_1 - 2R_3 \to R_3]{6R_3 + R_2 \to R_2} \begin{pmatrix} 1 & 1 & 0 \\ 0 & 1 & 0 \\ 0 & 0 & 1 \end{pmatrix} \xrightarrow{R_1 - R_2 \to R_1} \begin{pmatrix} 1 & 0 & 0 \\ 0 & 1 & 0 \\ 0 & 0 & 1 \end{pmatrix}$$

Consequently

$$\begin{pmatrix} 1 & 1 & -8\ 0 & -1 & 6 \\ 0 & 0 & 1 & & \end{pmatrix} \begin{pmatrix} 1 & 0 & 0 \\ 0 & 0 & 1 \\ 0 & 1 & 0 \end{pmatrix} \begin{pmatrix} 1 & 0 & 0 \\ -2 & 1 & 0 \\ 3 & 0 & 1 \end{pmatrix} \begin{pmatrix} 1 & 1 & 2 \\ 2 & 2 & 5 \\ -3 & -4 & 0 \end{pmatrix}$$

We find

$$\begin{pmatrix} 1 & 1 & -8\ 0 & -1 & 6 \\ 0 & 0 & 1 & & \end{pmatrix}^{-1} = \begin{pmatrix} 1 & 1 & 2 \\ 0 & -1 & 6 \\ 0 & 0 & 1 \end{pmatrix}$$

$$\begin{pmatrix} 1 & 0 & 0 \\ 0 & 0 & 1 \\ 0 & 1 & 0 \end{pmatrix}^{-1} = \begin{pmatrix} 1 & 0 & 0 \\ 0 & 0 & 1 \\ 0 & 1 & 0 \end{pmatrix}$$

$$\begin{pmatrix} 1 & 0 & 0 \\ -2 & 1 & 0 \\ 3 & 0 & 1 \end{pmatrix}^{-1} = \begin{pmatrix} 1 & 0 & 0 \\ 2 & 1 & 0 \\ -3 & 0 & 1 \end{pmatrix}$$

Multiplying by the inverses gives us

$$\begin{pmatrix} 1 & 1 & 2 \\ 2 & 2 & 5 \\ -3 & -4 & 0 \end{pmatrix} = \begin{pmatrix} 1 & 0 & 0 \\ 2 & 1 & 0 \\ -3 & 0 & 1 \end{pmatrix} \begin{pmatrix} 1 & 0 & 0 \\ 0 & 0 & 1 \\ 0 & 1 & 0 \end{pmatrix} \begin{pmatrix} 1 & 1 & 2 \\ 0 & -1 & 6 \\ 0 & 0 & 1 \end{pmatrix}$$

as our LDU-decomposition.

8.2 *QR*-Decomposition

The *LDU* decomposition requires our matrix be invertible. This means that some square matrices can't be put into this form—and *no* nonsquare matrix can be. This means we should consider other possible decompositions.

To create such an approach, note that row reduction relies on interpreting the matrix entries as coefficients of a system of linear equations. We can also interpret the matrix entries as column vectors.

Activity 8.4: Decomposition Using Gram-Schmidt

In the following, let $A = \begin{pmatrix} \mathbf{v}_1 & \mathbf{v}_2 & \dots & \mathbf{v}_n \end{pmatrix}$, where the $\mathbf{v}_i$s are column vectors in $\mathbb{F}^m$. Assume the $\mathbf{v}_i$s form an independent set of vectors.

A8.4.1 Let $Q = (\mathbf{w}_1 \quad \mathbf{w}_2 \quad \ldots \quad \mathbf{w}_n)$, where the $\mathbf{w}_i$s form a different basis for the column space of A.

a) Prove: There is a matrix N where $Q = AN$.
b) Prove: There is a matrix R where $QR = A$.
c) What are the dimensions of N and R?

A8.4.2 Let $Q = (\mathbf{w}_1 \quad \mathbf{w}_2 \quad \ldots \quad \mathbf{w}_n)$ as above, with $Q = AN$ and $QR = A$. However, this time let the $\mathbf{w}_i$s be orthogonal vectors produced using the Gram-Schmidt process.

a) What are the entries in the first column of N?
b) $\mathbf{w}_2$ is the second vector produced by the Gram-Schmidt orthogonalization process. What do you know about $\mathbf{w}_2$ in terms of the $\mathbf{v}_i$s?
c) What does this tell you about the entries in the second column of N?
d) What does this tell you about the form of the matrix N?
e) What does this tell you about the form of R?

A8.4.3 The QR-decomposition of A is defined by the matrix Q, an orthogonal basis for A produced by the Gram-Schmidt process, and a matrix R, where $QR = A$. Find the QR decomposition of the given matrices.

a) $\begin{pmatrix} 1 & 3 \\ 2 & 5 \end{pmatrix}$

b) $\begin{pmatrix} 1 & -3 \\ 0 & 2 \\ 2 & 1 \end{pmatrix}$

Example 8.4. *Find a QR-decomposition for* $A = \begin{pmatrix} 3 & 2 \\ 1 & 5 \\ 4 & -3 \end{pmatrix}$

Solution. *Treating our columns as vectors, we find the orthogonal basis*

$$\begin{pmatrix} 3 \\ 1 \\ 4 \end{pmatrix} = 1 \begin{pmatrix} 3 \\ 1 \\ 4 \end{pmatrix}$$

$$\begin{pmatrix} 55 \\ 131 \\ -74 \end{pmatrix} = 26 \begin{pmatrix} 3 \\ 1 \\ 4 \end{pmatrix} + 1 \begin{pmatrix} 2 \\ 5 \\ -3 \end{pmatrix}$$

Consequently

$$\begin{pmatrix} 3 & 55 \\ 1 & 131 \\ 4 & -74 \end{pmatrix} = \begin{pmatrix} 3 & 2 \\ 1 & 5 \\ 4 & -3 \end{pmatrix} \begin{pmatrix} 1 & 1 \\ 0 & 26 \end{pmatrix}$$

Right multiplication by the inverse of the triangular matrix gives

$$\begin{pmatrix} 3 & 55 \\ 1 & 131 \\ 4 & -74 \end{pmatrix} \begin{pmatrix} 1 & -\frac{1}{26} \\ 0 & \frac{1}{26} \end{pmatrix} = \begin{pmatrix} 3 & 2 \\ 1 & 5 \\ 4 & -3 \end{pmatrix}$$

8.3 Eigendecompositions

Both LU and QR decompositions rely on row reduction, which is a relatively "cheap" operation to apply. And while it's not true that paying more guarantees a better product, the fact that LU and QR decompositions *are* cheap suggests that they might be of limited use. The next step up in complexity are eigenvalue decompositions, which are significantly more expensive in terms of computational cost—but are significantly more useful as well.

To begin with, suppose A is a nondefective $n \times n$ matrix. Then we know the eigenvectors of A form a basis for $\mathbb{F}^n$. We can use this to provide a useful factorization of A.

Activity 8.5: Eigendecomposition

A8.5.1 Let $A = \begin{pmatrix} 5 & -6 \\ 4 & -5 \end{pmatrix}$.

a) Find the eigenvalues λ_1, λ_2 and corresponding eigenvectors $\vec{v}_1$, $\vec{v}_2$.

b) Let $\vec{x} = a_1\vec{v}_1 + a_2\vec{v}_2$. Find $A\vec{x}$.

c) Find $A^{100}\vec{x}$.

d) Find $A^{255}\vec{x}$.

A8.5.2 Suppose Λ is a diagonal matrix. Find an expression for Λ^n.

A8.5.3 Suppose M is a nondefective $n \times n$ matrix, where $\mathcal{V} = \{\vec{v}_1, \vec{v}_2, \ldots, \vec{v}_n\}$ is an eigenbasis for $\mathbb{F}^n$; assume $\vec{v}_i$ is an eigenvector corresponding to eigenvalue λ_i.

a) Prove: Every vector $\vec{x} \in \mathbb{F}^n$ can be expressed as a linear combination of the vectors in $\mathcal{V}$.

b) Suppose the coordinates of $\vec{x}$ with respect to $\mathcal{V}$ are $(x_1, x_2, \ldots, x_n)$. Find the coordinates of $M\vec{x}$ with respect to $\mathcal{V}$.

c) Let Λ be the diagonal matrix with $\Lambda_{ii} = \lambda_i$ (the eigenvalue corresponding to eigenvector $\vec{v}_i$), and let P be the matrix whose ith column is the eigenvector $\vec{v}_i$. Show

$$M\vec{x} = P\Lambda \begin{pmatrix} x_1 \\ x_2 \\ \vdots \\ x_n \end{pmatrix}$$

d) Why does this show $M = P\Lambda P^{-1}$?

A8.5.4 Let $A = \begin{pmatrix} 5 & -6 \\ 4 & -5 \end{pmatrix}$ (this is the same matrix from Activity A8.5.1).

a) Use the approach of Activity A8.5.3 to write A as a product $P\Lambda P^{-1}$, where Λ is the diagonal matrix of the eigenvalues and P is the matrix whose columns are the corresponding eigenvectors.

b) Use the factorization $A = P\Lambda P^{-1}$ to find an expression for A^n.

c) Find A^{100} and A^{255}

Since the equation $M\vec{x} = P\Lambda P^{-1}\vec{x}$ does not depend on $\vec{x}$, we invoke our aphorism that two things that *do* the same thing *are* the same thing, and conclude $M = P\Lambda P^{-1}$. Consequently $M^n = P\Lambda^n P^{-1}$. What makes this useful is that because Λ^n is a diagonal matrix, Λ^n is very easy to compute.

This makes matrices that can be expressed in this form very interesting. We define:

Definition 8.1 (Diagonalizable). *Let A be a matrix. A is diagonalizable if it can be expressed in the form $A = P\Lambda P^{-1}$, where Λ is a diagonal matrix.*

While diagonalizability is very useful, we might wonder how many matrices really are diagonalizable.

Activity 8.6: Diagonalizable Matrices

A8.6.1 Construct a matrix M with the given properties.

a) Eigenvalues $\lambda = 1, 2$ with eigenvectors $\langle 1, 2 \rangle$ and $\langle 3, 5 \rangle$.

b) Eigenvalues $\lambda = 1, 3$, where $\lambda = 1$ has eigenvectors $\langle 1, 2, -1 \rangle$ and $\langle 3, 1, -2 \rangle$, and $\lambda = 3$ has eigenvector $\langle -1, -1, 1 \rangle$.

A8.6.2 Suppose M is a nondefective matrix. Prove/disprove: The eigendecomposition $M = P\Lambda P^{-1}$ is unique.

A8.6.3 Suppose M is diagonalizable. Prove/disprove: M is invertible.

A8.6.4 Suppose M is a nondefective $n \times n$ matrix with a single eigenvalue. Prove/disprove: M is a scalar multiple of the identity.

A8.6.5 Suppose M is a nondefective matrix with a row of 0s (for convenience, assume it's the last row of M). Prove/disprove: M must have an eigenvalue of 0.

A8.6.6 Suppose M is a nondefective matrix with eigenvalue 0. Prove/disprove: M must have a row of 0s.

As Activity 8.6 suggests, many different types of matrices are diagonalizable. Thus it's useful to define:

Definition 8.2 (Eigendecomposition). *Let A be a nondefective square matrix. Let Λ be the diagonal matrix of eigenvalues, and P a matrix whose column vectors form an eigenbasis for the column space of A (with the columns corresponding to the eigenvalues in Λ). An* ***eigendecomposition of*** *A is the factorization $A = P\Lambda P^{-1}$.*

Example 8.5. *Find an eigendecomposition of* $A = \begin{pmatrix} 8 & -12 & -30 \\ -3 & 8 & 15 \\ 3 & -6 & -13 \end{pmatrix}$

Solution. *This matrix has eigenvalues $\lambda = 2$, with eigenvectors $\langle 5, 0, 1 \rangle$ and $\langle 2, 1, 0 \rangle$; and $\lambda = -1$, with eigenvector $\langle 2, -1, 1 \rangle$. Consequently*

$$\Lambda = \begin{pmatrix} 2 & 0 & 0 \\ 0 & 2 & 0 \\ 0 & 0 & -1 \end{pmatrix} \qquad P = \begin{pmatrix} 5 & 2 & 2 \\ 0 & 1 & -1 \\ 1 & 0 & 1 \end{pmatrix}$$

We also find

$$P^{-1} = \begin{pmatrix} 1 & -2 & -4 \\ -1 & 3 & 5 \\ -1 & 2 & 5 \end{pmatrix}$$

so

$$\begin{pmatrix} 8 & -12 & -30 \\ -3 & 8 & 15 \\ 3 & -6 & -13 \end{pmatrix} = \begin{pmatrix} 5 & 2 & 2 \\ 0 & 1 & -1 \\ 1 & 0 & 1 \end{pmatrix} \begin{pmatrix} 2 & 0 & 0 \\ 0 & 2 & 0 \\ 0 & 0 & -1 \end{pmatrix} \begin{pmatrix} 1 & -2 & -4 \\ -1 & 3 & 5 \\ -1 & 2 & 5 \end{pmatrix}$$

What if A is a defective matrix?

Activity 8.7: Eigendecompositions With Defective Matrices

In the following, let $\mathbf{x}_k$ be generalized eigenvector for λ of rank k, and define $(A - \lambda I)\mathbf{x}_{i+1} = \mathbf{x}_i$. Let

$$\mathcal{V} = \{\mathbf{x}_1, \mathbf{x}_2, \ldots, \mathbf{x}_{k-1}, \mathbf{x}_k\}$$

A8.7.1 Show $A\mathbf{x}_{i+1} = \mathbf{x_i} + \lambda \mathbf{x_{i+1}}$.

A8.7.2 Suppose $\mathbf{x}$ is in Span $\mathcal{V}$, with coordinates $(a_1, a_2, \ldots, a_k)$ with respect to $\mathcal{V}$.

a) Find the coordinates of $A\mathbf{x}$.
b) Find J so the product

$$J \begin{pmatrix} a_1 \\ a_2 \\ \vdots \\ a_k \end{pmatrix}$$

gives the coordinates of $A\mathbf{x}$ with respect to $\mathcal{V}$.
c) Find P so

$$\mathbf{x} = P \begin{pmatrix} a_1 \\ a_2 \\ \vdots \\ a_k \end{pmatrix}$$

d) Explain why P^{-1} must exist, where

$$P^{-1}\mathbf{x} = \begin{pmatrix} a_1 \\ a_2 \\ \vdots \\ a_k \end{pmatrix}$$

e) Find $A\mathbf{x}$ in terms of the matrices J and P.

Activity 8.7 leads to a decomposition of defective matrices known as the **Jordan canonical form**. We describe the Jordan canonical form as follows: let eigenvalue λ have generalized eigenvector rank k. A **Jordan block** is a $k \times k$ matrix

$$J_\lambda = \begin{pmatrix} \lambda & 1 & 0 & 0 & \ldots & 0 \\ 0 & \lambda & 1 & 0 & \ldots & 0 \\ 0 & 0 & \lambda & 1 & \ldots & 0 \\ \vdots & \vdots & \vdots & \vdots & \ddots & \vdots \\ 0 & 0 & 0 & & \ldots & \lambda \end{pmatrix}$$

You produced this in Activity A8.7.2e as a way to determine the *coordinates* of $A\mathbf{x}$ from the coordinates of $\mathbf{x}$, for vectors $\mathbf{x}$ in the span of a set of generalized eigenvectors for λ. Since we can recover the actual value of $\mathbf{x}$ by multiplying its coordinates by the matrix of basis vectors, we have

$$A\mathbf{x} = PJ_\lambda\mathbf{c}$$

where $\mathbf{c}$ is the column vector of the coordinates of $\mathbf{x}$, and each column of P corresponds to a generalized eigenvector of increasing rank. But again, we can recover the actual value of $\mathbf{x}$ by multiplying its coordinates by P, so:

$$AP\mathbf{c} = PJ_\lambda \mathbf{c}$$

"Things that do the same thing are the same thing," so we can conclude:

$$AP = PJ_\lambda$$

Since P must be invertible, it follows that

$$A = PJ_\lambda P^{-1}$$

What's to recognize is that this equality *only* holds for vectors $\mathbf{x}$ in the span of the generalized eigenvectors for λ.

Activity 8.8: The Jordan Normal Form

In the following, assume that A is a defective $n \times n$ matrix, with eigenbasis $\mathcal{L} = \{\mathbf{v}_1, \mathbf{v}_2, \ldots, \mathbf{v}_n\}$. Assume that the eigenvectors are ordered, so that the first few eigenvectors correspond to eigenvalue λ_1; the next correspond to eigenvalue λ_2, and so on.

A8.8.1 Let $\mathbf{c}$ be the coordinates of some vector $\mathbf{x}$ in $\mathbb{R}^n$ with respect to $\mathcal{L}$. Explain why the coordinates of $A\mathbf{x}$ can be found by the product $J\mathbf{c}$, where

$$J = \begin{pmatrix} J_1 & \mathbf{0} & \ldots & \mathbf{0} \\ \mathbf{0} & J_2 & \ldots & \mathbf{0} \\ \vdots & \vdots & \ddots & \vdots \\ \mathbf{0} & \mathbf{0} & \ldots & J_k \end{pmatrix}$$

where the $\mathbf{0}$s are appropriately sized zero matrices; and the J_is are appropriate constituted Jordan blocks.

A8.8.2 Prove: $A = PJP^{-1}$.

Example 8.6. *Find a Jordan decomposition for* $A = \begin{pmatrix} 9 & -3 & -1 \\ 10 & -3 & 0 \\ 14 & -5 & -2 \end{pmatrix}$

Solution. *In Example 7.8, we found this matrix has*

- $\lambda = 2$, *with eigenvector* $\langle 1, 2, 1 \rangle$,
- $\lambda = 1$, *with chain of generalized eigenvectors* $\langle -3, -5, -5 \rangle$ *(rank 2) and* $\langle -4, -10, -2 \rangle$ *(a "normal" eigenvector).*

We'll make the first Jordan block correspond to $\lambda = 2$, so the first column of P will just be the eigenvector $\langle 1, 2, 1 \rangle$.

The second Jordan block corresponds to $\lambda = 1$, so the corresponding columns of P will be $\langle -4, -10, -2 \rangle$ (a rank 1 generalized eigenvector) and $\langle -3, -5, -5 \rangle$ (a rank 2 generalized eigenvector); thus

$$J = \begin{pmatrix} 2 & 0 & 0 \\ 0 & 1 & 1 \\ 0 & 0 & 1 \end{pmatrix}, P = \begin{pmatrix} 1 & -4 & -3 \\ 2 & -10 & -5 \\ 1 & -2 & -5 \end{pmatrix}, P^{-1} = \begin{pmatrix} 20 & -7 & -5 \\ \frac{5}{2} & -1 & -\frac{1}{2} \\ 3 & -1 & -1 \end{pmatrix}$$

8.4 Singular Value Decomposition

As we've seen, linear algebra has many applications, ranging from modeling population growth, classifying documents, analyzing networks, navigating the internet, displaying and printing color images, and encrypting data. One of the most important uses is known as Singular Value Decomposition, or SVD for short. We'll begin by developing some intuition for SVD.

Activity 8.9: More Transformations

In the following activity, elementary geometry will more easily prove your results. We'll use $|PQ|$ to represent the length of the line segment between P and Q.

A8.9.1 Let $ABCD$ be the a rectangle with one vertex at the origin and sides along the x and y axes. Moreover, imagine the sides have a fixed length: this would be the case if, for example, the rectangle were a real rectangle made out of real materials (we say the sides of the rectangle $ABCD$ are **rigid objects**). Let transformation T pull vertex at C in the direction of the diagonal $\overrightarrow{AC}$, and remember we're assuming the sides stay the same length; thus $AD = AD'$, $AB = AB'$, etc.

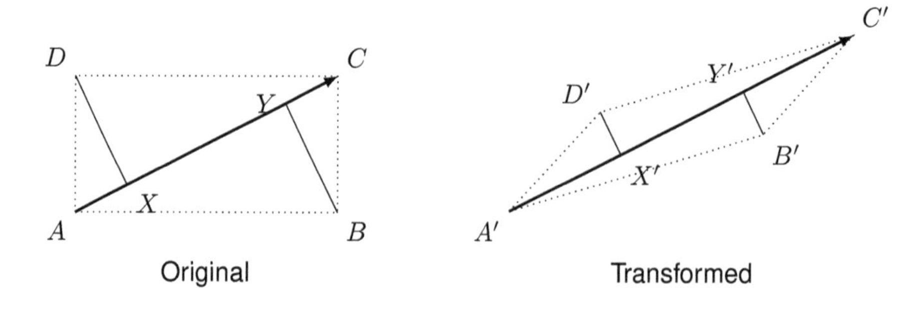

a) Explain why, under this transformation, $|A'C'| = \lambda_1|AC|$ for some λ_1.

b) Show that the distances to the diagonal DX, BY in the original will satisfy $|D'X'| = \lambda_2|DX|$ and $|B'Y'| = \lambda_2|BY|$ for some λ_2.

"Math ever generalizes," so let's consider transformations that can be described by stretching or compressing by *any* amount along perpendicular axes.

Activity 8.10: Stretching and Compressing

A8.10.1 Let Σ be the matrix for a transformation that stretches an object horizontally by a factor of λ_1, and vertically by a factor of λ_2. Find Σ.

A8.10.2 Let R_θ be the transformation that rotates an object counterclockwise through an angle of θ. Describe the geometric effect $R_{-\theta}\Sigma R_\theta$.

A8.10.3 From Activity 3.2, you determined $R_\theta = \begin{pmatrix} \cos\theta & -\sin\theta \\ \sin\theta & \cos\theta \end{pmatrix}$

a) Find T, the transformation that stretches an object by a factor of 3 in the direction of $\overrightarrow{v} = \langle 1, 1\rangle$, and by a factor of $\frac{1}{2}$ in the direction perpendicular to $\overrightarrow{v}$.

b) Find U, the transformation that stretches an object by a factor of 2 in the direction of $\overrightarrow{v} = \langle 1, 5\rangle$, and by a factor of 5 in the direction perpendicular to $\overrightarrow{v}$.

c) Find V, the transformation that stretches an object by a factor of -5 in the direction of $\overrightarrow{v} = \langle 3, 1\rangle$, and by a factor of 2 in the direction perpendicular to $\overrightarrow{v}$.

d) What do you notice about matrices T, U, V?

A8.10.4 In a **shear transformation**, one particular line remains unchanged, while all parallel lines are shifted by a distance proportional to their distance from the fixed line; an illustration of the shear transformation of the unit square $ABCD$ into a sheared square $A'B'C'D'$ is shown below:

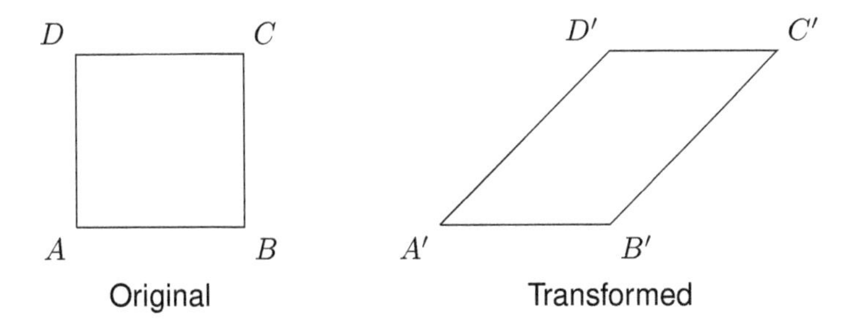

Explain why shear transformations *cannot* be expressed in the form $R_{-\theta}\Sigma R_{\theta}$. Suggestion: Suppose it can be represented by a matrix S. What would S have to look like?

Unfortunately, the results of Activity 8.10 suggest that any transformation that can be described in terms of a rotation, followed by a stretch/compression, followed by a counter-rotation, will be described using a symmetric matrix. Thus, while it appears we can always factor symmetric matrices this way, that seems too strict a requirement to provide a generally useful factorization.

Or is it? Let's consider the factorization more closely. If S is a symmetric matrix, we saw that we can describe it as a rotation, followed by a stretch/compression along perpendicular axes; followed by a counter-rotation. Since the rotation and counterrotation are inverses, we can express one as R and the other as R^{-1}; to align our notation with our eventual description, we'll write this factorization as $S = R\Sigma R^{-1}$.

Now remember R represents a rotation. Consequently R is an orthogonal matrix, and so $R^{-1} = R^T$, and our factorization can be expressed as $S = R\Sigma R^T$. Thus our symmetric matrix S is the product of three matrices:

- An orthogonal matrix R,
- A diagonal matrix Σ,
- Another orthogonal matrix R^T.

"Math ever generalizes." This usually applies to the concepts of mathematics, but it can also apply to the *questions*: If we can write a symmetric matrix as such a product, might we be able to write a nonsymmetric matrix as the product of

- An orthogonal matrix U,
- A diagonal matrix Σ,
- Another orthogonal matrix V^T

At this point, we'll introduce one of the most important strategies in mathematical research:

Strategy. *Try it and see where it takes you.*

Suppose $A = U\Sigma V^T$, where U, V are orthogonal matrices and Σ is a diagonal matrix. How could we find U, Σ, and V?

Activity 8.11: Singular Value Decomposition

In the following, let A be *any* matrix of *any* dimension, and suppose $A = U\Sigma V^T$, where U and V are orthogonal matrices and Σ is a diagonal matrix.

A8.11.1 There are two slightly different approaches to finding U, V, and Σ. Suppose A is a $n \times m$ matrix.

a) Suppose U, V are square matrices. What are the dimensions of U, V, and Σ?

b) Suppose instead that Σ is a square matrix. What are the dimensions of U, V, and Σ?

A8.11.2 Let A be a matrix, and suppose we can find orthogonal matrices U and V, and diagonal matrix Σ where $A = U\Sigma V^T$.

a) Find $A^T A$ and AA^T in terms of U, V, and Σ.

b) What is the connection between U, V, and the eigenvectors of $A^T A$ and AA^T? Explain why.

c) Remember we can either require Σ to be square, or U, V to be square. Based on your answers, which is preferable? Why?

A8.11.3 Let A be a $m \times n$ matrix, and suppose we can find orthogonal matrix U and V and diagonal matrix Σ where $A = U\Sigma V^T$, and Σ is a square matrix with nonzero diagonal entries.

a) Find the dimensions of U, V, Σ.

b) If it exists, find Σ^{-1}; if it does not exist, explain why not.

c) Find a right inverse of A, if it exists. In other words, find B where $AB = I$. Suggestion: Consider each factor of $A = U\Sigma V^T$ separately. What can you multiply by to eliminate the factors?

d) Find a left inverse of A, if it exists.

A8.11.4 Consider the problem of finding Σ where $\Sigma^T\Sigma = \Lambda$, where Λ and Σ are diagonal matrices.

a) "Every problem in linear algebra can be solved by setting up and solving a system of linear equations." Explain why, in this case, what is not true about this claim.

b) Find Σ.

c) Why does your work mean that Σ is not unique? (In other words: If $A = U\Sigma V^T$, there is more than one possible choice of Σ.)

d) Suppose you wanted to define a "standard" Σ. How might you define it?

e) How will this choice of Σ affect U, V? (You should reference your answer to Activity A8.11.2b.)

A8.11.5 If possible, find a diagonal (not necessarily square) matrix Σ with the indicated products. If not possible, explain why not.

a) $\Sigma\Sigma^T = (26)$ and $\Sigma^T\Sigma = \begin{pmatrix} 26 & 0 \\ 0 & 0 \end{pmatrix}$

b) $\Sigma\Sigma^T = \begin{pmatrix} 10 & 0 \\ 0 & 1 \end{pmatrix}$ and $\Sigma^T\Sigma = \begin{pmatrix} 10 & 0 & 0 \\ 0 & 1 & 0 \\ 0 & 0 & 0 \end{pmatrix}$

c) $\Sigma\Sigma^T = \begin{pmatrix} 5 & 0 \\ 0 & 1 \end{pmatrix}$ and $\Sigma^T\Sigma = \begin{pmatrix} 5 & 0 & 0 \\ 0 & 1 & 0 \\ 0 & 0 & 4 \end{pmatrix}$

From Activity 8.11, you found that if we can write $A = U\Sigma V^T$, where U, V are orthogonal matrices and Σ is a diagonal matrix, then

$$A^T A = V\Sigma^T \Sigma V^T \qquad AA^T = U\Sigma\Sigma^T U^T$$

However, there were two approaches. Either we could make Σ a square matrix, or we could make U, V square matrices. Which approach should we adopt?

Let's consider $A^T A$ for a moment. Since $A^T A$ is symmetric, it's nondefective, so it has an eigendecomposition. Moreover, because it's symmetric, its eigenvectors are orthogonal. We can go one step further and normalize all the eigenvectors, and write $A^T A = P\Lambda P^T$, where P is an orthonormal matrix (one whose column vectors form an orthonormal set) and Λ is the diagonal matrix of the eigenvalues of P.

Now let's compare our expressions for $A^T A$. On the one hand, $A^T A = V\Sigma^T\Sigma V^T$. On the other hand, $A^T A = P\Lambda P^T$. It seems both expedient and reasonable to let $P = V$, which means we want V (and, by a similar argument, U) to be square matrices.

What about Σ? If we take $P = V$, then $\Sigma^T\Sigma = \Lambda$, the diagonal matrix of the eigenvalues of $A^T A$. Thus we can find Σ directly: the entries of Σ will be the square roots of the eigenvalues of $A^T A$.

This last point should concern us, since if our eigenvalues are negative, we won't be able to take the square roots.

Activity 8.12: More Symmetric Matrices

In the following, let A be *any* matrix, not necessarily square.

A8.12.1 Explain why $A^T A = R\Sigma R^T$, where R is an orthogonal matrix and Σ is a diagonal matrix.

A8.12.2 Suppose λ, $\mathbf{v}$ is an eigenvalue-eigenvector pair for $A^T A$.

a) Find $||A\mathbf{v}||^2$. Suggestion: Remember the relationship between the dot product $\vec{u} \cdot \vec{v}$ and the matrix product $\mathbf{u}^T\mathbf{v}$.

b) What does this tell you about the eigenvalue λ?

c) Is the same result true for if μ, $\mathbf{u}$ is an eigenvalue-eigenvector pair for AA^T?

Activity 8.12 leads to a useful result:

Proposition 25. *Let A be any matrix with real entries. The eigenvalues of A^TA and AA^T are real and nonnegative.*

This leads us to one of the most remarkable theorems in linear algebra:

Theorem 8.2 (Singular Value Decomposition). *Let A be any matrix (not necessarily square). There exists orthogonal matrices U and V, and diagonal matrix Σ, where $A = U\Sigma V^T$.*

There's one more issue that must be resolved.

Activity 8.13: Choices and Ambiguities

A8.13.1 Let M be a symmetric matrix.

a) Explain why we can write $M = P\Lambda P^T$, where P is an orthogonal matrix and Λ is a diagonal matrix of the eigenvalues of M.

b) If possible, explain how you could find a P', Λ' different from P, Λ where $M = P'\Lambda'P'^T$. If not possible, explain why not.

c) Suppose you require the entries of Λ be the eigenvalues in nondecreasing order. Does this make Λ unique (in other words, given the eigenvalues of M, could you find two different diagonal matrices whose nonzero entries are the eigenvalues in nondecreasing order)? Why/why not?

d) Again, suppose you require the entries Λ to be the eigenvalues in nondecreasing order. Does this make P unique? Why/why not?

A8.13.2 Suppose $A = \begin{pmatrix} 1 & 3 \\ 5 & 3 \end{pmatrix}$, and assume $A = U\Sigma V^T$ for some orthogonal matrices U, V, and some diagonal matrix Σ.

a) Find the eigenvalues of AA^T, then use them to choose an appropriate diagonal matrix Σ.

b) Based on your choice of Σ, find U, an orthonormal basis for the eigenspace of AA^T.

c) Based on your choice of Σ, find U', a *different* orthonormal basis for the eigenspace of AA^T.

d) Based on your choice of Σ, how many other matrices U will *also* be an orthonormal basis for the eigenspace of AA^T? Find them.

e) Find the eigenvectors of $A^T A$.

f) Find *all* orthogonal matrices V whose columns form an orthonormal basis for the eigenspace of A^T.

g) What problem does this suggest you will encounter when trying to factor $A = U\Sigma V^T$?

A8.13.3 Suppose $A = U\Sigma V^T$, where U, V are orthogonal matrices and Σ is a diagonal matrix whose entries are in nondecreasing order. Consider U', whose columns are the same as those of U but the kth column has been multiplied by -1.

a) Explain why U' is still an orthogonal matrix whose columns form a basis for the eigenspace of AA^T.

b) Will $U'\Sigma V^T = A$? Why/why not?

Activity 8.13 identifies an important problem with our attempt to factor $A = U\Sigma V^T$. We can identify a unique Σ, because the entries of Σ are the square roots of the eigenvalues of $A^T A$, and we can put these entries in nondecreasing order.

But even given a choice of Σ, we still have choices for the eigenvectors, since any scalar multiple of an eigenvector is an eigenvector. In particular, multiplying an eigenvector by -1 will affect neither the orthogonality nor the magnitude, so it's possible to have U, U', and others, all of which correspond to an orthonormal basis for the eigenspace of AA^T; similarly for V, V', and so on, which correspond to an orthonormal basis for the eigenspace of $A^T A$. Since every column of U and V can have its sign changed, this means that if A is a $m \times n$, there are 2^m possibilities for U and 2^n possibilities for V. So, which U and V do we need?

Activity 8.14: Sign Ambiguity

In the following, suppose A is a $m \times n$ matrix, where $A = U\Sigma V^T$, where U, V are orthogonal and Σ is a diagonal matrix whose diagonal entries are σ_1, $\sigma_2, \ldots$.

A8.14.1 Suppose $A = U\Sigma V^T$.

a) Explain why finding U requires finding the eigenvectors of a $m \times m$ matrix, while finding V requires finding the eigenvectors of a $n \times n$ matrix.

b) Suppose you find U. How could you find some or all of the entries of V *without* finding the eigenvectors of a $n \times n$ matrix?

c) Suppose you could find V. How could you find some or all of the entries of U *without* finding the eigenvectors of a $m \times m$ matrix?

Activity 8.14 suggests the following approach: Once we have Σ, find *either* U or V. Then:

- If you've found V, find U by solving $AV = U\Sigma$.
- If you've found U, finding V by solving $UA = \Sigma V^T$.

The reason this is useful is because both $U\Sigma$ and ΣV^T have very simple forms.

Activity 8.15: Singular Value Decomposition

A8.15.1 Let $A = \begin{pmatrix} 1 & 1 \\ 2 & 4 \end{pmatrix}$

a) Suppose $A = U\Sigma V^T$, where U, V are orthonormal matrices and Σ is a diagonal matrix. Find the dimensions of U, V, and Σ.

b) Express $A^T A$ in terms of U, V, and Σ.

c) Find $A^T A$; its eigenvalues; and its eigenvectors; then find matrix V.

d) From $AV = U\Sigma$, find U.

e) Verify $A = U\Sigma V^T$.

A8.15.2 Let $B = \begin{pmatrix} 2 & 5 & 1 & 3 & 5 \\ 1 & 6 & -3 & -4 & 3 \end{pmatrix}$ and suppose $B = U\Sigma V^T$, where U, V are square orthogonal matrices and Σ is a diagonal matrix.

a) Explain why U is the orthonormal basis for the eigenspace of a 2×2 matrix, while V is the orthonormal basis for the eigenspace of a 5×5 matrix.

b) Find U and Σ.

c) Explain why the equation $AU = \Sigma V^T$ *can't* be used to find V.

d) Explain why you *can* find *some* of the entries of V. Which entries of V can you find?

e) Let V' be the matrix consisting of the entries of V you can find, with 0s in all the remaining places. Find $U\Sigma V'$, and round your final entries to the nearest whole number. What do you notice?

Example 8.7. *Find a SVD for* $A = \begin{pmatrix} 2 & 1 & 4 \\ 1 & 3 & 5 \end{pmatrix}$

Solution. *We'll need to find the eigenvectors and eigenvalues of either* $A^T A$

or AA^T*, but since* AA^T *is will be a* 2×2 *matrix and* A^TA *is a* 3×3 *matrix, we'll work with* AA^T*. We find*

$$AA^T = \begin{pmatrix} 21 & 25 \\ 25 & 35 \end{pmatrix}$$

This matrix has eigenvalues $\lambda_1 \approx 53.9616$ *and* $\lambda_2 \approx 2.0384$*. Finding and normalizing the corresponding eigenvectors give us*

$$U = \begin{pmatrix} 0.6043 & -0.7968 \\ 0.7968 & 0.6043 \end{pmatrix}$$

Σ *is the diagonal matrix whose entries are the square roots of the eigenvalues of* AA^T*, so*

$$\Sigma = \begin{pmatrix} 7.3458 & 0 \\ 0 & 1.4277 \end{pmatrix}$$

From $A = U\Sigma V^T$ *and the fact that* U *is orthonormal, we have* $U^TA = \Sigma V^T$*, which means the rows of* U^TA *will be scalar multiples of the rows of* V^T*. We find:*

$$U^TA = \begin{pmatrix} 2.0054 & 2.9946 & 6.4010 \\ -0.9892 & 1.0162 & -0.1655 \end{pmatrix}$$

where the first row will be 7.3458 *times the corresponding entries of* V^T*, and the second row will be* 1.4277 *times the entries of* V^T*. Thus*

$$V^T = \begin{pmatrix} 0.2730 & 0.4077 & 0.8714 \\ -0.6929 & 0.7118 & -0.1159 \end{pmatrix}$$

Thus

$$\begin{pmatrix} 2 & 1 & 4 \\ 1 & 3 & 5 \end{pmatrix} = \begin{pmatrix} 0.6043 & -0.7968 \\ 0.7968 & 0.6043 \end{pmatrix} \begin{pmatrix} 7.3458 & 0 \\ 0 & 1.4277 \end{pmatrix} \begin{pmatrix} 0.2730 & 0.4077 & 0.8714 \\ -0.6929 & 0.7118 & -0.1159 \end{pmatrix}$$

It seems like SVD is a tremendous amount of work. So why bother? We close our journey through elementary linear algebra with *one* application of SVD that has a profound impact on life in the 21st century.

Activity 8.16: Compressing Matrices

You may want to consult Activity 7.8 for ways to find eigenvalues numerically.

A8.16.1 Let $C = \begin{pmatrix} 1 & 7 & 2 & 5 & 3 \\ 2 & 15 & 5 & 9 & 7 \\ 1 & 6 & 3 & 4 & 3 \end{pmatrix}$.

a) Using a numerical method, find *only* the largest eigenvalue and corresponding eigenvector of CC^T.

b) Let U' be the matrix whose first column is the normalized eigenvector, and whose remaining columns are 0s. Find U'.

c) Let Σ' be the diagonal matrix whose first entry is the square root of the eigenvalue, and whose remaining columns are 0s. Find Σ'.

d) Use the relation $U^T A = \Sigma V^T$ to find V', assuming any value you can't find to be 0.

e) Find $U'\Sigma'V'$. Round your final values to the nearest whole number. What do you notice?

A8.16.2 Let $A = \begin{pmatrix} 2 & 5 & 1 & 3 & 5 \\ 1 & 6 & -3 & -4 & 3 \end{pmatrix}$ and suppose $A = U\Sigma V^T$, where U, V are square orthogonal matrices and Σ is a diagonal matrix.

a) Explain why U is the orthonormal basis for the eigenspace of a 2×2 matrix, while V is the orthonormal basis for the eigenspace of a 5×5 matrix.

b) Find U and Σ.

c) Write a matrix equation to find V.

d) Explain why you can't use the matrix equation to find all columns of V.

e) Let V' be the matrix formed using the columns of V you *can* find from the matrix equation, and set the remaining columns equal to 0. Find $U\Sigma V'$.

f) Use A, U, and Σ to find V.

A8.16.3 Let $A = \begin{pmatrix} 5 & 5 & 2 & 7 & 6 \\ 8 & 9 & 3 & 13 & 11 \\ 6 & 6 & 2 & 8 & 8 \end{pmatrix}$

a) Find the largest eigenvalue of $A^T A$ and AA^T.

b) Find the corresponding eigenvectors.

c) Assume the remaining eigenvalues are 0. Find Σ', an approximation to Σ.

d) Your normalized eigenvectors will give you the first column of U and V. Assume the other columns are 0, to give you U' and V', the approximations to U and V.

e) Suppose you know the entries of A are integers. Round the entries of $A' = U'\Sigma' V'^T$ to the nearest integer. How accurate an approximation to A can you find using U', Σ', and V'?

As Activity 8.16 shows, if A is a $n \times m$ matrix where $A = U\Sigma V^T$, then $A' = U'\Sigma' V'^T$ can be a good approximation to A, where U' consists of the first column of U with the remaining columns 0; Σ' is a diagonal matrix whose only nonzero entry is the square root of the largest eigenvalue of $A^T A$; and V' consists of the first column of V with the remaining columns 0s.

What's important here is that since we know most of the entries of U', Σ', and V' are zero, and we know where those 0s occur, the relevant information consists of:

- The values of a single $n \times 1$ column vector, namely the first column of U
- A single eigenvalue, namely the largest eigenvalue of $A^T A$,
- The values of a single $m \times 1$ column vector, namely the first column of V.

This means we can get an approximation of the matrix A, which has nm entries, using just $n+1+m$ values. If A is a modest sized matrix—say a 1000×1000 matrix that records the "red" intensity values of a small digital image—the *exact* values of A would require keeping track of $1000 \times 1000 = 1,000,000$ numbers, but by using the largest eigenvalue, we can get an approximation A' that only requires $1000 + 1 + 1000 = 2001$ numbers, a savings of almost 98%.

In actual practice, using just a single eigenvalue won't get a very good reproduction. However, if we use the largest few eigenvalues (which we can find using the method of Activity 7.8), we can generally get a reasonably good approximation to the matrix, and achieve a significant savings in the amount of data we must record in order to store a digital image.

9

Extras

This chapter includes material that is, strictly speaking, not part of linear algebra. However, it plays an important role in some of the major applications of linear algebra.

9.1 Properties of Polynomials

In general, it is impossible to factor a polynomial to recover its roots. However, we can often say something about the roots of a polynomial, even without knowing its factorization.

Activity 9.1: Properties of Polynomials

A9.1.1 Suppose

$$f(x) = (x - r_1)(x - r_2) \cdots (x - r_n)$$

a) Find the degree of f.
b) Find $f(0)$. What does this tell you about the *constant* term in the expansion of $f(x)$?
c) Find the coefficient of the x^{n-1} term in the expansion of $f(x)$.

9.2 Complex Numbers

In college algebra, you might want to solve an equation for x: Find out what x *is*. In higher mathematics, we're often more interested in how things behave: Given that x is a solution to an equation, what can we say about $2x$ or x^3?

Activity 9.2: Complex Numbers

A9.2.1 Suppose x is a number where $x^2 + 1 = 0$. Starting with the equation $x^2 + 1 = 0$, find and simplify an expression for:

a) x^2
b) x^4. Suggestion: $x^4 = (x^2)^2$.
c) x^3. Suggestion: $x^3 = x^2 x$, and you found an expression for x^2.
d) x^{13}. Suggestion: Write x^{13} in terms of powers of x whose values you know.

A9.2.2 Suppose z is a solution to $z^2 + z + 8 = 0$. **WITHOUT** solving for z, find and simplify an expression for:

a) z^2
b) z^3
c) $z^4 + 3z^3 + 2z + 8$.

This leads to the definition:

Definition 9.1 (Complex Unit). *The solution to the equation $x^2 + 1 = 0$ is designated i.*

The behavior of the solutions to the equation $x^2 + 1 = 0$ gives us the following theorem:

Theorem 9.1 (Powers of i). *If $x = i$ is the solution to $x^2 + 1 = 0$, then:*

$$i^2 = -1 \qquad i^3 = -i \qquad i^4 = 1$$

We define:

Definition 9.2 (Complex Number). *Let a, b be real numbers, and let i be the solution to $x^2 + 1 = 0$. Then $a + bi$ is a* ***complex number*** *with* ***real part*** *a and* ***imaginary part*** *bi.*

Activity 9.3: Complex Arithmetic

A9.3.1 Simplify the following complex expressions. Remember to use the fact that $i^2 = -1$.

a) $(4 + 3i) + (2 + i)$
b) $(8 - 3i) - (7 - 6i)$
c) $i(3 - 4i)$
d) $(2 + i)(8 - i)$

A9.3.2 Suppose we have a complex number z_1. We want to find a complex number z_2 where $z_1 + z_2$ is a *real* number.

a) Let $z_1 = i$. Find a complex number z_2 where $z_1 + z_2$ is a real number. Is there more than one solution?
b) Let $z_1 = 3+4i$. Find a complex number z_2 where z_1+z_2 is a real number. Is there more than one solution?

A9.3.3 Suppose we have a complex number z_1. We want to find a complex number z_2 where $z_1 z_2$ is a *real* number.

a) Let $z_1 = i$. Find a complex number z_2 where $z_1 z_2$ is a real number. Is there more than one solution?
b) Let $z_1 = 3i$. Find a complex number z_2 where $z_1 z_2$ is a real number. Is there more than one solution?
c) Let $z_1 = 3 + 4i$. Find a complex number z_2 where $z_1 z_2$ is a real number. Suggestion: Let $z_2 = a + bi$. Then find an expression for $z_1 z_2$ in terms of a, b. What do you *want* to be true?

A9.3.4 You should have found that if z_1 is a complex number, there are many complex numbers that we can add to z_1 to get a real number; and many complex numbers we can multiply by z_1 to get a real number.

a) Let $z_1 = i$. Find z_2 where *both* $z_1 + z_2$ and $z_1 z_2$ are real numbers. Is there more than one solution?
b) Let $z_1 = 3 + i$. Find z_2 where *both* $z_1 + z_2$ and $z_1 z_2$ are real numbers. Is there more than one solution?
c) Let $z_1 = a + bi$, where a, b are real numbers. Find z_2 where *both* $z_1 + z_2$ and $z_1 z_2$ are real numbers. Is there more than one solution?

Activity A9.3.4 leads to the following: Given a complex number z_1, there is a unique complex number z_2 where *both* $z_1 + z_2$ *and* $z_1 z_2$ are real numbers. This suggests:

Definition 9.3 (Complex Conjugate). *If a, b are real numbers, then $a + bi$ and $a - bi$ are* ***complex conjugates****.*

We often say that one number is the complex conjugate of the other, and indicate the conjugate of a complex number z as $\overline{z}$.

Example 9.1. *Find $\overline{3 + 4i}$ and $\overline{5 - 12i}$.*

Solution. *The complex conjugate has the same terms, but the operation on the imaginary part is changed:*

$$\overline{3 + 4i} = 3 - 4i$$
$$\overline{5 - 12i} = 5 + 12i$$

Activity 9.4: Conjugates and Polynomials

A9.4.1 Suppose z is a real number. What is $\overline{z}$?

A9.4.2 Let $u = a + bi$.

a) Find $\overline{u}$.
b) Find u^2.
c) Find $\overline{u}^2$. (This is the square of $\overline{u}$.)
d) Find $\overline{(u)^2}$. (This is the conjugate of u^2.)
e) True or false: $\overline{u^2} = \overline{(u^2)}$. On what is your conclusion based?

A9.4.3 Let $u = a + bi$, $v = c + di$.

a) Find $\overline{u}$, $\overline{v}$.
b) Find $\overline{uv}$. (This is the conjugate of uv.)
c) Find $\overline{u}\overline{v}$. (This is the product of $\overline{u}$ with $\overline{v}$.)
d) Prove/disprove: $\overline{uv} = \overline{u}\overline{v}$.

A9.4.4 In the following, we'll build an induction proof that $\overline{z}^n = \overline{z^n}$.

a) Prove: $\overline{(z^2)} = \overline{z}^2$. (This is known as the *base step*.)
b) Show that if $\overline{(z^k)} = \overline{z}^k$, then $\overline{(z^{k+1})} = \overline{z}^{k+1}$. Suggestion: Apply the result of Activity A9.4.3 to find $\overline{(z^k z)}$, and remember you've assumed $\overline{(z^k)} = \overline{z}^k$. (This is known as the *induction step*.)
c) Explain why your work means that $\overline{(z^n)} = \overline{z}^n$ is true for all whole numbers n.

A9.4.5 Suppose $f(x) = a_n x^n + a_{n-1}x^{n-1} + \ldots + a_1 x + a_0$, where the a_is are real numbers.

a) Find an expression for $f(z)$.
b) Find an expression for $f(\overline{z})$.
c) Find an expression for $\overline{f(z)}$. (This is the conjugate of $f(z)$.)
d) Compare $f(\overline{z})$ to $\overline{f(z)}$. What do you notice?
e) Suppose z is a root of $f(z)$. What can you say about $\overline{z}$?

Complex numbers are properly algebraic, in the sense that they represent solutions to systems of equations. However, it's useful to construct a geometric view of them.

Activity 9.5: The Complex Plane

A9.5.1 Consider a number line, like that shown below, with origin O. Let the number P correspond to the point shown.

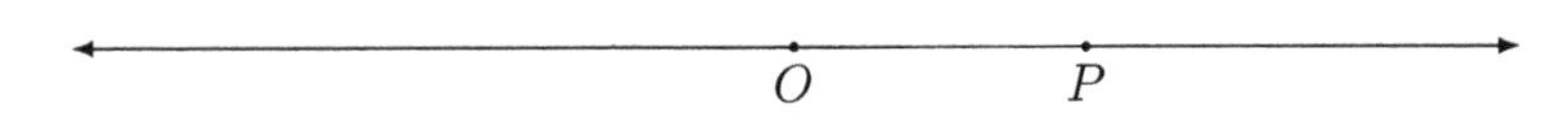

a) Where will the point corresponding to $2P$ be located, relative to the point P? Defend your placement.

b) Where will the point $\frac{1}{2}P$ be located, relative to the point P? Defend your placement.

c) Where will the point $-1P$ be located, relative to the point P? Defend your placement.

A9.5.2 Again, refer to the number line.

a) Where will the point $P + 3$ be located, relative to the point P? Defend your placement.

b) Where will the point $P - 7$ be located, relative to the point P? Defend your placement.

A9.5.3 Consider some number P, and let a, b be real numbers.

a) Describe the transformation aP. In particular: How is the location of aP related to the location of P?

b) Describe the transformation bP.

c) Describe the transformation $(ab)P$. **Note**: The notation $(ab)P$ means that ab is computed first to obtain a real number; P is then multiplied by this real number.

d) Describe the transformation $a(bP)$. **Note**: The notation $a(bP)$ means that bP is computed first; the result is then multiplied by a.

e) How do $(ab)P$ and $a(bP)$ compare?

A9.5.4 Consider the transformation $-1P$. There are two ways to interpret this transformation.

a) Interpret this transformation as a reflection across the origin.

b) Interpret this transformation as a rotation around the origin.

A9.5.5 Consider the transformations x^2P.

a) Explain why this can be interpreted as "Apply x twice in succession to P."

b) Of the two possible interpretations for $-1P$, which can be expressed as "Apply a transformation twice to P?" What is the transformation that would be applied twice to P to obtain the same result as $-1P$?

c) What does this suggest about a possible interpretation for iP (where $i^2 = -1$)?

A9.5.6 Based on your graph in Activity A9.5.5 and your work in Activity A9.5.2, place the following.

a) $i + 3$

b) $-i + 1$

c) Where would you place $3i$? (Suggestion: What is aP, based on your work in Activity A9.5.1?)

d) Where would you place $5 + 3i$?

e) Where would you place $x + iy$, where x, y are real numbers?

Activity 9.5 leads to the following *geometric* representation of complex numbers: Multiplying a number (real or complex) by i produces a rotation by $90°$.

Actually, there is one issue we must address: based on the algebra alone, there's no obvious preference for which direction (clockwise or counterclockwise) we should rotate. However, since angles are generally measured counterclockwise, we conventionally identify multiplication by i with a $90°$ *counterclockwise* rotation. This leads to the following: The complex number $x+iy$ can be represented in the **complex plane** by the point with coordinates (x, y). This is sometimes called an **Argand diagram**, after the French mathematician Jean-Robert Argand (1768–1822).

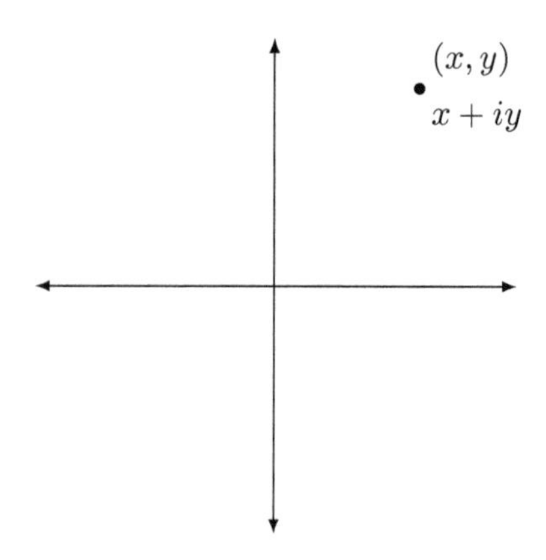

9.3 Mod-N Arithmetic

Another important feature of higher mathematics is asking *why* something happens. Thus in arithmetic, we might "What is $5+1$?" In higher mathematics, we ask "Why is $5+1$ equal to 6?" Answering this question allows us to create new areas of mathematics.

Activity 9.6: Introduction to Mod n Arithmetic

In higher arithmetic, we say that $5+1=6$ because 6 is "the number after" 5. More generally, $n+1$ is "the number after n." The significance of this change in viewpoint is that it allows us to extend arithmetic to *any* set where "the object after" is defined.

A9.6.1 Consider the days of the week. By interpreting $n+1$ as "the object after," find:

a) Monday $+1$
b) Wednesday $+1$
c) Interpret, then find, Wednesday $+3$.
d) Interpret, then find, Monday $+4$.
e) Interpret, then find, Friday $+7$.
f) Interpret, then find, Thursday $+7$.
g) In what way is 7 "like" 0?

A9.6.2 If we consider $+1$ to be "the next one after," then we might consider -1 to be "the one right before."

a) Find Sunday -1?
b) Interpret, then find, Tuesday -5.
c) Interpret, then find, Tuesday $+2$.
d) In what way is -5 "like" 2?

A9.6.3 Find the following.

a) Tuesday $+17$.
b) Friday $+25$.
c) Sunday -30.
d) Monday $+593$.
e) Wednesday -2185.

In Activity 9.6, you found that in some sense, adding or subtracting 7 to a weekday is like adding or subtracting 0: it doesn't change the weekday. It's tempting to write something like $7 = 0$; however, this looks a little peculiar, so mathematicians write $7 \equiv 0$ to indicate that, in some ways, 7 is like 0, but we recognize that 7 is not in fact 0. We read this as "7 is *equivalent to* 0."

It takes only a little creativity to see that we could do the same thing with other numbers: for example, if we consider time on a 12-hour clock, then we can write statements like $12 \equiv 0$, since adding or subtracting 12 to a time gives the same time. However, to avoid confusion, we need to specify whether we're working with time or the days of the week, so we define:

Definition 9.4 (Modulus). *The modulus of an arithmetic system is the least positive value that is equivalent to* 0.

Thus, if we're talking time, our modulus is 12; if we're talking days of the week, our modulus is 7. "Math ever generalizes," so we could imagine taking other numbers as our modulus: 23, 9, or $2^{16} + 1$.

Once we've chosen our modulus N, we can reduce a mod N by finding the least nonnegative value b which is equivalent.

Example 9.2. *Find* 10 mod 7.

Solution. *Our modulus is* 7*, so adding or subtracting* 7 *is like adding or subtracting* 0*. We can write:*

$$10 = 3 + 7$$

where we can use equality, because the two expressions are in fact equal. But since $7 \equiv 0$*, we can continue:*

$$\equiv 3 + 0$$

where we change to $\equiv$ *in recognition of the fact that* 7 *is not equal* ($=$) *to* 0*, but rather equivalent* ($\equiv$)*. Finally, since* $3 + 0 = 3$*, we can complete our reduction:*

$$10 \equiv 3$$

Since the values don't make sense unless we're working mod 7, we *should* write $10 \bmod 7 \equiv 3 \bmod 7$. Alternatively, since the mod 7 applies to both, we *could* write $10 \equiv 3, \bmod 7$, where the comma indicates the "mod 7" applies to both 10 and 3. Unfortunately, we *usually* write $10 \equiv 3 \bmod 7$. It's important to understand that in a statement like this (known as a **congruence**), the modulus applies to *both* sides of the $\equiv$.

Activity 9.7: Arithmetic mod N

A9.7.1 Suppose the modulus of our system is 7. Find the least nonnegative number equivalent to the give number.

a) 10.
b) 35
c) -2. Suggestion: $-2+0 \equiv -2+7$,
d) -12

A9.7.2 Suppose a is a positive integer, and r is the least nonnegative value for which $a \equiv r \bmod N$.

a) What are the possible values of r?
b) What is the relationship between a, r, N?
c) Find $53 \bmod 11$.
d) Find $389 \bmod 17$.

One of the joys of working mod N is that we can (and *should*) reduce a number so that it is between 0 and $N-1$: in other words, $N-1$ is the largest number we're ever required to work with. Since working with small numbers is easier than working with large numbers, this suggests we'll want to reduce as we compute.

Activity 9.8: Multiplication and Powers Mod N

A9.8.1 Compute, then reduce.

a) $(7 \times 3) \bmod 8$
b) $(8 \times 9) \bmod 11$
c) $(12 \times 5) \bmod 20$.

A9.8.2 Reduce the factors, then find the product.

a) $(23 \times 54) \bmod 7$.
b) $(198 \times 49) \bmod 11$.
c) $(13049871 \times 1297431343) \bmod 2$.

A9.8.3 Find the following.

a) $9^2 \bmod 23$.
b) $9^4 \bmod 23$. Suggestion: $9^4 = (9^2)^2$.
c) $9^8 \bmod 23$. Suggestion: $9^8 = (9^4)^2$.

d) $9^{16} \bmod 23$.

e) $9^{25} \bmod 23$. Suggestion: $a^m a^n = a^{m+n}$.

A9.8.4 Find the following.

a) $5^{22} \bmod 23$.

b) $5^{28} \bmod 29$.

c) $5^{30} \bmod 31$.

d) $5^{20} \bmod 21$.

e) "One of these things is not like the other." What do you observe about your results?

What about division?

Activity 9.9: Division mod N

A9.9.1 One way to define division is "The number of times you can subtract the divisor before obtaining 0."

a) Find the *nonnegative* value equivalent to $(3 - 6) \bmod 7$.

b) Find the *nonnegative* value equivalent to $(1 - 9) \bmod 11$.

c) Find the *nonnegative* value equivalent to $(2 - 7) \bmod 13$.

A9.9.2 Find the following. Defend your conclusion based on the idea that $a \div b$ is the number of times you can subtract b before ending with 0.

a) $(3 \div 6) \bmod 7$

b) $(5 \div 8) \bmod 9$

c) $(4 \div 6) \bmod 11$

A9.9.3 We can avoid dealing with negative numbers (and then finding nonnegative numbers equivalent to them) by remembering we can add any multiple of the modulus.

a) Explain why $3 \equiv 10 \bmod 7$.

b) Find four other positive numbers 3 is equivalent to, mod 7.

c) Which of these is exactly divisible by 6? (If none of them are, find more positive numbers equivalent to $3 \bmod 7$).

d) How does this fit with your earlier conclusion about $(3 \div 6) \bmod 7$?

A9.9.4 The standard approach to division mod N is to consider the quotient $a \div b$ as $a \times b^{-1}$, where b^{-1} is the multiplicative inverse of b.

a) Explain why if $5x \equiv 1 \bmod 17$, then $x \equiv 5^{-1} \bmod 17$.

b) Solve: $5x \equiv 1 \bmod 17$. Suggestion: Find 5×1, 5×3, 5×3, and so on, until you find a product where $5x \equiv 1 \bmod 17$.

c) Find: $(11 \div 5) \bmod 17$. Remember: $11 \div 5 = 11 \times 5^{-1}$.

A9.9.5 Activity 2.15 introduces a method of finding integer solutions to certain types of linear equations.

a) Suppose $5x = 1 + 17y$. Explain why $5x \equiv 1 \bmod 17$.

b) Use Bezout's method (Activity 2.15) to solve: $5x \equiv 1 \bmod 17$.

c) Solve: $11x \equiv 1 \bmod 23$. Then use your value to find $37 \times 11^{-1} \bmod 23$.

9.4 Polar Coordinates

The most natural way to describe the location of an object is to point at the object and give the distance (range). This leads to the use of polar coordinates.

Activity 9.10: Polar Coordinates

Any coordinate system relies on the existence of a center (the origin) and a principal direction. As a rule, we take the principal direction to be horizontally to the right.

A9.10.1 In rectangular coordinates, we describe the location of a point by how far we travel from the origin in the principal direction, followed by how far we move perpendicular to the principal direction.

a) Locate the point we arrive at if we travel 4 units along the principal direction, and 3 units perpendicular ("upward") to the principal direction.

b) Could we have arrived at this point by moving 3 units perpendicular ("upward"), then 4 units parallel to the principal direction?

A9.10.2 In polar coordinates, we describe the location of a point by specifying an angle θ measured from the principal direction, and a distance r along the terminal side of that angle.

a) Locate the point we arrive at if we rotate an angle of $\frac{\pi}{4}$ from the principal direction, then travel 3 units in the direction we're facing.

b) Could we have arrived at the same point if we traveled 3 units in the direction we're facing, then rotated by an angle of $\frac{\pi}{4}$?

A9.10.3 Let (x, y) be the rectangular coordinates of a point. Find $[r, \theta]$, the polar coordinates of the point.

A9.10.4 Let $[r, \theta]$ be the polar coordinates of a point. Find (x, y), the rectangular coordinates of the point.

Bibliography

[1] B. Acharya, D. Jena, Patra, S. K., and G. Panda. Invertible, Involutory, and Permutation Matrix Generation Methods for Hill Cipher Systems. *International Conference on Advanced Computer Control*, pages 410–414, 2008.

[2] S. Axler. Down with Determinants! *American Mathematical Monthly*, pages 139–154, 1995.

[3] S. Axler. *Linear Algebra Done Right*. Springer-Verlag, 1997.

[4] R. B. Bapat. *Graphs and Matrices*. Springer-Verlag, 2010.

[5] M. Berry, Z. Drmac, and E. Jessup. Matrices, Vector Spaces, and Information Retrieval. *SIAM Review*, pages 335–362, 1945.

[6] Kurt Bryan and Tanya Leise. The $25,000,000,000 Eigenvector: The Linear Algebra Behind Google. *SIAM Review*, 48:569–581, 2006.

[7] G. Canright and K. Engø-Monsen. Roles in Networks. *Science of Computer Programming*, pages 195–214, 2004.

[8] H. Dogan-Dunlap. Linear Algebra Students' Modes of Reasoning: Geometric Representations. *Linear Algebra and its Applications*, pages 2141–2159, 2010.

[9] W. P. Galvin. Matrices with "Custom-Built" Eigenspaces. *The American Mathematical Monthly*, pages 308–309, 1984.

[10] J. Hannah. A Geometric Approach to Determinants. *The American Mathematical Monthly*, pages 401–409, 1996.

[11] T. Hern. Gaussian Elimination in Integer Arithmetic: An Application of the L-U Factorization. *The College Mathematics Journal*, pages 67–71, 1993.

[12] K. J. Heuvers. Symmetric Matrices with Prescribed Eigenvalues and Eigenvectors. *Mathematics Magazine*, pages 106–111, 1982.

[13] L. S. Hill. Cryptography in an Algebraic Alphabet. *American Mathematical Monthly*, pages 306–312, 1929.

[14] L. S. Hill. Concerning Certain Linear Transformation Apparatus of Cryptography. *American Mathematical Monthly*, pages 135–154, 1931.

[15] J. Hoffstein, J. Pipher, and J. H. Silverman. *An Introduction to Mathematical Cryptography*. Springer Verlag, 2008.

[16] D. Kalman. A Singularly Valuable Decomposition: The SVD of a Matrix. *The College Mathematics Journal*, pages 2–23, 1996.

[17] J. P. Keeners. *SIAM Review*, pages 80–93, 1993.

[18] A. Langville and C. D. Meyer. A Survey of Eigenvector Methods for Web Information Retrieval. *SIAM Review*, pages 135–161, 2005.

[19] A. N. Langville and C. D. Meyer. *Google's PageRank Algorithm and Beyond*. Princeton University Press, 2012.

[20] A. N. Langville and C. D. Meyer. *Who's #1? The Science of Rating and Ranking*. Princeton University Press, 2012.

[21] P. H. Leslie. On the Use of Matrices in Certain Population Mathematics. *Biometrika*, pages 183–212, 1945.

[22] A. McAndrew. Using the Hill Cipher to Teach Cryptographic Principles. *International Journal of Mathematical Education in Science and Technology*, pages 967–979, 2008.

[23] W. A. McWorter Jr. and L. F. Meyers. Computing Eigenvalues and Eigenvectors without Determinants. *Mathematics Magazine*, pages 24–33, 1998.

[24] T. Muir. *The Theory of Determinants in the Historical Order of Development*. Dover, 1906, 1911.

[25] J.-C. Renaud. Matrices with Integer Entries and Integer Eigenvalues. *The American Mathematical Monthly*, pages 202–203, 1984.

[26] G. Salton. *A Theory of Indexing*. Cornell University Press, 1974.

[27] G. Salton and C. Buckley. Term-Weighting Approaches in Automatic Text Retrieval. *Information Processing and Management*, pages 513–523, 1988.

[28] G. Salton, A. Wong, and C. S. Yang. A Vector Space Model for Automatic Indexing. *Communications of the ACM*, pages 613–620, 1975.

[29] G. Schay. *A Concise Introduction to Linear Algebra*. Birkhäuser, 2012.

[30] T. S. Shores. *Applied Linear Algebra and Matrix Analysis*. Springer-Verlag, 2007.

[31] J. Suzuki. *Constitutional Calculus.* Johns Hopkins University Press, 2015.

[32] J. Suzuki. *Patently Mathematical.* Johns Hopkins University Press, 2018.

[33] J. Suzuki. Eigenvalues and Eigenvectors: Generalized and Determinant Free. *Mathematics Magazine*, pages 200–212, 2020.

[34] M. Trigueros and E. Possani. Using an Economics Model for Teaching Linear Algebra. *Linear Algebra and its Applications*, pages 1779–1792, 2013.

[35] P. D. Turney and P. Pantel. From Frequency to Meaning: Vector Space Models of Semantics. *Journal of Artificial Intelligence Research*, pages 141–188, 2010.

Index

For Product Safety Concerns and Information please contact our EU
representative GPSR@taylorandfrancis.com
Taylor & Francis Verlag GmbH, Kaufingerstraße 24, 80331 München, Germany

www.ingramcontent.com/pod-product-compliance
Ingram Content Group UK Ltd.
Pitfield, Milton Keynes, MK11 3LW, UK
UKHW021057080625
459435UK00003B/35